BACTERIAL BIOSURFACTANTS

Isolation, Purification, Characterization, and Industrial Applications

BACTERIAL BIOSURFACTANTS

Isolation, Purification, Characterization, and Industrial Applications

Bolin Kumar Konwar, PhD

APPLE ACADEMIC PRESS

First edition published 2022

Apple Academic Press Inc.
1265 Goldenrod Circle, NE,
Palm Bay, FL 32905 USA

4164 Lakeshore Road, Burlington,
ON, L7L 1A4 Canada

CRC Press
6000 Broken Sound Parkway NW,
Suite 300, Boca Raton, FL 33487-2742 USA

4 Park Square, Milton Park,
Abingdon, Oxon, OX14 4RN UK

Library and Archives Canada Cataloguing in Publication

CIP data on file with Canada Library and Archives

Library of Congress Cataloging-in-Publication Data

CIP data on file with US Library of Congress

ISBN: 978-1-77463-056-3 (hbk)
ISBN: 978-1-77463-931-3 (pbk)
ISBN: 978-1-00318-813-1 (ebk)

About the Author

Bolin Kumar Konwar, PhD
Dean, School of Science and Technology and Professor, Department of Molecular Biology and Biotechnology, Tezpur University (Central), and former Vice Chancellor, Nagaland University (Central), India

Bolin Kumar Konwar, PhD, was Dean at the School of Science and Technology, Tezpur University, India, in 2008. In 2011, he was appointed as Vice-Chancellor, Nagaland University (Central), Nagaland, India. He was formerly an Associate Professor in the Department of Plant Breeding and Genetics at Assam Agricultural University, Jorhat, Assam, India. Other positions included Senior Scientist (Botany and Biotechnology) at Tocklai Experimental Station, Tea Research Association; Professor and Head of Department of Molecular Biology and Biotechnology at Tezpur University (Central). He also headed the ONGC-sponsored Centre for Petroleum Biotechnology, DBT (Govt of India), which sponsored MSc Biotechnology program and Bioinformatics Infrastructure Facility (BIF).

Dr. Konwar and his research group so far have published 143 research papers in peer-reviewed international journals (impact factor 1–7), 80 in seminars and conference proceedings, and 91 research papers at different seminars/conferences all over the country and abroad. He has carried out 12 research projects of national importance as Principal Investigator. A total of 46 MSc students carried out research projects under him both at Assam Agricultural University and Tezpur University. To date, 15 students have obtained PhD degree under his supervision. He and his research group so far have deposited 11 genes (DNA) sequences in gene banks.

Dr. Konwar published more than 130 articles on popular sciences, environment, biotechnology, history, national integration, higher education, research needs, and other articles in various Assamese magazines and newspapers, as well as more than 40 scientific articles in national English magazines. He has also published five books, three booklets, and 21 book chapters. His five books are *Lipase: An Industrial Enzyme Through Metagenomics; Prospects of Microbe and Medicinal Plant Resources; Medicinal Plant Repertoire: A Perspective of Biogeographical Gateway of India; Joimati-Gadapani: Ek Amar Dampatya Premgatha; and Deshapremi Pariyal: Barbaruah Barphukan* (History: Assamese).

Dr. Konwar earned his MSc in Agriculture from Assam Agricultural University, Jorhat, Assam, India, where he earned a gold medal. He obtained his MSc (Agriculture) degree in Plant Breeding and Genetics from the same university with the first-class distinction (above 80% marks). He served Assam Agricultural University (AAU) as a Lecturer in the Department of Plant Breeding and Genetics and soon became Assistant Professor. With a fellowship from the Govt of India, he studied at the Imperial College of Science, Technology, and Medicine, University of London, and obtained a DIC in Microbiology and PhD in Plant Biotechnology.

Contents

Abbreviations

AGCL	Assam Gas Company Limited
BATH	bacterial adhesion to hydrocarbons
BS	biosurfactant
BS	biosurfactant solution
CF	cystic fibrosis
CFU	colony-forming unit
CGMA	chemical gradient motility agar
CLP	cyclic depsipeptide
CMC	critical micelle concentration
CMD	critical micelle dilution
CTAB	cetyl trimethyl ammonium bromide
DMSO	dimethyl sulfoxide
DSC	differential scanning calorimetry
ES	electrospray
FTIR	Fourier transform infrared spectroscopy
GC	gas chromatography
GI	germination index
HBE	hyperbranched epoxy
HLB	hydrophilic-lipophilic balance
HTS	high-throughput screening
IFT	interfacial tension
IOCL	Indian Oil Corporation Limited
ION-RL	iron oxide nanocrystal-rhamnolipid
KNO_3	potassium nitrate
LB	Luria Bertani
LC-MS	liquid chromatography and mass spectroscopy
LPS	lipopolysaccharide
MAM	motility agar medium
MDR	multidrug resistance
MEL	mannosylerythritol lipid
MEOR	microbial enhanced oil recovery
MHA	Mueller Hinton agar
MHBE	modified hyperbranched epoxy

MIC	minimum inhibitory concentration
MSM	mineral salt medium
Na_2HPO_4	disodium hydrogen phosphate
$NaHCO_3$	sodium bicarbonate
NAPLs	non-aqueous phase liquids
NH_4Cl	ammonium chloride
NRL	Numaligarh Refinery Limited
OIL	Oil India Limited
OMMT	organically modified montmorillonite clay
ONGC	oil and natural gas corporation
ORFs	open reading frames
PAH	polycyclic aromatic hydrocarbons
PHA	polyaromatic hydrocarbons
PTSA	p-toluene sulfonic acid
RL	rhamnolipid
RT	room temperature
SD	sabouraud dextrose
SDA	sabouraud dextrose agar
SDS	sodium dodecyl sulfate
SEM	scanning electron microscopy
SNP	silver nanoparticles
SNP-RL	silver nanoparticle rhamnolipid
TEM	transmission electron microscopy
TGA	thermogravimetric analysis
TLC	thin layer chromatography
TPH	total petroleum hydrocarbon
UPLC	ultra-pure liquid chromatography
w/o	water-in-oil

Preface

Surfactants are a group of commonly used surface-active agents capable of lowering surface and interfacial tensions (IFTs) of liquids. Characteristically, they are organic amphipathic compounds containing the hydrophilic (polar) moiety usually referred to as the 'head,' and the nonpolar hydrophobic moiety known as the 'tail.' Because of their amphiphilic nature, surfactant molecules accumulate at the interfaces such as solid-liquid, liquid-liquid, or vapor-liquid. The hydrophobic portion concentrates at the surface with a strong attraction to the surrounding solution, while the hydrophilic portion is oriented towards the solution with weak to non-attraction forces. At the interface, these molecules reduce the free energy of the system by reducing the forces of repulsion between unlike phases or surfaces and allow the two phases to mix more easily. They can increase solubility, detergency power, wettability, and foaming capacity. Virtually, all surfactants are chemically synthesized from petroleum hydrocarbons. Nevertheless, in recent years, biosurfactants (BSs) have been receiving attention due to their diversity, selectivity, environmentally friendly nature, performance under extreme conditions, the possibility of large-scale production, and potential applications in environmental protection. Surfactants are used in detergents and cleaners, textiles, leather and paper, chemical processes, cosmetics, and pharmaceuticals, the food industry, and agriculture. World production of surfactants is estimated at 15 M ton per year, of which about half are soaps. The market is expected to grow further, with an annual average of 4.5%.

Surfactants have varying chemical structures according to ionic charge in the polar part. Anionic, cationic, nonionic, and zwitterionic surfactants exist. The hydrophobic moiety is usually a hydrocarbon chain of varying lengths and mainly synthesized from chemical-based materials such as petroleum-derived hydrocarbons, lignosulfonates, or triglycerides. Most of the synthetic surfactants are comprised of linear alkylbenzene sulphonates, alcohol sulfates, alcohol ether sulfates, alcohol glyceryl ether sulphonates, α-olefin sulphonates, alcohol ethoxylates, and alkylphenol ethoxylates.

For reducing hazardous wastes, greener processes with reduced waste products are being progressively integrated with modern developments.

Since most of the conventional surfactants are synthetic and of petroleum origin, they cost not only high but also pose potential threats to the environment due to their recalcitrant and persistent nature. The surfactant molecules at higher concentrations affect enzymes and other cellular proteins necessary for the basic functions of microorganisms. Prolonged exposure of skin to surfactants can cause chaffing because surfactants disrupt the lipid coating that protects skin and cells. With the increased environmental awareness among consumers and new stringent legislation, the environmental compatibility of surfactants has become an important factor in their application for various uses. As a result, attention has been paid to the alternative environment-friendly processes for the production of different types of surfactants from microorganisms. The use of biosurfactants in place of chemical surfactants can minimize the threats caused by synthetic surfactants.

Biosurfactants are produced by a large number of microorganisms such as bacteria, actinomycetes, fungi, yeast, etc. They are amphipathic molecules comprised of a hydrophilic portion represented by monosaccharides, oligosaccharides or polysaccharides; amino acids or peptides; carboxylate or phosphate groups; and a hydrophobic portion made of saturated or unsaturated (hydroxy) fatty acids or fatty alcohols. These molecules are partitioned at the interface between fluid phases with different degrees of polarity and hydrogen bonding, such as in oil/water or air/water interfaces. Accumulation of biosurfactant molecules at the interfaces ultimately reduces the surface and interfacial tension of the system. They impart better wetting, spreading, foaming, detergency, and emulsifying traits, rendering the most versatile process.

Attention towards biosurfactants has been gradually increasing in recent years due to the possibility of their production through fermentation technology and their potential applications in environmental protection. In spite of numerous advantages over synthetic chemical surfactants, biosurfactants are unable to compete with chemically synthesized surfactants. This could be due to their high production costs in relation to the inefficient bioprocessing techniques, poor strain productivity, and use of costly substrates. The success of biosurfactant production depends on the development of cheaper processes and the use of low-cost raw materials. Factors that need to be focused on to reduce the cost of production of biosurfactants are microorganisms, process, growth substrate or process feedstock, and the process byproduct.

CHAPTER 1

Introduction

Surfactants are surface-active agents with broad range properties including the lowering of surface and interfacial tensions (IFTs) of liquids (Christofi and Ivshina, 2002; Bordoloi and Konwar, 2007). These surface-active compounds are a group of commonly used chemicals in everyday life. Characteristically, these are organic amphipathic compounds containing the hydrophilic (polar) moiety usually referred to as the 'head,' and the nonpolar hydrophobic moiety known as the 'tail.' Because of their amphiphilic nature, surfactant molecules accumulate at the interfaces such as solid-liquid, liquid-liquid or vapor-liquid. The hydrophobic portion concentrates at the surface with a strong attraction to the surrounding solution, while the hydrophilic portion is oriented towards the solution with weak to non-attraction forces. At interface, these molecules reduce the free energy of the system by reducing the forces of repulsion between unlike phases or surfaces and allow the two phases to mix more easily. Surfactant molecules can lower the surface tension, increase solubility, detergency power, wettability, and foaming capacity (Mulligan, 2005). The unique surface-active properties of surfactants have been exploited in various areas such as detergency, emulsification, adhesion, coatings, wetting, foaming, soil, and water remediation, paints, chromatographic separations, medicine, agriculture, cosmetics, personal care and almost every sector of modern industry (Mulligan and Gibbs, 1993). Surfactants are also the key ingredients found in detergents, shampoos, toothpaste, oil additives, and several other consumer products.

Virtually all surfactants are chemically synthesized chiefly from petroleum hydrocarbons. Nevertheless, in the recent years, biosurfactants (BSs) have been receiving much more attention due to their diversity, selectivity, environmentally friendly nature, performance under extreme conditions, possibility of large-scale production, and potential applications in environmental protection (Urum and Pekdemir, 2004). Further, increasing environmental concern had also led to consider the biological surfactants in various applications, especially in bioremediation-related technologies.

1.1 SURFACTANTS MARKET

Surfactants are used in detergents and cleaners (54%); as auxiliaries for textiles, leather, and paper (13%); in chemical processes (10%); in cosmetics and pharmaceuticals (10%); in the food industry (3%); in agriculture (2%) and in others (8%) (Rahman and Gakpe, 2008). World production of surfactants is estimated at 15 M ton per year, of which about half are soaps. The other surfactants produced on a large scale are linear alkylbenzene sulfonates (1700 k ton per year), lignin sulfonates (600 k ton per year), fatty alcohol ethoxylates (700 k tons per year), alkylphenol ethoxylates (500 k ton per year) (Kosswig, 2005). The total production of the surfactant exceeded around 10 million tons in 2007 for their increasing demands in various industries such as polymers, lubricants, and solvents (Van Bogaert et al., 2007; Makkar et al., 2011). The market is expected to grow over the US $41 billion in 2018, with an average annual growth of 4.5% (Reznik et al., 2010). Such an increase in growth rate is related to the world demand for detergents since this sector uses over 50% of surfactant production (Deleu and Paquot, 2004). Out of the total production of surfactants, about 54% are consumed as household or laundry detergents, and only 32% are destined for industrial use (Singh et al., 2007; Franzetti et al., 2008; Develter and Lauryssen, 2010).

1.2 CHEMICAL NATURE OF SURFACTANT

Surfactants are classified with varying chemical structures according to their ionic charge residing in the polar part of the molecule. Anionic, cationic, nonionic, and zwitterionic (combined presence of anionic and cationic charges) surfactants exist (Christofi and Ivshina, 2002; Urum and Pekdemir, 2004) in nature. The hydrophobic moiety is usually a hydrocarbon chain of varying lengths in different surfactants (Mulligan and Gibbs, 1993). Surfactants are mainly synthesized from chemical-based materials such as petroleum-derived hydrocarbons, lignosulfonates, or triglycerides (Mulligan and Gibbs, 1993). The majority of the synthetic surfactants include linear alkylbenzene sulfonates, alcohol sulfates, alcohol ether sulfates, alcohol glyceryl ether sulfonates, α-olefin sulfonates, alcohol ethoxylates, and alkylphenol ethoxylates (Desai and Banat, 1997). Surfactants are potentially useful in every industry dealing with multiphase systems such as sodium dodecyl sulfate (SDS, $C_{12}H_{25}SO_4^-$ Na^+), a widely used anionic surfactant. It contains a straight-chain aliphatic hydrocarbon with a sulfate group.

1.3 ACTIONS OF SURFACTANTS

Surfactants enhance the aqueous solubility of non-aqueous phase liquids (NAPLs) by reducing their surface/interfacial tension (IFT) at air-water and water-oil interfaces (Urum and Pekdemir, 2004). The effectiveness of a surfactant is determined by its ability to lower the surface tension, which is a measure of the surface free energy per unit area or the work required to bring a molecule from the bulk phase to the surface (Rosen, 1978). Another distinguishing character of surfactants is their ability to self-assemble in solution into dynamic aggregates called micelles. As the IFT is reduced at the air-liquid interface and the aqueous surfactant concentration increased, the monomers aggregate to form micelles. The minimum concentration of the surfactant at which micelles first initiate to form is referred to as critical micelle concentration (CMC) (Becher, 1965). The concentration corresponds to the point where the surfactant molecules reduce the maximum surface tension of the system. The CMC of a surfactant is influenced by pH, temperature, and ionic strength (Mulligan and Gibbs, 1993). The physical properties used to characterize surfactants that depend on CMC include emulsion formation, oil solubilization, foaming, and detergency, interfacial, and surface tensions. Generally, the dispersions of the complex fluids form and stabilize by absorbing surfactants onto air-liquid interface, known as foams. These dispersed fluids contain small bubbles with large surface areas, which can be stabilized using surfactants. Formation of heavy foams signifies the better detergency character which indicates the effectiveness of the surfactant in separating oily material from a medium.

The dynamics of the surfactant adsorption is of vast significance for practical applications such as foaming, emulsifying or coating processes, where bubbles or drops are rapidly generated and need to be stabilized. The dynamics of surfactant adsorption depend on its diffusion coefficient. As the interface is created, the adsorption of the surfactant is limited by the diffusion at the interface. In certain cases, there is an existence of energy barrier for the adsorption or desorption of the surfactants at the interface, then the adsorption dynamics is known as 'kinetically limited.' Such energy barrier is due to the steric or electrostatic repulsions between the surfactant molecules. The surface rheology of surfactant layers, including their elasticity and viscosity plays a very vital role in foam or emulsion stability. The immiscible liquids such as oil and water, when mixed, one liquid is dispersed into the other, and small droplets form emulsion. An emulsion is defined as a "heterogeneous system," consisting of at least one immiscible liquid dispersed in another in the form of droplets, whose diameters, in general, exceed 0.1 mm (Christofi and Ivshina, 2002; Bordoloi and Konwar, 2007). The behavior of the emulsion is related to the equilibrium phase behavior of the oil/water/surfactant system from which it is made. Formation of such small droplets provides a large amount of interfacial surface area and hence greater interfacial free energy in the system. In normal condition, the droplets may rapidly coalesce, and two separate phases will form to minimize the interfacial area of the system (Becher, 1965). With the addition of surfactants, emulsion formed may stabilize by reducing the IFT and decreasing the rate of coalescence (Rosen, 1978). Usually there are two types of emulsion, i.e., water-in-oil (w/o) or oil-in-water (o/w). The term hydrophilic-lipophilic balance (HLB) is used to classify which type of emulsion the emulsifier will favor.

1.4 DISADVANTAGES OF SYNTHETIC SURFACTANTS

As a part of the global effort to reduce hazardous wastes, greener processes with reduced waste products are being progressively integrated with modern developments (Kiran et al., 2010). Since all the conventional surfactants commercially available today are synthetic and of petroleum origin, they not only cost high but also pose potential threats to the environment due to their recalcitrant and persistent nature (Salihu et al., 2009; Aparna et al., 2012). There is a concern on the possible toxic effect of the synthetic surfactants on aquatic organisms, especially if they are used in nearshore waters (Otitoloju and Popoola, 2009). Besides these, the effects of synthetic surfactants on biostimulation of indigenous microorganisms in enhancing the removal of organic pollutants yielded inconsistent results. Such a decrease in the rate of biodegradation of organic pollutants, especially at higher concentrations of surfactant could be due to the interaction of surfactant with the lipid membrane. It is a well-known fact that surfactant molecules at higher concentration affect enzymes and other cellular proteins necessary for basic functions of the microorganisms (Sun et al., 2008). Most anionic and nonionic surfactants are nontoxic, having LD50 comparable to sodium chloride. Prolonged exposure of skin to surfactants can cause chaffing because surfactants disrupt the lipid coating that protects skin and cells (Kosswig, 2005). Moreover, most of synthetic surfactants during the manufacturing process and the byproducts cause serious environmental hazards (Makkar and Cameotra, 2002). With the increased environmental awareness among consumers and new stringent legislation, environmental compatibility of surfactants has become an important factor in their application for various uses (Maier and Soberon-Chavez, 2000; Sarubbo et al., 2007). As a result, with advances in biotechnology, attention has been paid to the alternative environment friendly processes to produce different types of surfactants from microorganisms (Lotfabad et al., 2009). Hence, the use of biosurfactants (BSs) in place of chemical surfactants can minimize the threats caused by the synthetic surfactant.

1.5 GREEN SURFACTANT OR BIO-SURFACTANT

In ancient times some natural surfactants such as soap (fatty acid salt), lecithin (phospholipid) and saponins (glycolipid) were extracted from plants or animals and widely used in households and industry. Natural surfactants are usually present in lesser quantity in their natural sources and the cost involvement in their extraction procedure exceeds the cost of chemical synthesis (Kitamoto et al., 2002). Therefore, the investigation for an alternative source of natural surfactant is significant.

Investigation of the literature indicates that the capability to produce natural surfactant, well-known as BSs is prevalent among the bacterial and archaeal domains (Bodour et al., 2003). BSs are heterogeneous groups of surface-active microbial surfactants produced by a wide variety of microorganisms such as bacteria, actinomycetes, fungi, yeast, etc. (Xia et al., 2011). These surface-active molecules have been reported as being produced on the microbial cell surfaces and excreted extracellularly (Sim et al., 1997). Basically, they are amphipathic molecules that comprise a hydrophilic portion, which might consist of monosaccharides, oligosaccharides, or polysaccharides, amino acids or peptides, or carboxylate or phosphate groups, and a hydrophobic portion, which is composed of saturated or unsaturated (hydroxy) fatty acids or fatty alcohols. These molecules are partitioned at the interface between fluid phases with different degrees of polarity and hydrogen bonding, such as in oil/water or air/water interfaces (Kulkarni et al., 2007). Such accumulation of BS molecules at the interfaces ultimately reduces the surface and IFT of the system. They impart better wetting, spreading, foaming, detergency, and emulsifying traits, rendering them the most versatile process chemicals (Banat et al., 2000). These qualities make them more competitive and suitable to various application needs (Cameotra and Makkar, 2004; Nitschke et al., 2005; Van Hamme et al., 2006).

1.6 BIOSURFACTANTS (BSS) VS. SYNTHETIC SURFACTANTS

The most attractive aspects of BS use are their biodegradability and ecological acceptance (Bafghi and Fazaelipoor, 2012). BSs could retain their surface-active properties even under extreme conditions of temperature, pH, salinity, and metal salts (Xia et al., 2011; Ron and Rosenberg, 2002; Joshi et al., 2008). Due to the low irritancy and compatibility with human skin (Pornsunthorntawee et al., 2009), they are constantly used in the sector of cosmetics and pharmaceutical industries. Because of these properties, BSs have a broad range of potential applications, including in detergent, pharmaceuticals, agriculture, cosmetics, food, cleanser, paint, and petroleum-based industries (Lotfabad et al., 2009; Banat et al., 2000). Other main advantages of BSs include bioavailability, activity under diverse conditions, ecological acceptability, low toxicity, their capacity to be modified by biotechnological techniques, and their capability of increasing the bioavailability of poorly soluble organic compounds, such as polyaromatic hydrocarbons (PHA) (Jain et al., 2012). From an environmental standpoint, BSs have promising applications in various fields, including bioremediation of contaminated environments, tertiary oil recovery such as MEOR (microbial enhanced oil recovery), flotation process, detergent formulations, etc. BSs could also be easily produced from renewable resources through microbial fermentation. In recent years, BSs have been gaining much attention in the field of nanobiotechnology because of their unique chemical composition (Palanisamy and Raichur, 2009; Koopmans and Aggeli, 2010). Various aspects of BSs, such as their biomedical and therapeutic properties as well as natural roles, have been recently reviewed (Rodrigues et al., 2006; Gudiana et al., 2010).

1.7 PROBLEMS OF COMMERCIALIZATION OF BIOSURFACTANT (BS)

Attention towards BSs has been gradually increasing in recent years due to the possibility of their production through fermentation technology and their potential applications in specific areas such as environmental protection. Despite numerous advantages over the synthetic chemical surfactants, BSs are still unable to compete with the chemically synthesized surfactants in the surfactant market. This could be due to their high production costs in relation to the inefficient bioprocessing techniques, poor strain productivity and the need to use costly substrates (Deleu and Paquot, 2004; Pornsunthorntawee et al., 2009). This has led to concerted efforts during the last decade, focused on minimizing production costs in order to facilitate wider commercial use. The success of BS production depends on the development of cheaper processes and the use of low-cost raw materials, which account for 10–30% of the overall cost. Four factors need to be focused to reduce the cost of production of BSs: the microbes, process, microbial growth substrate or process feedstock, and the process byproducts (Kosaric et al., 1984).

KEYWORDS

- **critical micelle concentration**
- **hydrophilic-lipophilic balance**
- **microbial enhanced oil recovery**
- **non-aqueous phase liquids**
- **polyaromatic hydrocarbons**
- **sodium dodecyl sulfate**
- **water-in-oil**

REFERENCES

Aparna, A., et al., (2012). Production and characterization of biosurfactant produced by a novel *Pseudomonas* sp. 2B. *Colloids Surf. B. Biointerfcaes, 95*, 2–29.

Bafghi, M. K., & Fazaelipoor, M. H., (2012). Application of rhamnolipid in the formulation of a detergent. *J. Surfact. Deterg., 15*, 679–684.

Banat, I. M., et al., (2000). Potential commercial applications of microbial surfactants. *Appl. Microbiol. Biotechnol., 53*, 495–508.

Becher, P., (1965). In: *Emulsions, Theory and Practice* (2nd edn.). Reinhold Publishing, New York.

Bodour, A. A., et al., (2003). Distribution of biosurfactant-producing bacteria in undisturbed and contaminated arid southwestern soils. *Appl. Environ. Microbiol., 69*, 3280–3287.

Bordoloi, N. K., & Konwar, B. K., (2007). Microbial surfactant-enhanced mineral oil recovery under laboratory conditions. *Colloids Surf. B. Biointerfaces, 63*, 73–82.

Cameotra, S. S., & Makkar, R. S., (1998). Synthesis of biosurfactants in extreme conditions. *Appl. Microbiol. Biotechnol., 50*, 520–529.

Cameotra, S. S., & Makkar, R. S., (2004). Recent application of biosurfactants as biological and immunological molecules. *Curr. Opin. Microbiol., 7*, 262–266.

Cameotra, S. S., & Makkar, R. S., (2010). Biosurfactant-enhanced bioremediation of hydrophobic pollutants. *Pure Appl. Chem., 82*, 97–116.

Christofi, N., & Ivshina, I. B., (2002). Microbial surfactants and their use in field studies of soil remediation. *J. Appl. Microbiol., 93*, 915–929.

Deleu, M., & Paquot, M., (2004). From renewable vegetables resources to microorganisms: New trends in surfactants. *Comptes. Rendus. Chimie., 7*, 641–646.

Desai, J. D., & Banat, I. M., (1997). Microbial production of surfactants and their commercial potential. *Microbiol. Mol. Bio. Rev., 61*, 47–64.

Develter, D. W. G., & Lauryssen, L. M. L., (2010). Properties and industrial applications of sophorolipids. *Eur. J. Lipid Sci. Technol., 112*, 628–638.

Franzetti, A., et al., (2008). Surface-active compounds and their role in bacterial access to hydrocarbons in *Gordonia* strains. *FEMS Microbiol. Ecol., 63*, 238–248.

Gudiana, E. J., et al., (2010). Isolation and functional characterization of a biosurfactant produced by *Lactobacillus paracasei*. *Colloids Surf. B. Biointerfcaes, 16*, 298–304.

Jain, R. M., et al., (2012). Isolation and structural characterization of biosurfactant produced by an alkaliphilic bacterium *Cronobacter sakazakii* isolated from oil-contaminated wastewater. *Carbohydrate Polym., 87*, 2320–2326.

Joshi, S., et al., (2008). Biosurfactant production using molasses and whey under thermophilic conditions. *Bioresour. Technol., 99*, 195–199.

Kiran, S. G., et al., (2010). Optimization and characterization of a new lipopeptide biosurfactant produced by marine *Brevibacterium aureum* MSA 13 in solid-state culture. *Bioresour. Technol., 101*, 2389–2396.

Kitamoto, D., et al., (2002). Functions and potential applications of glycolipid biosurfactants-from energy-saving materials to gene delivery carriers. *J. Biosci. Bioeng., 94*, 187–201.

Koopmans, R. J., & Aggeli, A., (2010). Nanobiotechnology-quo Vadis? *Curr. Opinion Microbiol., 13*, 327–334.

Kosaric, N., et al., (1984). The role of nitrogen in multiorganism strategies for biosurfactant production. *J. Am. Oil. Chem. Soc., 61*, 1735–1743.

Kosswig, K., (2005). Surfactants. In: *Ullmann's Encyclopedia of Industrial Chemistry*. Wiley-VCH, Weinheim.

Kulkarni, M., et al., (2007). Novel tensio-active microbial compounds for biocontrol applications. In: Ciancio, A., & Mukherji, K. G., (eds.), *General Concept Integrated Pest and Disease Management* (pp. 61–70). Springer, Netherlands.

Lotfabad, T. B., et al., (2009). An efficient biosurfactant-producing bacterium *Pseudomonas aeruginosa* MR01, isolated from oil excavation areas in south of Iran. *Colloids Surf. B. Biointerfaces, 69*, 183–193.

Maier, R. M., & Soberon-Chavez, G., (2000). *Pseudomonas aeruginosa* rhamnolipids: biosynthesis and potential applications. *Appl. Microbiol. Biotechnol., 54*, 625–633.

Makkar, R. S., & Cameotra, S. S., (2002). An update on the use of unconventional substrates for biosurfactant production and their new applications. *Appl. Microbiol. Biotechnol., 58*, 428–434.

Makkar, R. S., et al., (2011). Advances in utilization of renewable substrates for biosurfactant production. *AMB Express, 1*, 5–10.

Mulligan, C. N., & Gibbs, B. F., (1993). Factors influencing the economics of biosurfactants. In: Kosaric, N., (ed.), *Biosurfactants, Properties, Applications* (pp. 329–371). Marcel Dekker, New York.

Mulligan, C. N., (2005). Environmental applications for biosurfactants. *Environ. Pollut., 133*, 183–198.

Nitschke, M., et al., (2005). Rhamnolipid Surfactants: An update on the general aspects of these remarkable biomolecules. *Biotechnol. Prog., 21*, 1593–1600.

Otitoloju, A. A., & Popoola, T. O., (2009). Estimation of "environmentally sensitive" dispersal ratios for chemical dispersants used in crude oil spill control. *Environmentalist, 29*, 371–380.

Palanisamy, P., & Raichur, A. M., (2009). Synthesis of spherical NiO nanoparticles through a novel biosurfactant mediated emulsion technique. *Mater. Sci. Eng. C., 29*, 199–204.

Pornsunthorntawee, O., et al., (2009). Solution properties and vesicle formation of rhamnolipid biosurfactants produced by *Pseudomonas aeruginosa* SP4. *Colloids Surf. B. Biointerfaces, 72*, 6–15.

Rahman, P. K. S. M., & Gakpe, E., (2008). Production, characterization and application of biosurfactants-review. *Biotechnol., 7*, 360–370.

Reznik, G. O., et al., (2010). Use of sustainable chemistry to produce an acyl amino acid surfactant. *Appl. Microbiol. Biotechnol., 86*, 1387–1397.

Rodrigues, L. R., et al., (2006). Biosurfactants: Potential applications in medicine. *J. Antimicrob. Chemother., 57*, 609–618.

Ron, E., & Rosenberg, E., (2002). Biosurfactants and oil bioremediation. *Curr. Opinion Biotechnol., 13*, 249–252.

Rosen, M. J., (1978). *Surfactants and Interfacial Phenomena* (pp. 149–171). John Wiley & Sons, New York.

Salihu, A., et al., (2009). An investigation for potential development on biosurfactants. *Biotechnol. Mol. Biol. Rev., 3*, 111–117.

Sarubbo, L. A., et al., (2007). Co-utilization of canola oil and glucose on the production of a surfactant by *Candida lipolytica*. *Curr. Microbiol., 54*, 68–73.

Sim, L., et al., (1997). Production and characterization of a biosurfactant isolate from *Pseudomonas aeruginosa* UW-1. *J. Ind. Microbiol. Biotechnol., 19*, 232–238.
Singh, A., et al., (2007). Surfactants in microbiology and biotechnology: Part 2. Application aspects. *Biotechnol. Adv., 25*, 99–121.
Sun, N., Wang, H., Chen, Y., Lu, S., & Xiong, Y., (2008). Effect of surfactant SDS, tween 80, triton X-100 and rhamnolipid on biodegradation of hydrophobic organic pollutants. In: *2nd International Conf. Bioinfor. and Biomed. Engg.* (pp. 4730–4734). Shanghai, China.
Urum, K., & Pekdemir, T., (2004). Evaluation of biosurfactant for crude oil contaminated soil washing. *Chemosphere, 57*, 1139–1150.
Van, B. I., et al., (2007). Microbial production and application of sophorolipids. *Appl. Microbiol. Biotechnol., 76*, 23–34.
Van, H. J. D., et al., (2006). Physiological aspects: Part 1: In a series of papers devoted to surfactants in microbiology and biotechnology. *Biotechnol. Adv., 24*, 604–620.
Xia, W., et al., (2011). Comparative study of biosurfactant produced by microorganisms isolated from formation water of petroleum reservoir. *Colloids Surf. A Physicochem. Eng. Aspects, 392*, 124–130.

FURTHER READING

Bharali, P., & Konwar, B. K., (2011). Production and physicochemical characterization of a biosurfactant produced by *Pseudomonas aeruginosa* OBP1 isolated from petroleum sludge. *Appl. Biochem. Biotechnol., 164*, 1444–1460.
Bharali, P., Das, S., Konwar, B. K., & Thakur, A. J., (2011). Crude biosurfactant from thermophilic *Alcaligenes faecalis*: Feasibility in Petro-spill bioremediation. *Inter. Biodeter. Biodegr., 65*, 682–690.
Das, S., Kalita, S. J., Bharali, P., Konwar, B. K., Das, B., & Thakur, A. J., (2013). Organic reactions in "green surfactant": An avenue to bisuracil derivative. *ACS Sustainable Chem. Eng.*, (p. 301). doi: 10.1021/sc4002774. Publication Date (Web).
Pranjal, B., (2015). *Bioremediation of Crude Oil Contaminated Soil.* (Thesis: Supervisor B K Konwar), Dept. of Mol. Biol. and Biotechnology, Tezpur University (Central), Napaam-784028, Assam, India.

CHAPTER 2

Classification of Biosurfactants

Most commonly, surfactants are generally categorized according to the type of the polar group present. Based on the chemical composition of the polar head group, they are classified as nonionic, anionic, cationic, and zwitterionic (Christofi and Ivshina, 2002; Urum and Pekdemir, 2004). Rosenberg and Ron (2002) suggested that biosurfactants (BSs) could be divided into low molecular weight molecules and high molecular weight polymers. The lower molecular weight BSs which include glycolipids, lipopeptides, flavolipids, corynomycolic acids, and phospholipids, efficiently lower the surface and interfacial tensions (IFTs) at the air/water interfaces. The high molecular weights polymers also known as bioemulsans which includes emulsan, alasan, liposan, polysaccharides, and protein complexes, are highly efficient emulsifiers that work at low concentrations, exhibit considerable substrate specificity and are more effective in stabilizing oil-in-water emulsions (Franzetti et al., 2008; Salihu et al., 2009). However, general classification of BS is based on the parent chemical structure and their surface properties and the major classes of BSs are: (i) glycolipids, (ii) phospholipids and fatty acids, (iii) lipopeptide/lipoproteins, (iv) particulate surfactants, and (v) polymeric surfactants.

2.1 MAJOR CLASSES OF BIOSURFACTANTS (BSS)

2.1.1 GLYCOLIPIDS

Glycolipids are carbohydrate groups in combination with long-chain aliphatic acids or hydroxy aliphatic acids. The connection is by means of either an ether or ester group.

Glucosyl diglycerides present in the cell membrane of a wide variety of bacteria are the most common glycolipids; these are discussed in subsections.

2.1.1.1 TREHALOSE LIPIDS

Microbial trehalolipid, a glycolipid type biosurfactant (BS) produced by most of the sp. of *Mycobacterium*, *Corynebacterium*, and *Nocardia*. Trehalolipid consists of disaccharide trehalose linked to C-6 and C-6′ to mycolic acid. Mycolic acids are the long-chain, α-branched, and β-hydroxy fatty acids. Trehalolipids from diverse organisms differ in the size and structure of mycolic acid, the number of carbon atoms present and the degree of unsaturation (Asselineau and Asselineau, 1978). *Rhodococcus erythropolis* and *Arthrobacter* sp. were reported to produce trehalolipid that reduces the surface tension and interfacial tension (IFT) of the culture broth (Figure 2.1) (Kretschmer et al., 1949).

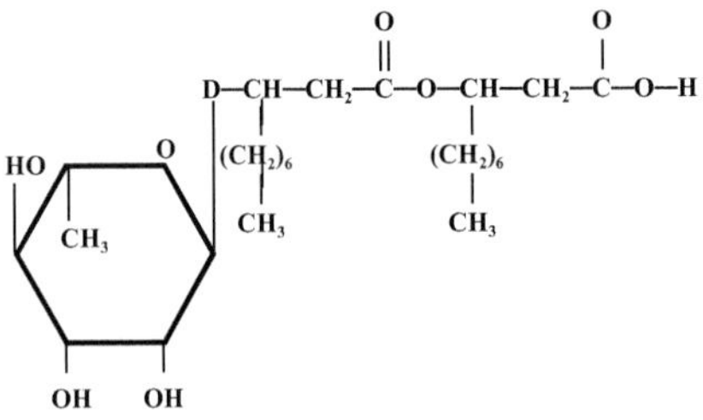

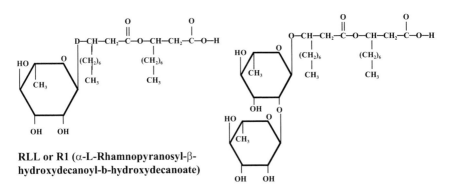

RLL or R1 (α-L-Rhamnopyranosyl-β-hydroxydecanoyl-b-hydroxydecanoate)

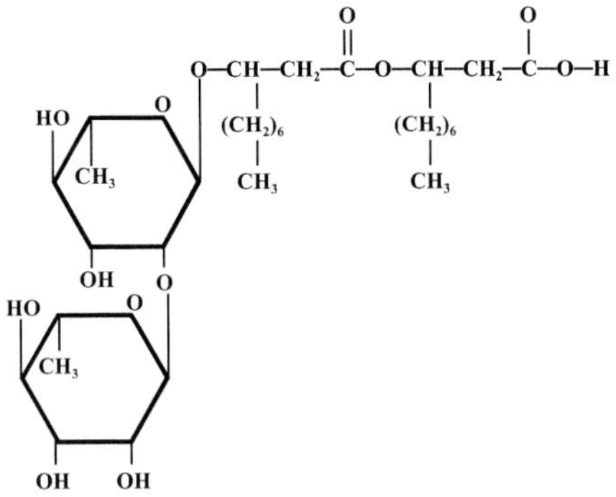

RRLL or R3 (2-O-α-L-Rhamnopyranosyl-α-L-rhamnopyranosyl-β-hydroxydecanoyl-β-hydroxydecanoate)

FIGURE 2.1 Chemical structure of rhamnolipids.

2.1.1.2 RHAMNOLIPIDS (RLS)

Rhamnolipids (RLs) are composed of hydrophilic head formed by one or two rhamnose molecules, known as mono- and di-rhamnolipid, and a hydrophobic tail which contains up to three molecules of hydroxyl fatty acids of varying chain length from 8 to 14 of which β-hydroxydecanoic acid is predominant (Banat et al., 2000; Monterio et al., 2007; Silva et al., 2010). Almost all RL species possess similar chemical structure and have an average molecular weight of 577 (Wei et al., 2005; Dos Santos et al., 2010). The bacterial genus *Pseudomonas* has been highlighted for its ability to use diverse hydrophobic or hydrophilic substrates such as hydrocarbons, vegetable oils, carbohydrates, or even wastes from the food industry as carbon source to produce RL-type BSs (Makkar et al., 2011; Bafghi and Fazaelipoor, 2012; Dos Santos et al., 2010).

Jarvis and Johnson (1949) reported the production of RL from *Pseudomonas aeruginosa*, an important BS with tremendous applications both in industrial and environmental sector. RL usually contains one or two molecules of rhamnose, which are connected to one or two molecules of β-hydroxydecanoic acid. The -OH group of one of the acids is involved in glycosidic linkage with the reducing end of the rhamnose disaccharide, the -OH group of the second acid is occupied in ester formation (Karanth et al., 1999). L-Rhamnosyl-Lrhamnosyl-β-hydroxydecanoyl-β-hydroxydecanoate and L-rhamnosyl-β-hydroxide canoyl-β-hydrtocyde canote, known as RL 1 and 2, respectively, are principal glycolipids synthesized by *P. aeruginosa* (Edward and Hayashi, 1965) that can lower the IFT against n-hexadecane to about 1 mNm^{-1} and had a CMC of 10 ± 30 $mg.l^{-1}$ depending on pH and salt concentration. Bharali et al. (2011, 2013) used RL impregnated nanocomposite as a potential antibiotic. Saikia et al. (2013) highlighted the application of RLs in the synthesis and protection of nanocomposites

The production of RL is known to be regulated by quorum sensing mechanism and it depends on various environmental and nutritional factors like pH, temperature, phosphates, and iron content, as well as the nature of the carbon source (Lebron-Paler, 2008).

2.1.1.3 SOPHOROLIPIDS

Sophorolipids are mainly produced by yeast such as *Torulopsis bombicola* (Cooper and Paddock, 1984; Hommel, 1987). These molecules mainly composed of a dimeric carbohydrate sophorose attached to a long-chain hydroxyl fatty acid by a glycosidic linkage. Generally, they are found as a mixture of free acid and macro-lactones. These BSs are a combination of at least 6–9 varied hydrophobic sophorolipids. Hu and Ju (2001) reported the uses of lactone of sophorolipid in various applications. The sophorolipids lower surface and IFTs, although they are not effective emulsifying agents (Poremba et al., 1991). The pure lactonic sophorolipid (10 $mg.l^{-1}$) lowered the IFT between n-hexadecane and water from 40 mNm^{-1} to about 5 mNm^{-1}, relatively independently of pH (6–9), salt concentration and temperature (20 ± 90°C) (Ron and Rosenberg, 2002).

The component consisting of four β-hydroxydecanoic acids linked together by ester bonds is coupled glycosidically with C-1 of glucose (Abraham et al., 1998). *Alcanivorax borkumensis*, a marine bacterium was also reported to produces an anionic glucose lipid type BS consisting of a tetrameric oxyacyl side chain with N-terminal esterified with glycine. Schulz et al. (1991) reported a marine *Arthrobacter* sp. SI1 that produced *disaccharide* trehalose (trehalose tetraester) and trehalose dicorynomycolates (trehalose diester) when grown on mihagol-S and ethanol separately as carbon sources. The minimal IFTs (between aqueous salt solutions and n-hexadecane) achieved with corynomycolic acids, trehalose monocorynomycolates, and trehalose dicorynomy-colates were 6, 16, and 17 mNm^{-1} respectively (Ron and Rosenberg, 2002). Suzuki et al. (1969) identified trehalose dimycolates, a type of glycolipid, present in the emulsion layer of culture broths of *Arthrobacter paraffineus* during their growth on hydrocarbon substrates. Wagner and co-workers have carried out extensive studies of trehalose dimycolates produced by *Rhodococcus erythropolis* with special reference to their interfacial activities and possible application in enhanced oil recovery (Kim et al., 1990). Mannosylerythritol lipids (MELs), extracellular microbial surfactants, have several interesting and biological properties such as they inhibit growth of human promyelocytic leukemia cell lines and induce monocytic differentiation (Isoda et al., 1997).

2.1.2 PHOSPHOLIPIDS AND FATTY ACIDS

Certain hydrocarbon-degrading bacteria and yeast produce appreciable amounts of phospholipids and fatty acids when grown on *n*-alkanes (Cirigliano and Carman, 2015). These surfactants are to produce optically clear micro-emulsions of alkanes in water. The important candidates are saturated fatty acids in the range of C_{12} to C_{14} and the complex fatty acids containing hydroxyl groups and alkyl branches (MacDonald and Chandler, 1981). Phospholipids produced by *Thiobacillus thioxidan* could help in wetting of elemental sulfur. Miyazima et al. (1985) reported that *Aspergillus* sp. could produce appreciable amounts of phospholipids when grown on hydrocarbons. Kretschmer et al. (1982) observed that certain hydrocarbon-degrading microbes could produce extracellular free fatty acids when grown on alkanes and exhibited good surfactant activity. Wayman et al. (1984) and Robert et al. (1989) reported that *Arthobacter* strain AK-19 and *P. aeruginosa* Ti could accumulate up to 40–80% (w/w) of such lipids when cultivated on hexadecane and olive oil, respectively.

2.1.3 LIPOPEPTIDE/LIPOPROTEINS

Decapeptide (gramicidins) and lipopeptide (polymyxins) antibiotics produced by *B. brevis* (Marahiel et al., 1977) and *B. polymyxa* (Suzuki et al., 1969) respectively possess remarkable surface-active properties. Similarly, peptides containing lipids exhibit BS activity. Surfactin, a cyclic lipopeptide, reported by Arima et al. (1968) in *B. subtilis* ATCC-21332 is one of the most effective BSs known so far. The ability of surfactin to lyse red blood cells is of limited use, but this discovery has led to the development of a quick method for the screening of BS producing microbes (Mulligan, 2005). Surfactant BL-86 was reported to be a mixture of lipopeptides with the major components ranging from 979–1091 Daltons with varying increments of 14 Daltons. There are seven amino acids per molecule while, lipid portion is composed of 8–9 methylene groups, a mixture of linear and branched tails (Horowitz and Griffin, 1991).

McInerney et al. (1990) observed that lichenysin is a lipopeptide surfactant produced by *B. licheniformis* JF2 has similar structure and physiochemical properties like that of surfactin. This species could produce several other surface-active agents, which could act synergistically and exhibit excellent temperature, pH, and salt stability. Horowitz (1991) observed that BL-86, a surfactant produced by *B. licheniformis* BL-86, lowered surface tension of water to 27 dynes.cm^{-1} and IFT between water and n-hexadecane to 0.36 dynes.cm^{-1}. Horowitz and Currie (1990) reported stability of the surfactant over a wide range of pH, temperature, and NaCl concentrations and could promote dispersion of colloidal 3-silicon carbide and aluminum nitrite slurries much more efficiently then the commercial agents.

The lipopeptide and lipoprotein BSs are mostly produced by bacterial species belonging to Bacillus genus, and they are:

1. **Surfactin:** *B. subtilis* produces cyclic lipopeptide BS known as surfactin that has various potential applications. It is composed of seven amino-acid rings joined to a fatty acid chain by means of lactone linkage. It reduces the surface tension from 72 to 27.9 mNm^{-1} at concentration as low as 0.005% and shows a minimum IFT against hexadecane up to 1 mNm^{-1} (Arima et al., 1968).
2. **Iturin:** Iturin A was isolated from a *B. subtilis* obtained from soil (Delcambe and Devignat, 1957). The Iturin groups of compounds are cyclic lipo-heptapeptides which contain β-amino fatty acid in

its side chain and reported to have potent antifungal property which can be used as biopesticide (Vater et al., 2002). The molecules have remarkable efficacy against a broad variety of clinically important pathogenic fungi and yeast strains. However, their application in medicine is limited because of possible toxicity (Bonmatin et al., 2003).

3. **Fengycin:** It is a type of lipopeptide BS, consists of lipodecapeptide with β-hydroxy fatty acid in its side chain (Vater et al., 2002). The compound comprises of C_{15}-C_{17} variants, which have characteristic Ala-Val dimorphy at position 6 of the peptide ring (Vanittanakom et al., 1986).
4. **Lichenysin:** *B. licheniformis* produces lichenysin which acts synergistically and exhibit stability at extreme temperature, pH, and salinity. These molecules are similar in their structural and physicochemical properties to surfactin (McInerney et al., 1990). Lichenysin BS produced by *B. licheniformis* can reduce the surface tension of water to 27 mNm^{-1} and the IFT between water and n-hexadecane to 0.36 mNm^{-1}. Several lipopeptides, including gramicidin and polymyxin show both antibiotic and surface-active properties (Rosenberg and Ron, 1999). The polymixins are a group of closely related lipopeptide antibiotics produced by *B. polymyxa*, *B. brevis* and other related bacilli (Marahiel et al., 1977). Polymyxin B is composed of deca-peptide in which amino acids 3 ± 10 form a cyclic octapeptide. A branched-chain fatty acid is connected to the terminal 2, 4-diaminobutyric acid. The synthesis of one or more peptide antibiotic during the early stages of sporulation is specific characteristics common to some members of Bacillus (Katz and Demain, 1977).

2.1.4 PARTICULATE BIOSURFACTANTS (BSS)

The extracellular-membrane vesicles tend to partition hydrocarbons to form a microemulsion which plays an important role in alkane uptake by microbial cells (Rosenberg et al., 1979). *Acinetobacter* sp. HO1-N was reported to secrete extracellular vesicles having a diameter of 20–50 nm and a buoyant density of 1.158 cubic g cm^{-1}. Such vesicles are mainly consisting of protein, phospholipids, and lipopolysaccharide (LPS) (Kappeli and Finnerty, 1979). The vesicles tend to partition hydrocarbons in the form of microemulsion and play an important role in alkane uptake by the cells. The purified vesicles are composed of proteins, phospholipids, and LPSs. The vesicles have phospholipid five times higher and polysaccharide content, 360-fold higher than that observed in the outer membrane of the same organism. The surface components that contribute to the surfactant activity include M-protein and lipoteichoic acid on *Streptococci* group-A, protein-A of *Staphylococcus aureus,* layer-A of *Aeromonas salmonicids*, prodigiosin of *Serratia* spp., gramicidins in *B. brevis* spores, and thin fimbriae in *A. calcoaceticus* RAG-1 (Rosenberg et al., 1988; Fattom and Shilo, 1985).

The cellular lipid content of *Pseudomonas nautical* reported to be increased in eicosane grown cells up to 3.2-fold, compared with acetate-grown cells. Zinjarde and Pant (2002) reported that phospholipids, mainly the phosphatidyl-ethanolamines and phosphatidylglycerides, were accumulated in eicosanes-grown cells of *Pseudomonas nautical.* For *Sphingomonas* sp., the cell surface of bacteria of this strain was covered with extracellular vesicles when grown on polyaromatic hydrocarbons (PHA). Nevertheless, the surfaces were smooth when cells were grown on a hydrophilic substrate such as acetate (Maneerat, 2005).

2.1.5 POLYMERIC SURFACTANTS

These are high molecular weight biopolymers which exhibit useful properties, such as high viscosity, tensile strength and resistance to shear, and are known to have a variety of industrial applications. The best-studied polymeric BSs are emulsan, biodispersan, liposan, and other polysaccharide-protein complexes (Desai and Desai, 1993; Maneerat, 2005).

The composition of emulsanosol to be 70–75% oil. Emulsanosols remain stable for months and have the ability to withstand enormous shear without any inversion. Shoham et al. (1983) isolated the enzyme responsible for the depolymerization of emulsion by transelimination. Navonvenezia et al. (1995) first described alasan as an anionic alanine-containing hetero-polysaccharide protein produced by *Acinetobacter radioresistens* KA-53. Sar and Rosenberg (1983) demonstrated that polysaccharide had no emulsification activity alone but become a potent emulsifier when combined with some proteins released during growth on ethanol. Cirigliano and Carman (1985) reported that *Candida lipolytica* could produce an extracellular water-soluble emulsifier called liposan, which is composed of 83% carbohydrate and 17% protein. The carbohydrate portion is a heteropolysaccharide consisting of glucose, galactose, galactosamine, and galacturonic acid.

The production of large amounts of mannoprotein by *S. cerevisiae* exhibiting excellent emulsifier activity toward several oils, alkanes, and organic solvents was reported by Cameron et al. (1998). The purified emulsifier contained 44% mannose and 17% protein. Kappeli et al. (1984) isolated a mannan-fatty acid complex from alkane grown *Candida tropicalis*. This complex stabilized hexadecane-in-water emulsion. *Schizonella malanogramma* and *Ustilago maydis* produced BSs as erythritol and mannose-containing lipid. Kitamoto et al. (1993) demonstrated the production of two kinds of mannosylerythhritol lipids from *Candida Antarctica* T-34. Raddy et al. (2009) and Chameotra and Singh (1990) isolated an emulsifying and solubilizing factor containing protein and carbohydrate from *Pseudomonas* sp. grown on hexadecane. Desai et al. (1988) reported the production of bioemulsifier, composed of 50% carbohydrate, 19.6% protein, and 10% lipid by *P. fluorescens*.

2.2 EMULSAN

Acinetobacter calcoaceticus RAG-1 produces an extracellular potent polyanionic amphipathic heteropolysaccharide bioemulsifier (Rosenberg et al., 1979). The heteropolysaccharide backbone contains repeating trisaccharide of N-acetyl-Dgalactosamine, N-acetyl-galactosamine uronic acid, and an unidentified N-acetyl amino sugar (Zukerberg et al., 1979). Based on dry weight, 10–15% of fatty acids are shown to be linked to the polysaccharide through O-ester linkages (Belsky et al., 1979). Emulsan does not appreciably reduce the surface tension, but it is an effective emulsifying agent for hydrocarbons in water (Zosim et al., 1982) even at a concentration as low as 0.001 to 0.01%.

2.3 BIODISPERSAN

Acinetobacter calcoaceticus A2 was reported to produce an extracellular, non-dialyzable dispersing agent called biodispersion. It is an anionic heteropolysaccharide having an average molecular weight of 51 and four reducing sugars, namely glucosamine, 6-methyl aminohexose, galactosamine uronic acid, and an unidentified amino sugar (Rosenberg et al., 1988).

2.4 LIPOSAN

Liposan is an extracellular water-soluble emulsifier produced by *Candida lipolytica* and composed mainly of 83% carbohydrate and 17% protein (Cirigliano and Carman, 1985). Husain et al. (1997) reported a polymeric type BS produced by *Pseudomonas nautica* which consists of proteins, carbohydrates, and lipid at the ratio of 35:63:2, respectively. Zinjarde and Pant (2002) also reported that *Yarrowia lipolytica*, a tropical marine strain produces an emulsifier (lipid-carbohydrate-lipid) complex associated with the cell wall in the earlier stages of growth but displayed the extracellular emulsifier activity towards the stationary phase during their growth on alkanes or crude oil.

2.5 LIPOPEPTIDES AND LIPOPROTEINS

2.5.1 SURFACTIN

Bacillus subtilis produces a cyclic lipopeptide called surfactin or subtilisin (Arima et al., 1968; Cooper and Zajic, 1980). These groups of BS contain a lipid-linked to a polypeptide chain. *Bacillus subtilis* produces cyclic lipopeptide type BS known as surfactin that has various potential applications. It is composed of a seven amino-acid ring structure joined to a fatty acid chain by means of lactone linkage. It reduces the surface tension from 72 to 27.9 mNm^{-1} at a concentration as low as 0.005% and shows a minimum IFT against hexadecane up to 1 mNm^{-1} (Delcambe and Devignat, 1957).

2.5.2 ITURIN

Iturin A was isolated from a *Bacillus subtilis* strain taken from the soil in Iturin (Vater et al., 2002). The Iturin groups of compounds are cyclic lipo-heptapeptides which contain a β-amino fatty acid in its side chain and reported to have potent antifungal agents which can be used as biopesticides for plants protection (Bonmatin et al., 2003). These molecules have remarkable efficacy against a broad variety of clinically important pathogenic fungi and yeast strains. However, their application in medicine is limited because of possible toxicity.

2.5.3 FENGYCIN

Fengycin, a type of lipopeptide BS consists of lipodecapeptide with β-hydroxy fatty acid in its side chain (Vanittanakom et al., 1986). This group of compounds comprises of C15 to C17 variants, which have characteristic Ala-Val dimorphy at position 6 of the peptide ring.

2.5.4 LICHENYSIN

Bacillus licheniformis produces several types of biosurfactants such as lichenysin that act synergistically and exhibit stability at extreme temperature, pH, and salinity. Several lipopeptide including decapeptide antibiotics (gramicidins) and lipopeptide antibiotics (polymyxins) shows both the antibiotics as well as the potent surface-active properties (Ron and Rosenberg, 2002). The polymixins are a group of closely related lipopeptide antibiotics produced by *Bacillus polymyxa*, *Bacillus brevis* and other related bacilli (Marahiel et al., 1977). Polymyxin B is composed of deca-peptide in which amino acids 3 ± 10 form a cyclic octapeptide. A branched-chain fatty acid is connected to the terminal 2, 4-diaminobutyric acid (Dab) (Ron and Rosenberg, 2002).

The synthesis of one or more peptide antibiotic during the early stages of sporulation is specific characteristics common to the selected members of the genus Bacillus (Katz and Demain, 1977). *Bacillus brevis* produces the cyclosymmetric decapeptide antibiotic known as gramicidin S. In solution, gramicidin S exists in the form of a rigid ring with the two positively charged ornithine side-chains constrained to one side of the ring and the side-chains of the remaining hydrophobic residues oriented toward the opposite side of the ring (Krauss and Chan, 1983). *Pseudomonas* strains produce viscosin, a peptidolipid BS that lowers surface tension to 27 mNm^{-1} (Neu and Poralla, 1990).

2.6 FATTY ACIDS, PHOSPHOLIPIDS, AND NEUTRAL LIPIDS

Various bacteria and yeasts secrete large amounts of fatty acids and phospholipid surfactants during their growth on n-alkanes (Cirigliano and Carman, 1985). The hydrophilic and lipophilic balance (HLB) is directly proportional to the length of the hydrocarbon chain in their structures.

These surfactants can produce optically clear microemulsions of alkanes in water. Kappeli and Finnerty (1979) reported that *Acinetobacter* sp. When grown on n-alkane produces phosphatidylethanolamine-rich vesicles that form optically clear microemulsions of alkanes in water and lowers the IFT between hexadecane and water to less than 1 mNm^{-1} with a critical micelle concentration (CMC) of 30 $mg.l^{-1}$. *Myroides* sp. SM1 was found to produce bile acids, cholic acid, deoxycholic acid and their glycine conjugate when cultivated in Marine broth (Maneerat et al., 2005). *Thiobacillus thiooxidans* produces a measurable quantity of phospholipids and has a role in the wetting of element sulfur (Beeba and Umbreit, 1971). Miyazima et al. (1985) reported the production of phospholipids by the fungus *Aspergillus* sp. that grown on hydrocarbons. The extracellular free fatty acids are in the range of C12 to C14 and the complex fatty acid contains hydroxyl groups and alkyl branches (Kretschmer et al., 1982). Such fatty acids are produced by various microorganisms during their growth on alkanes and exhibit the properties of surfactants. *Arthrobacter* AK-19161 and *P. aeruginosa* 44T1162 have been shown to accumulate up to 40–80% w/w lipid when cultivated on hexadecane and olive oil, respectively.

2.7 POLYSACCHARIDE-PROTEIN COMPLEXES

The surface-active properties of *Acinetobacter calcoaceticus* BD4 is due to the production of heteropolysaccharide-containing capsules (Kaplan and Rosenberg, 1982). These capsules are composed of repeating units of heptasaccharide and are released in the medium during the growth on hydrocarbons. Sar and Rosenberg (1983), polysaccharides alone showed no emulsification activity, but polysaccharides released with protein during the growth of a parent strain on ethanol or by a mutant strain BD-413 showed potent emulsification activity. Cameron et al. (1988) reported the production of mannoprotein emulsifier from *Saccharomyces cerevisiae* that emulsifies many oils, alkanes, and organic solvents, and the emulsions were reported to be stable at extreme temperature, pH, and salt concentrations. Kappeli et al. (1984) had isolated a mannan-fatty acid complex from alkane-grown *Candida tropicalis* that stabilized hexadecane in water emulsions. *Shizonella malanogramma* and *Ustilago maydis* were reported to produce a BS that was characterized as erythritol- and mannose-containing lipid (Fluharty and O'Brien, 1969). Cameotra and Singh (1990) had isolated, purified, and characterized an emulsifying and solubilizing factor from hexadecane grown *Pseudomonas* sp. Desai et al. (1988) reported the production of bioemulsifier which is composed of trehalose (50% carbohydrate) and lipid-*o*-dialkyl monoglycerides (10% lipid and 19.6% protein) by *Pseudomonas fluorescens* during growth on gasoline. *Bacillus subtilis* FE-2 had reported to produce a glycolipopeptide that was capable of emulsifying water-immiscible organophosphorus pesticides (Patel and Gopinathan, 1986).

2.8 NATURAL ROLES OF BIOSURFACTANT (BS)

When taking the account of BS's role in microbial physiology, it is important to highlight their production by the various groups of microorganisms, also possessing diverse chemical structures and surface properties. Thus, it is rational to believe that diverse groups of BSs have different natural roles which are specific to the physiology and ecology of the producing microorganisms (Ron and Rosenberg, 2002). Therefore, it will not be correct to draw any generalization for assigning common functions to all microbial surfactants. Surfactants are surface-active agents with a broad range of properties, including lowering of surface and IFTs of liquids. The immiscible liquids such as oil and water when mixed, liquid is dispersed into the other, and small droplets form emulsion, a heterogeneous system, consisting of at least one immiscible liquid dispersed in another in the form of droplets, whose diameters, in general, exceed 0.1 mm (Christofi and Ivshina, 2002; Bordoloi and Konwar, 2007). Characteristically, these are organic amphipathic compounds containing the hydrophilic (polar) moiety usually referred to as the 'head,' and the nonpolar hydrophobic moiety known as the 'tail.' While lowering the surface tension they increase solubility, detergency power, wettability, and foaming capacity (Mulligan, 2005). These properties are exploited in various areas such as detergency, emulsification, adhesion, coatings, wetting, foaming, soil, and water remediation, paints, chromatographic separations, medicine, agriculture, cosmetics, personal care, and almost every sector of modern industry (Mulligan and Gibbs, 1993).

2.9 MODE OF ACTION OF SURFACTANTS

Surfactant molecules tend to accumulate at the interfaces of solid-liquid, liquid-liquid or vapor-liquid. The hydrophobic portion concentrates at the surface with a strong attraction to the surrounding solution, while the hydrophilic portion is oriented towards the solution with weak to non-attraction forces. At interface, these molecules reduce the free energy of the system by reducing the forces of repulsion between unlike phases or surfaces and allow the two phases to mix more easily. They enhance the aqueous solubility of non-aqueous phase liquids (NAPLs) by reducing their surface/IFT at air-water and water-oil interfaces (Urum and Pekdemir, 2004). Surfactants tend to self-assemble in solution into dynamic aggregates called micelles. On reduction of IFT at the air-liquid interface and the aqueous surfactant concentration increased, the monomers aggregate to form micelles. The minimum concentration of the surfactant at which micelles first initiate to form is referred to as CMC. The concentration corresponds to the point where the surfactant molecules reduce the maximum surface tension of the system. The CMC is influenced by pH, temperature, and ionic strength (Mulligan and Gibbs, 1993; Desai and Banat, 1997).

The dynamics of the surfactant adsorption is of vast significance for practical applications such as foaming, emulsifying or coating processes, where bubbles or drops are rapidly generated and needed to be stabilized. As the interface is created, the adsorption of the surfactant is limited by the diffusion at the interface. The surface rheology of surfactant layers, including their elasticity and viscosity plays a very vital role in foam or emulsion stability. Formation of heavy foams signifies the better detergency character which indicates the effectiveness of the surfactant in separating oily material from a medium. With the addition of surfactants, emulsion formed may stabilize by reducing the IFT and decreasing the rate of coalescence. Usually, emulsions use to be water-in-oil (w/o) or oil-in-water. The hydrophilic-lipophilic balance (HLB) is used to classify emulsifiers (Mulligan and Gibbs, 1993; Rosen, 1978). Bharali et al. (2013) isolated crude BS from thermophilic *Alcaligenes faecalis* and applied in petro-spill bioremediation.

2.10 BIOSURFACTANT (BS) AND ITS ROLE IN THE PRODUCING MICROBE

BSs possess diverse properties and physiological functions like increase in the surface area and bioavailability of hydrophobic water-insoluble substrates, heavy metal binding, bacterial pathogenesis, and quorum sensing and biofilm formation (Rahman and Gakpe, 2008). One of the most essential survival strategies of microorganisms is their ability to establish themselves in an ecological niche where they can propagate. Most of the bacteria isolated from sites having a history of contaminations by hydrocarbons and derivatives are gram-negative, and it might be a characteristic that contributes to the survival of the populations in such harsh environments (Kosswig, 2005). Most of them produce BSs, which improve the ability of microbial cells to utilize hydrophobic compounds as growth substrates (Van Bogaert et al., 2007). BSs are either produced on microbial cell surfaces or excreted extracellularly (Makkar et al., 2011). The capacity of bacteria to produce BS specifically with antimicrobial property could be a survival strategy in competitive environments (http://www.primaryinfo.). The mode of action of BSs is the modification of the cell surface hydrophobicity Zhang and Miller (1990) and/or in promoting emulsification and/or solubilization of substrates (Haferburg et al., 1987). In the case of *Acinetobacter* strains, the cell-surface hydrophobicity was reduced in the presence of its cell-bound emulsifier (Rosenberg, 1988). Such observations suggest that microorganisms can use their BSs to regulate their cell-surface properties in order to attach or detach from surfaces according to the need. He found that emulsan, an extracellular polymeric heteropolysaccharide capsule, is used by *Acinetobacter calcoaceticus* to facilitate detachment from crude oil droplets with the exhaustion of carbon source.

KEYWORDS

- ***Acinetobacter calcoaceticus***
- **critical micelle concentration**
- **glycolipids**
- **hydrophilic-lipophilic balance**
- **hydrophobic compounds**
- **non-aqueous phase liquids**

REFERENCES

Abalos, A., et al., (2001). Physicochemical and antimicrobial properties of new rhamnolipids produced by *Pseudomonas aeruginosa* AT10 from soybean oil refinery wastes. *Langmuir, 17*, 1367–1371.

Abraham, W. R., et al., (1998). Novel glycine containing glycolipids from the alkane using bacterium *Alcanivorax borkumensis*. *Biochim. Biophys. Acta, 1393*, 57–62.

Al-Tahhan, R. A., et al., (2000). Rhamnolipid induced removal of lipopolysaccharide from *Pseudomonas aeruginosa*: Effect on cell surface properties and interaction with hydrophobic substrates. *Appl. Environ. Microbiol., 66*, 3262–3268.

Andersen, J. B., et al., (2003). Surface motility in *Pseudomonas sp.* DSS73 is required for efficient biological containment of the root-pathogenic micro fungi *Rhizoctonia solani* and *Pythium ultimum*. *Microbiol., 149*, 37–46.

Arima, K., et al., (1968). Surfactin, a crystalline peptide lipid surfactant produced by *Bacillus subtilis*: isolation, characterization and its inhibition of fibrin clot formation. *Biochem. Biophys. Res. Commun., 31*, 488–494.

Asselineau, C., & Asselineau, J., (1978). Trehalose containing glycolipids. *Prog. Chem. Fats Lipids, 16*, 59–99.

Beal, R., & Betts, W. B., (2000). Role of rhamnolipid biosurfactants in the uptake and mineralization of hexadecane in *Pseudomonas aeruginosa*. *J. Appl. Microbiol., 89*, 158–168.

Bechard, J., et al., (1998). Isolation and partial chemical characterization of an antimicrobial peptide produced by a strain of *Bacillus subtilis*. *J. Agric. Food. Chem., 46*, 5355–5361.

Beeba, J. L., & Umbreit, W. W., (1971). Extracellular lipid of *Thiobacillus thiooxidans*. *J. Bacteriol., 108*, 612–615.

Belsky, I., et al., (1979). Emulsifier of *Arthrobacter* RAG-1: Determination of emulsifier bound fatty acids. *FEBS Lett., 101*, 175–178.

Bharali, P., et al., (2011). Crude biosurfactant from thermophilic *Alcaligenes faecalis*: Feasibility in Petro-spill bioremediation. *Inter. Biodeter. Biodegr., 65*, 682–690.

Bharali, P., et al., (2013). Colloidal silver nanoparticles/rhamnolipid (SNPRL) composite as novel chemotactic antibacterial agent. *Inter. J. Biol. Macromol., 61*, 238–242.

Bonmatin, J. M., et al., (2003). Diversity among microbial acyclic lipopeptides: Iturin and surfactin activity-structure relationship to design new bioactive agents. *Combinator. Chem. Throughput Screen, 6*, 541–556.

Boonchan, S., et al., (1998). Surfactant-enhanced biodegradation of high molecular weight polycyclic aromatic hydrocarbons by *Stenotrophomonas maltophilia*. *Biotechnol. Bioengg., 59*, 482–494.

Cameotra, S. S., & Singh, H. D., (1990). Purification and characterization of alkane solubilizing factor produced by *Pseudomonas* PG-1. *J. Ferment. Bioengg., 69*, 341–344.

Cameron, D. R., et al., (1988). The mannoprotein of *Saccharomyces cerevisiae* is an effective emulsifier, *Appl. Environ. Microbiol., 54*, 1420–1425.

Carrillo, C., et al., (2003). Molecular mechanism of membrane permeabilization by the peptide antibiotic surfactin. *Biochim. Biophys. Acta, 1611*, 91–97.

Cirigliano, M. C., & Carman, G. M., (1985). Purification and characterization of liposan, a emulsifier from *Candida lipolytica*. *Appl. Environ. Microbiol., 50*, 846–850.

Cooper, D. G., & Paddock, D. A., (1984). Production of biosurfactants from *Torulopsis bombicola*. *Appl. Environ. Microbiol, 47*, 173–176.

Cooper, D. G., & Zajic, J. E., (1980). Surface-active compounds from microorganisms. *Adv. Appl. Microbiol., 26*, 229–253.

Delcambe, L., & Devignat, R., (1957). L'iturine, nouvelantibiotique d'origine congolaise. *Acad. Sci. Coloniales, 6,* 1–77.

Desai, A. J., et al., (1988). Emulsifiers production by *Pseudomonas fluorescens* during the growth on hydrocarbon. *Curr. Sci., 57*, 500–501.

Desai, J. D., & Desai, A. J., (1993). Production of biosurfactants. In: Kosaric, N., (ed.), *Biosurfactants, Production, Properties, Applications* (pp. 65–97). Marcel Dekker, New York.

Dos, S. S. C., et al., (2010). Evaluation of substrates from renewable-resources in biosurfactants production by *Pseudomonas* strains. *Afr. 1. Biotechnol., 9*, 5704–5711.

Edward, J. R., & Hayashi, J. A., (1965). Structure of a rhamnolipid from *Pseudomonas aeruginosa*. *Arch. Biochem. Biophys., 111*, 415–421.

Fattom, A., & Shilo, M., (1985). Production of emulcyan by phormidium J-1: Its activity and function. *FEMS Microbiology Letters, 31*(1), 3–9.

Fluharty, A. L., & O'Brien, J. S., (1969). A mannose and erythritol containing glycolipid from *Ustilago maydis*. *Biochem., 8*, 2627–2632.

Gerard, J., et al., (1997). Massetolides A-H, antimycobacterial cyclic depsipeptides produced by two pseudomonads isolated from marine habitats. *J. Nat. Prod., 60*, 223–229.

Grangemard, I., et al., (2001). Lichenysin: A more efficient cation chelator than surfactin. *Appl. Biochem. Biotechnol., 90*, 199–210.

Haba, E., et al., (2003). Physicochemical characterization and antimicrobial properties of rhamnolipids produced by *Pseudomonas aeruginosa* 47T2 NCBIM 40044. *Biotechnol. Bioengg., 81*, 316–322.

Haferburg, D., et al., (1987). Antiviral activity of rhamnolipids from *Pseudomonas aeruginosa*. *Acta Biotech., 1*(7), 353–356.

Harshada, K., (2014). Biosurfactant: A potent antimicrobial agent. *J. Microbiol. Expl.*, 1–5.

Hauler, S., et al., (1998). Purification and characterization of a cytotoxic exolipid of *Burkholderia pseudomallei*. *Infect. Immun., 66*, 1588–1593.

Hisatsuka, K., Nakahara, T., Sano, N., & Yamada, K., (1971). Formation of rhamnolipid by *Pseudomonas aeruginosa*: Its function in hydrocarbon fermentations. *Agric. Biol. Chem., 35*, 686–692.

Holakoo, L., & Mulligan, C. N., (2002). On the capability of rhamnolipids for oil·spill control of surface water. *Proc. Ann. Confe. Can. Soc. Civil. Engg.*, 5–8.

Hommel, R., et al., (1987). Production of water-soluble surface-active exolipids by *Torulopsis apicola*. *Appl. Microbiol. Biotechnol., 26*, 199–205.

Horowitz, S., & Currie, J. K., (1990). Novel dispersants of silicon carbide and aluminum nitride. *J. Dispersion Sci. Tech., 11*(6), 637–659.

Horowitz, S., & Griffin, W. M., (1991). Structural analysis of *Bacillus* licheniformis 86 surfactant. *J. Indus. Microbiol., 7*(1), 45–52.

Hu, Y., & Ju, L. K., (2001). Purification of lactonic sophorolipids by crystallization. *J. Biotechnol., 87*, 263–272.

Husain, D. R., et al., (1997). The effect of temperature on eicosane substrate uptake modes by a marine bacterium *Pseudomonas Nautica* strain 617: Relationship with the biochemical content of cells and supernatants. *World J. Microbiol. Biotechnol., 13*, 587–590.

Hutchison, M. L., & Gross, D. C., (1997). Lipopeptide phytotoxins produced by *Pseudomonas syringae*pv. *syringae*: Comparison of the biosurfactant and ion channel-forming activities of syringopeptin and syringomycin. *Mol. Plant Microb. Interact., 10*, 347–54.

Isoda, H., et al., (1997). Differentiation of human promyelocytic leukemia cell line HL60 by microbial extracellular glycolipids. *Lipids, 32*, 263–271.

Itoh, S., et al., (1971). Rhamnolipids produced by *Pseudomonas aeruginosa* grown on n-paraffin (mixture of C_{12}, C_{13} and C_{14} fractions). *J. Antibiot., 24*, 855–859.

Jacques, P., Hbid, C., Destain, J., Razafindralambo, H., Paquot, M., De Pauw, E., & Thonart, P., (1999). Optimization of biosurfactant lipopeptide production from *Bacillus subtilis* S499 by Plackett-Burman design. In: *Twentieth Symposium on Biotechnology for Fuels and Chemicals* (pp. 223–233). Humana Press, Totowa, NJ.

Jarvis, F. G., & Johnson, M. J., (1949). A glycolipid produced by *Pseudomonas aeruginosa. J. Am. Oil Chem. Soc., 71*, 4124–4126.

Johnson, M. K., & Boese-Marrazzo, D., (1980). Production and properties of heat-stable extracellular hemolysin from *Pseudomonas aeruginosa. Infect. Immun., 29*, 1028–1033.

Kaplan, N., & Rosenberg, E., (1982). Exopolysaccharide distribution and bioemulsifier production in *Acinetobacter calcoaceticus* BD4 and BD413. *Appl. Environ. Microbiol., 44*, 1335–1341.

Kappeli, O., & Finnerty, W. R., (1979). Partition of alkane by an extracellular vesicle derived from hexadecane grown Acinetobacter. *J. Bacteriol., 140*, 707–712.

Kappeli, O., et al., (1984). Structure of cell surface of the yeast *Candida tropicalis* and its relation to hydrocarbon transport. *Arch. Microbiol., 138*, 279–282.

Karanth, N. G. K., et al., (1999). Microbial production of biosurfactants and their importance. *Curr. Sci., 77*, 116–126.

Katz, E., & Demain, A. L., (1977). The peptide antibiotics of *Bacillus*: Chemistry, biogenesis and possible functions. *Bacteriol. Rev., 441*, 449–474.

Kim, B. S., et al., (2000). *In vivo* control and *in vitro* antifungal activity of rhamnolipid B, a glycolipid antibiotic, against *Phytophthora capsici* and *Colletotrichum orbiculare. Pest Manage. Sci., 56*, 1029–1035.

Kim, J. S., et al., (1990). Microbial glycolipid production under nitrogen limitation and resting cell conditions. *J. Biotechnol., 13*, 257–266.

Kitamoto, D., et al., (1993). Surface active properties and antimicrobial activities of mannosylerythritol lipids as biosurfactants produced by Candida Antarctica. *J. Biotech., 29* (1/2), 91–96.

Krauss, E. M., & Chan, S. I., (1983). Complexation and phase transfer of nucleotides by gramicidin S. *Biochem., 22*, 4280–4285.

Kretschmer, A., et al., (1982). Chemical and physical characterization of interfacial active lipid from *Rhodococcus erythropolis* grown on n-alkane. *Appl. Environ. Microbiol., 44*, 864–870.

Lang, S., & Wagner, F., (1987). Structure and properties of biosurfactants, In: Kosaricet, N., et al., (eds.), *Biosurfactants and Biotechnology* (pp. 21–47). Dekker, New York.

Lebron-Paler, A., (2008). *Solution and Interfacial Characterization of Rhamnolipid Biosurfactant from P. aeruginosa ATCC·9027*. PhD Thesis, University of Arizona Graduate College, Arizona, USA.

Lourith, N., & Kanlayavattanakul, M., (2009). Natural surfactants used in cosmetics: Glycolipids. *Int. J. Cosmet. Sci., 31*, 255–261.

MacDonald, R. M., & Chandler, M. R., (1981). Bacterium-like organelles in the vesicular-arbuscular mycorrhizal fungus glomus caledonius. *New Phytologies, 89*(2), 241–246.

Maneerat, S., (2005). Production of biosurfactants using substrates from renewable resources. *Songklanakarin J. Sci. Technol., 27*, 675–683.

Maneerat, S., et al., (2005). Bile acids are new products of a marine bacterium. *Myroides sp.* strain SM1. *Appl. Microbiol. Biotechnol., 67*, 679–683.

Marahiel, M., et al., (1977). Biological role of gramicidin S in spore function. Studies on gramicidin-S negative mutants of *Bacillus brevis* 999. *Eur. J. Biochem., 99*, 49–52.

Mata-Sandoval, J. C., et al., (1999). High-performance liquid chromatography method for the characterization of rhamnolipid mixtures produced by *Pseudomonas aeruginosa* UG2 on com oil. *Chromatogr., 864*, 211–220.

Matsuyama, T., & Nakagawa, Y., (2009). Surface-active exolipids: Analysis of absolute chemical structures and biological functions. *J. Microbiol. Methods, 5*, 165–175.

McInerney, M. J., et al., (1990). Properties of the biosurfactant produced by *Bacillus liqueniformis* strain JF-2. *Ind. J. Microbiol. Biotechnol., 5*, 95–102.

Miller, R. M., & Bartha, R., (1989). Evidence from liposome encapsulation for transport limited microbial metabolism of solid alkanes. *Appl. Environ. Microbiol., 55*, 269–274.

Miyazima, M., et al., (1985). Phospholipid derived from hydrocarbons by fungi. *J. Ferment. Technol., 63*, 219–224.

Monterio, S. A., et al., (2007). Molecular and structural characterization of the biosurfactant produced by *Pseudomonas aeruginosa* DAUPE 614. *Chem Phys Lipids, 147*, 1–13.

Mulligan, C. N., et al., (1999). Metal removal from contaminated soil and sediments by the biosurfactant surfactin. *Environ. Sci. Technol., 33*, 3812–3820.

Navon-Venezia, et al., (1995). Alasan, a new bioemulsifier from *Acinetobacter* radioresistens. *Appl. Environ. Microbiol., 61*(9), 3240–3244.

Neu, T. R., & Poralla, K., (1990). Emulsifying agent from bacteria isolated during screening for cells with hydrophobic surfaces. *Appl. Microbiol. Biotechnol., 32*, 521–525.

Nielsen, T. H., & Sørensen, J., (2003). Production of cyclic lipopeptides by *Pseudomonas fluorescens* strains in bulk soil and in the sugar beet rhizosphere. *Appl. Environ. Microbiol., 69*, 861–868.

Nielsen, T. H., et al., (1999). Viscosinamide, a new cyclic depsipeptide with surfactant and antifungal properties produced by *Pseudomonas fluorescens* DR54. *J. Appl. Microbiol., 87*, 80–90.

Noordman, W. H., & Janssen, D. B., (2002). Rhamnolipid stimulates uptake of hydrophobic compounds by *Pseudomonas aeruginosa*. *Appl. Environ. Microbiol., 68*, 4502–4508.

Patel, M. N., & Gopinathan, K. P., (1986). Lysozyme sensitive bioemulsifier for immiscible organophosphorous pesticides. *Appl. Environ. Microbiol., 52*, 1224–1226.

Poremba, K., et al., (1991). Marine biosurfactants, III. Toxicity testing with marine microorganisms and comparison with synthetic surfactants. *Z. Naturforsch, 46C*, 210–216.

Read, R. C., et al., (1992). The effect of *Pseudomonas aeruginosa* rhamnolipids on guineapig tracheal mucociliary transport and ciliary beating. *J. Appl. Physiol., 72*, 2271–2277.

Reddy, S. A., et al., (2009). Synthesis of silver nanoparticles using surfactin: A biosurfactant as stabilizing agent. *Mat. Lett., 63*, 1227–1230.

Robert, M., et al., (1989). Effect of the carbon source on biosurfactant production by *Pseudomonas aeruginosa* 44T1. *Biotechnol. Lett., 11*, 871–874.

Rosen, M. J., (1978). *Surfactants and Interfacial Phenomena* (pp. 149–171). John Wiley & Sons, New York.

Rosenberg, E., & Ron, E. Z., (1999). High-and low-molecular-mass microbial surfactants. *Appl. Microbiol. Biotech., 52*(2), 154–162.

Rosenberg, E., (1988). Production of biodispersan by *Acientobacter calcoaceticus* A2. *Appl. Environ. Microbiol., 54*, 317–322.

Rosenberg, E., et al., (1979). Emulsifier *Arthrobacter* RAG-1: Isolation and emulsifying properties. *Appl. Environ. Microbiol., 37*, 402–408.

Rosenberg, E., et al., (1988). Purification and chemical properties of *Acinetobacter calcoaceticus* A2 biodispersan. *Appl. Environ. Microbiol., 54*, 323–326.

Saikia, J. P., et al., (2013). Possible protection of silver nanoparticles against salt by using rhamnolipid. *Colloids Surf. B. Biointerfaces, 104*, 330–332.

Sar, N., & Rosenberg, E., (1983). Emulsifier production by *Acinetobacter calcoaceticus* strains. *Current Microbiol., 9*(6), 309–314.

Schulz, D., et al., (1991). Marine biosurfactants, I. Screening for biosurfactants among crude oil-degrading marine microorganisms from the North Sea. *Z. Naturforsch, 46*C, 197–203.

Shaw, A., (1994). Lipid composition as a guide to the classification of bacteria. *Adv. Appl. Microbiol., 17*, 63–108.

Shoham, et al., (1983). Bacterial degradation of emulsan. *Appl. Environ. Microbiol., 46*(3), 573–579.

Shoham, Y. U. V. A. L., & Rosenberg, E., (1983). Enzymatic depolymerization of emulsan. *Bacteriology, 156*(1), 161–167.

Silva, S. N. R. L., et al., (2010). Glycerol as substrate for the production of biosurfactant by *Pseudomonas aeruginosa* UCP 0992. *Colloids Surf. B. Biointerfaces, 79*, 174–183.

Sotirova, A., et al., (2009). Effects of rhamnolipid-biosurfactant on cell surface of *Pseudomonas aeruginosa. Microbiol. Res., 164*, 297–303.

Sposito, G., (1989). *The Chemistry of Soils* (p. 277). Oxford University Press, Oxford.

Suzuki, T., et al., (1969). Trehalose lipid and a-branched-b-hydroxy fatty acids formed by bacteria grown on n-alkanes. *Agric. Biol. Chem., 33*, 1619–1625.

Syldatk, C., & Wagner, F., (1987). Production of biosurfactants. In: Kosaric, N., et al, (eds.), *Biosurfactants and Biotechnology* (pp. 89–120). Marcel Dekker, New York.

Tan, H., et al., (1994). Complexation of cadmium by a rhamnolipid biosurfactant, *Environ. Sci. Technol., 28*, 2402–2406.

Vanittanakom, N., et al., (1986). Fengycin-a novel antifungal lipopeptide antibiotic produced by *Bacillus subtilis* F-29-3. *J. Antibiot., 39*, 888–901.

Vater, J., et al., (2002). Matrix-assisted laser desorption ionization-time of flight mass spectrometry of lipopeptide biosurfactants in whole cells and culture filtrates of *Bacillus subtilis* C-1 isolated from petroleum sludge. *Appl. Environ. Microbiol., 68*, 6210–6219.

Volkering, F., et al., (1998). Microbiological aspects of surfactant use for biological soil remediation. *Biodegr., 8*, 401–417.

Wayman, M., et al., (1984). Biotechnology for oil and fat industry. *J. Am. Oil Chem. Soc., 61*, 121–131.

Wei, Y. R., et al., (2005). Rhamnolipid production by indigenous *Pseudomonas aeruginosa* J4 originating from petrochemical wastewater. *Biochem. Eng., 127*,146–154.

Wicken, A. J., & Knox, K. W., (1970). Studies on the group F antigen of lactobacilli: Isolation of a teichoic acid-lipid complex from *Lactobacillus fermenti* NCTC 6991. *J. Gen. Microbiol., 60*, 303–313.

Yakimov, M. M., Timmis, K. N., Wray, V., & Fredrickson, H. L., (1995). Characterization of a new lipopeptide surfactant produced by thermotolerant and halotolerant subsurface *Bacillus licheniformis* BAS50. *Appl. Environ. Microbiol., 61*(5), 1706–1713.

Zajic, J. E., Guignard, H., & Gerson, D. F., (1977). Properties and biodegradation of a bio-emulsifier from *Corynebacterium* hydrocarboclastus. *Biotechn. Bioengg., 19*(9), 1303–1320.

Zhang, Y., & Miller, R. M., (1994). Effect of a *Pseudomonas* rhamnolipid biosurfactant on cell hydrophobicity and biodegradation of octadecane. *Appl. Environ. Microbiol., 60*, 2101–2106.

Zhao, B., et al., (2005). Solubilization and biodegradation of phenanthrene in mixed anionic-nonionic surfactant solutions. *Chemosphere, 58*, 33–40.

Zinjarde, S. S., & Pant, A., (2002). Emulsifier from tropical marine yeast. *Yarrowia lipolytica* NCIM 3589. *J. Basic Microbiol., 42*, 67–73.

Zosim, Z., et al., (1982). Properties of hydrocarbon-in-water emulsions stabilized by *Acinetobacter* RAG-1 emulsan. *Biotechnol. Bioengg., 24*, 281–292.

Zukerberg, A., et al., (1979). Emulsifier of *Arthrobacter* RAG-1: Chemical and physical properties. *Appl. Environ. Microbiol., 37*, 414–420.

Zulianello, L., et al., (2006). Rhamnolipids are virulence factors that promote early infiltration of primary human airway epithelia by *Pseudomonas aeruginosa. Infect. Immun., 74*, 3134–3147.

FURTHER READING

Bharali, P., & Konwar, B. K., (2011). Production and physicochemical characterization of a biosurfactant produced by *Pseudomonas aeruginosa* OBP1 isolated from petroleum sludge. *Appl. Biochem. Biotechnol., 164*, 1444–1460.

Das, S., Kalita, S. J., Bharali, P., Konwar, B. K., Das, B., & Thakur, A. J., (2013). Organic reactions in "green surfactant": An avenue to bisuracil derivative. *ACS Sustainable Chem. Eng.,* (p. 301). doi: 10.1021/sc4002774, Publication Date (Web).

Pranjal, B., (2015). *Bioremediation of Crude Oil Contaminated Soil.* (Thesis: Supervisor B K Konwar), Dept. of Mol. Biol. and Biotechnology, Tezpur University (Central), Napaam – 784028, Assam, India.

CHAPTER 3

Interaction of Microorganisms with Hydrophobic Water-Insoluble Substrates

Volkering et al. (1993) suggested that the microorganisms may access the hydrophobic substrate via direct contact or by contact with pseudo-solubilized substrate in biosurfactant (BS) micelles or emulsion droplets. Rosenberg (1988) found that emulsan; an extracellular polymeric hetero-polysaccharide is used by *Acinetobacter calcoaceticus* to facilitate detachment from crude oil droplets exhausted of substrate. Once the utilizable substrates are consumed the emulsan coat is shed off, and then it converts the hydrophobic oil surface to a hydrophilic one. In *Acinetobacter radioresistens* KA53 the bioemulsifiers, alasan was found to increase the solubility of PAHs up to 27-fold. For pseudo-solubilization, addition of exogenous BS can enhance a noticeable aqueous solubility of organic compounds and modify the bioavailability (Boonchan et al., 1998; Zhao et al., 2005). The micelles or other aggregates are formed that partition hydrophobic substrates and may enhance biodegradation by allowing for closer cell-substrate interactions or may fuse directly with microbial membranes resulting in direct substrate delivery.

Rhamnolipid (RL) has been found to remove LPS in a dose-dependent manner from *P. aeruginosa,* resulting in increased cell surface hydrophobicity and enhanced uptake of hydrophobic substrates (Noordman and Janssen, 2002). Some hydrocarbon-degrading microbes respond to these in-soluble carbon sources by producing surface-active compounds, as well as by changing cell surface properties such as cell surface hydrophobicity (Al-Tahhan et al., 2000; Beal and Betts, 2000).

3.1 METAL CHELATION BY BIOSURFACTANTS (BSS)

Biosurfactants (BSs) were also been explored for metal chelation. Rhamnolipids (RLs) can remove metals and ions, forming stable complexes with metals. The mechanism of reduced toxicity was apparently via RL complexation of cadmium as well as by RL-induced lipopolysaccharide (LPS) removal from the cell surface (Van Hamme et al., 2006). Tang et al. (2007) studied the effect of mono-rhamnolipid produced by *P. aeruginosa* ATCC 9027 on the formation of metal complexes and reported that surfactant-metal interactions are rapid and stable. Mulligan et al. (1999) reported the use of surfactin from *Bacillus subtilis* ATCC 21332, ATCC 9027 and *Torulopsis bombicola* ATCC 22214 to treat soil and sediments contaminated with Zn, Cu, Cd, oil, and grease. In sorption, metal-ligand complexation, also with soil constituents and cation exchange processes are involved (Sposito, 1989) affecting access of the metal to the microorganisms.

3.2 ANTIMICROBIAL ACTIVITY OF BIOSURFACTANTS (BSS)

BSs tend to influence the physiological behavior of microbes in plant and animal pathogenesis; act as a "dispersing agent" in pathogenic microorganisms also as a "wetting agent" for the surface of host cell (Matsuyama and Nakagawa, 1996). *P. syringae* produces two necrosis-inducing lipopeptide toxins, syringopeptin, and syringomycin, types of pore-forming cytotoxins that form ion channels permeable to divalent cations during plant pathogenesis. *Serratia marcescens*, an opportunistic pathogenic bacterium produces serrawettin a type of non-ionic cyclic depsipeptide (CLP) that helps in wetting of the host cell surface and dispersion of bacteria.

P. aeruginosa produces a heat-stable extracellular glycolipid called hemolysin that has hemolytic activity (Johnson and Boese-Marrazzo, 1980). The di-rhamnolipid type BS from *Burkholderia pseudomallei* is similarly hemolytic for erythrocytes of various species and cytotoxic at high concentrations for non-phagocytic and phagocytic cell lines (Hauler et al., 1998). Zulianello et al. (2006) stated that *P. aeruginosa* requires the production of RLs to invade respiratory epithelia reconstituted with primary human respiratory cells.

A *Pseudomonas* sp. derived from marine alga produces eight types of massetolides A-H, novel CLPs. These massetolides A-H were found to exhibit *in vitro* anti-microbial activity against *Mycobacterium tuberculosis* and *M. avium-intracellulare* (Gerard et al., 1997). RLs from *Pseudomonas* sp PS-17 interact with the producing bacteria causing a reduction in LPS content, changing the outer membrane proteins, and had a direct impact on bacterial cell surface morphology. Kim et al. (2000) studied the effects of RL B on a range of plant pathogenic fungi, including *Phytophthora capsici, Colletotrichum orbiculare* and observed to cause zoospore lysis, inhibition, spore germination and hyphal growth inhibition. Andersen et al. (2003) reported isolation of *Pseudomonas* sp. DSS73 from the rhizoplane of sugar beet seedlings showing antagonism towards the root-pathogenic micro-fungi *Pythium ultimum* and *Rhizoctonia solani*.

Viscosinamide, an antibiotic isolated from *P. fluorescens* with BS properties was found to have antifungal properties (Nielsen et al., 1999). Bechard et al. (1998) isolated an anti-microbial lipopeptide from a strain of *B. subtilis* and demonstrated a broad spectrum of activity against Gram-negative bacteria, lesser activity against gram-positive ones. Grangemard et al. (2001) reported the chelating properties of lichenysin, a cyclic lipopeptide produced by *B. licheniformis*, which might explain the membrane disrupting effect of lipopeptides. Carrillo et al. (2003) studied the molecular mechanism of antibiotic and other important biological actions of surfactin produced by *Bacillus subtilis*. Nielsen and Sorensen (1999) found three cyclic lipopeptides (viscosinamide, tensin, amphisin) produced by *P. fluorescens* in the rhizosphere of germinating sugar beet seeds and considered that such lipopeptides confer a competitive advantage to the organism during colonization.

KEYWORDS

- ***Acinetobacter calcoaceticus***
- **biosurfactants**
- **hetero-polysaccharide**
- **lipopolysaccharide**
- **pseudo-solubilized**
- **rhamnolipids**

REFERENCES

Al-Tahhan, R. A., et al., (2000). Rhamnolipid induced removal of lipopolysaccharide from *Pseudomonas aeruginosa*: Effect on cell surface properties and interaction with hydrophobic substrates. *Appl. Environ. Microbiol., 66*, 3262–3268.

Andersen, J. B., et al., (2003). Surface motility in *Pseudomonas sp.* DSS73 is required for efficient biological containment of the root-pathogenic micro fungi *Rhizoctonia solani* and *Pythium ultimum. Microbiol., 149*, 37–46.

Beal, R., & Betts, W. B., (2000). Role of rhamnolipid biosurfactants in the uptake and mineralization of hexadecane in *Pseudomonas aeruginosa. J. Appl. Microbiol., 89*, 158–168.

Bechard, J., et al., (1998). Isolation and partial chemical characterization of an antimicrobial peptide produced by a strain of *Bacillus subtilis. J. Agric. Food. Chem., 46*, 5355–5361.

Boonchan, S., et al., (1998). Surfactant-enhanced biodegradation of high molecular weight polycyclic aromatic hydrocarbons by *Stenotrophomonas maltophilia. Biotechnol. Bioengg., 59*, 482–494.

Carrillo, C., et al., (2003). Molecular mechanism of membrane permeabilization by the peptide antibiotic surfactin. *Biochim. Biophys. Acta, 1611*, 91–97.

Gerard, J., et al., (1997). Massetolides A-H, antimycobacterial cyclic depsipeptides produced by two pseudomonads isolated from marine habitat. *J. Nat. Prod., 60*, 223–229.

Grangemard, I., et al., (2001). Lichenysin: A more efficient cation chelator than surfactin. *Appl. Biochem. Biotechnol., 90*, 199–210.

Hauler, S., et al., (1998). Purification and characterization of a cytotoxic exolipid of *Burkholderia pseudomallei, Infect. Immun., 66*, 1588–1593.

Johnson, M. K., & Boese-Marrazzo, D., (1980). Production and properties of heat-stable extracellular hemolysin from *Pseudomonas aeruginosa. Infect. Immun., 29*, 1028–1033.

Kim, B. S., et al., (2000). *In vivo* control and *in vitro* antifungal activity of rhamnolipid B, a glycolipid antibiotic, against *Phytophthora capsici* and *Colletotrichum orbiculare. Pest Manage. Sci., 56*, 1029–1035.

Matsuyama, T., & Nakagawa, Y., (2009). Surface-active exolipids: Analysis of absolute chemical structures and biological functions. *J. Microbiol. Methods, 5*, 165–175.

Mulligan, C. N., et al., (1999). Metal removal from contaminated soil and sediments by the biosurfactant surfactin. *Environ. Sci. Technol., 33*, 3812–3820.

Nielsen, T. H., et al., (1999). Viscosinamide, a new cyclic depsipeptide with surfactant and antifungal properties produced by *Pseudomonas fluorescens* DR54. *J. Appl. Microbiol., 87*, 80–90.

Noordman, W. H., & Janssen, D. B., (2002). Rhamnolipid stimulates uptake of hydrophobic compounds by *Pseudomonas aeruginosa. Appl. Environ. Microbiol., 68*, 4502–4508.

Rosenberg, E., et al., (1988). Purification and chemical properties of *Acinetobacter calcoaceticus* A2 biodispersan. *Appl. Environ. Microbiol., 54*, 323–326.

Sposito, G., (1989). *The Chemistry of Soils* (p. 277). Oxford University Press, Oxford.

Tang, et al., (2007). Enhanced crude oil biodegradability of *Pseudomonas aeruginosa* ZJU after preservation in crude oil-containing medium. *World Microbiol. Biotechnol., 23*, 7–14.

Van, H. J. D., et al., (2006). Physiological aspects: Part 1 in a series of papers devoted to surfactants in microbiology and biotechnology. *Biotechnol. Adv., 24*, 604–620.
Volkering, F., et al., (1998). Microbiological aspects of surfactant use for biological soil remediation. *Biodegr., 8*, 401–417.
Zhao, B., et al., (2005). Solubilization and biodegradation of phenanthrene in mixed anionic-nonionic surfactant solutions. *Chemosphere, 58*, 33–40.
Zulianello, L., et al., (2006). Rhamnolipids are virulence factors that promote early infiltration of primary human airway epithelia by *Pseudomonas aeruginosa. Infect. Immun., 74*, 3134–3147.

CHAPTER 4

Biochemistry of Biosurfactants

Microbial surfactants are commonly differentiated on the basis of their biochemical nature. Edward and Hayashi (1965) reported the formation of glycolipid, type R-1 containing two rhamnose and two β-hydroxydecanoic units in *P. aeruginosa*. Wiken and Knox (1970) reported micelle formation by glycosyl-diglycerides isolated from *L. fermenti*. According to Hisatsuka et al. (1971) the rhamnolipid (RL) biosurfactant (BS) could emulsify alkanes and stimulate the growth of *P. aeruginosa* in hexadecane. Itoh et al. (1971) reported a second kind of RL (R-2) containing one rhamnose. Shaw reported that diglycosyl-diglycerides are glycolipids and are present in the cell membrane of a wide variety of bacteria. The structure of this class of molecule contains a polar, water-soluble head and two lipophilic alkyl tails. Cooper and Paddock (1984) reported the production of sophorolipids by *T. petrophilum* on water-insoluble substrates alkanes and vegetable oil. These surface-active compounds were chemically identical to those produced by *T. bombicolam* but could not emulsify alkanes or vegetable oils. A protein containing alkane emulsifying agent was formed when *T. petrophilum* was grown on a glucose-yeast extract medium. Lang and Wagner (1987); Syldatk and Wagner (1987); and Cooper et al. (1981) reported trehalolipids from organisms differed in size and structure of mycolic acid, number of carbon atoms and degree of unsaturation. Hommel et al. (1987) reported the production of a mixture of water-soluble sophorolipids from yeasts.

The initial classification of the BS was on the molecular weight, chemical properties and cellular localization. The low molecular weight BSs such as glycolipids, lipopeptides, flavolipids, corynomycolic acids and phospholipids lower the surface and interfacial tensions (IFTs) at the air/water interfaces. The high molecular weight BSs are called bioemulsans, such as emulsan, alasan, liposan, polysaccharides, and protein complexes. They are efficient emulsifiers at low concentrations and exhibit considerable substrate specificity in stabilizing oil-in-water emulsions (Holakoo and Mulligan, 2002). Therefore, it can be expected to have diverse properties and physiological functions of BSs such as increasing the surface area and bioavailability of hydrophobic water-insoluble substrates, heavy metal binding, bacterial pathogenesis, quorum sensing and biofilm formation (Harshada, 2014).

4.1 BIOSURFACTANTS (BSS) VS. SYNTHETIC SURFACTANTS

Biosurfactants (BSs) possess lower toxicity; higher biodegradability (Zajic et al., 1977); better environmental compatibility (Maneerat, 2005) and higher foaming (Jacques et al., 1999) as compared to synthetic surfactants. The advantages of BSs as compared to the synthetic products are their biodegradability, low toxicity, and simple production by microbial fermentation processes. BSs show a wide range of applications, such as cleanup of oil spills, secondary, and tertiary oil recovery. It is also possible to use BSs as additives in cosmetics, foodstuffs, beverages, and pharmaceutical products (Lang and Wagner, 1987). The most attractive aspect of BS is their biodegradability and ecological acceptance (Bafghi and Fazaelipoor, 2012). BSs could retain their surface-active properties even under extreme conditions of temperature, pH, salinity, and metal salts (Joshi et al., 2008). Due to the low irritancy and compatibility with human skin (Pornsunthorntawee et al., 2009) they are constantly used in cosmetics and pharmaceuticals (Lourith and Kanlayavattanakul, 2009).

BSs also cause bioavailability under diverse conditions, ecological acceptability, low toxicity, capacity to be modified and their capability of increasing the bioavailability of poorly soluble organic compounds like polyaromatic hydrocarbons (PHA) (Jain et al., 2012). They are potential for bioremediation of contaminated environments, tertiary oil recovery such as microbial enhanced oil recovery (MEOR), flotation process, detergent formulations, etc., (Bafghi and Fazaelipoor, 2012; Cameotra and Makkar, 2010).

Rhamnolipids (RLs) with high biodegradability possess an additional advantage over the synthetic surfactants in soil washing and bioremediation processes (Mata-Sandoval et al., 1999). They are not only efficient but also exhibit excellent antimicrobial activity against several other microorganisms (Rodrigues et al., 2006) and disrupt host defense during infections (Read et al., 1992). RLs could be potential as bacteriocide (Haba et al., 2003), fungicide (Yoo et al., 2005), wound healer (Stipcevic et al., 2005) and others (Lourith and Kanlayavattanakul, 2009). They also assist in cell surface motility and influence the architecture of biofilm, especially in the formation and maintenance of fluid channels within the exo-polymeric matrix after bacteria adhere irreversibly on a substratum (Figure 4.1) (Deziel et al., 2003; Davey et al., 2003).

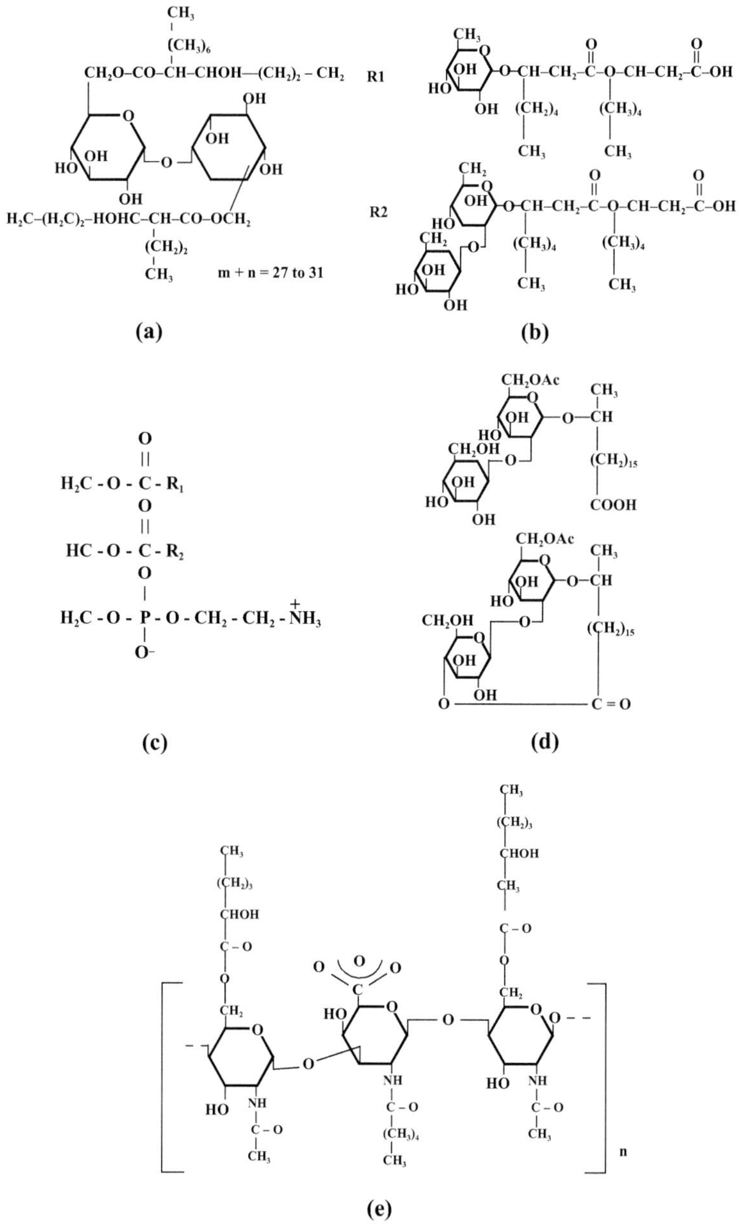

FIGURE 4.1 Structure of (a) trehalose lipids; (b) R1 and R2 rhamnolipids; (c) sophorolipid; (d) phosphatidylethanolamine; and (e) emulsan.

4.2 BIOSYNTHETIC PATHWAYS OF BIOSURFACTANT (BS)

BSs are synthesized by two primary metabolic pathways viz hydrocarbon and carbohydrate pathways (Edward and Hayashi, 1965) reported formation of glycolipid, type R-1 containing two rhamnose and two β-hydroxydecanoic units in *P. aeruginosa*. Wiken and Knox (1970) reported micelle formation by glycosyl-diglycerides isolated from *L. fermenti*. According to Hisatsuka et al. (1971) the RL BS could emulsify alkanes and stimulate the growth of *P. aeruginosa* in hexadecane. The action by *T. petrophilum* on water-insoluble substrates like alkanes and vegetable oil cause the production of sophorolipids. Syldatk and Wagner (1987); and Cooper et al. (1981) reported trehalolipids from organisms differed in size and structure of mycolic acid, number of carbon atoms and degree of unsaturation. Hommel et al. (1987) reported the production of a mixture of water-soluble sophorolipids from yeasts.

The initial classification of the BS was on the molecular weight, chemical properties and cellular localization. The low molecular weight BSs such as glycolipids, lipopeptides, flavolipids, corynomycolic acids and phospholipids lower the surface and IFTs at the air/water interfaces.

4.3 PRODUCTION AND EFFICACY OF BIOSURFACTANT (BS)

Haferberg et al. (1987); and Guerra Santos et al. (1984) reported that the majority of known BSs were synthesized by microorganisms grown on water-immiscible hydrocarbons, but some had been produced on water-soluble substrates such as glucose, glycerol, and ethanol (Palejwala and Desai, 1989). According to Roongsawang et al. (2002); Vater et al. (2002); and Bordoloi and Konwar (2009) glucose is the best carbon source for the production BS by the bacterial isolate *P. aeruginosa* MTCC7815 (31.1 $mN.m^{-1}$). According to them, saccharose and fructose to be good carbon sources, but glycerol severely decreased surfactin production. Cooper et al. (1981) reported that surfactin biosynthesis did not follow stimulation by hexadecane in contrast to other surfactants. Makkar and Cameotra (2002) described the ability of *Bacillus* strains to use starch and sucrose as the preferred carbon source for the maximum growth and BS production.

The bacterial isolate *P. aeruginosa* MTCC7815 exhibited higher reduction in surface tension of culture medium (29.1 $mN.m^{-1}$) irrespective of the carbon source used. Glycerol was found to be the second-best carbon source for the growth and BS production by *P. aeruginosa* MTCC7815 (Bharali et al., 2011). Turkovskaya et al. (2001) observed that agitation helped in intense aeration which was crucial for the growth of bacteria and BS synthesis by *Pseudomonas aeruginosa*. The agitation at 200 rpm was observed as the optimum for all the bacterial isolates (Sim et al., 1997; Cameotra and Makkar, 1998). An increase of agitation from 250 to 500 rpm caused a decrease in BS production by *Nocardia erythropolis* due to a shear rate effect on the growth kinetics of the microorganism (Syldatk and Wagner, 1987). Chayabutra and Ju (2000) observed that *P. aeruginosa* ATCC 10145 could grow optimally at a hexadecane concentration of 8% (v/v). RL production is influenced by the nutrients used in the culture media and on the applied culture parameters. Guerra-Santos et al. (1986) reported better yield of RLs, produce by *P. aeruginosa*, when the concentration of magnesium, calcium, potassium, sodium, and trace elements were minimized.

4.4 ALTERNATIVE CARBON SOURCE FOR BIOSURFACTANT (BS) PRODUCTION

The success of commercial BS production depends on the development of cheaper processes and raw materials, accounting for 50% of the final product cost (Cameotra and Makkar, 1998). Potato process effluents were used to produce BS by *B. subtilis* (Noah, 2005). Cassava wastewater generated during the preparation of cassava flour is a potential substrate to produce surfactin and RL by *B. subtilis* (Nitschke and Pastore, 2006) and *P. aeruginosa* (Costa et al., 2006), respectively. George and Jayachandran (2008) analyzed the RL BSs produced through submerged fermentation using orange fruit peelings as the sole carbon source. Patel and Desai (1997) reported the use of molasses and corn-steep liquor as the primary carbon and nitrogen source to produce RL using *P. aeruginosa* GS3. Dubey and Juwarkar (2004) studied the production of BS using industrial waste from distillery using *P. aeruginosa* BS2. Kitchen waste oils generated from domestic uses, vegetable oil refineries, or the soap industries have been reported to be suitable to produce BS through microbial fermentation (Yalcin and Ergene, 2009; Benincasa, 2009). Lima et al. (2009) reported the use of residual waste of soybean oils to produce BS by submerged fermentation in stirred tank reactors using *P. aeruginosa* PACL. Hazra et al. (2013) reported the utilization of de-oiled cakes of mahua (*Madhuca indica*), Karanja (*Pongamia pinnata*), jatropha (*Jatropha curcus*) and neem (*Azadirachta indica*) to produce RL using *P. aeruginosa* AB4. Daniel et al. (1998) used dairy wastes as carbon substrates and achieved production of high concentrations of sophorolipids using a two-stage cultivation process for the yeast *Cryptococcus curvatus* ATCC 20509, whereas Deshpande and Daniels (1995) used animal fat to produce sophorolipids BS using the yeast, *Candida bombicola*.

4.5 INTERACTION OF BIOSURFACTANT (BS) WITH POLYCYCLIC AROMATIC HYDROCARBONS (PAH)

Polycyclic aromatic hydrocarbons (PAH) are ubiquitous pollutants occurring mostly because of fossil fuel combustion and also as a by-product of industrial activities are suspected carcinogens, and as such, exposure may cause health risks. PAHs like phenanthrene, pyrene, fluorene, and crude oil being supplemented as the sole source of carbon and energy in the culture media can increase the yield of bacterial dry biomass and protein. *P. aeruginosa* strains MTCC7815 followed by MTCC7812 and MTCC7814 could utilize phenanthrene reducing the content to 70, 85, and 87 μg from the initial application of 180 μg in the medium in 12 days. The strains MTCC7814 and MTCC8165 on the other hand exhibited better utilization of pyrene with 89 μg and 93 μg, respectively from the initial application of 180 μg with increased biomass and protein production.

The growth of the bacterial isolates at the expense of fluorene as the sole source of carbon suggested utilization of 89, 90, and 92 μg of fluorene respectively from the initial application of 180 μg. Non-actinomycete bacteria such as *P. aeruginosa*, *P. pudita* and *Flavobacterium* species were reported to utilize pyrene, when supplemented with other forms of organic carbons (Trzesicka-Mlynarz and Ward, 1995). Soil-derived *Pseudomonas* species could degrade PAHs but failed to utilize them as the sole source of carbon and energy (Foght and Westlake, 1988). According to Bodour et al. (2003) BS producing bacteria are present in higher concentration in hydrocarbon-contaminated soils. BS production by bacterial strains increased significantly on the addition of phenanthrene along with pyrene and fluorine. The yield of BS in the culture supernatant with acid precipitation increased dramatically after 96 h of culture. String Fellow and

Aitken (1995) observed that *P. saccharophila* P15 could degrade pyrene on being induced by either phenanthrene or salicylate. In media having the combined addition of fluorene and phenanthrene caused better BS yield of 0.45 and 0.38 $g.l^{-1}$ by *P. aeruginosa* strains MTCC7815 and MTCC7814, respectively for the entire growth period of 96 h. In the same medium, the bacterial biomass increased from 0.6–1.2 $g.l^{-1}$ at 48 h of inoculation to a maximum of 1.0–1.5 $g.l^{-1}$ in 96 h of culture. Bouchez et al. reported that the addition of fluorene as a co-substrate could increase utilization of phenanthrene, however it might be a poor inducer of its own degradation, but fluorene could enhance phenanthrene biodegradation, possibly by a positive analog effect on enzyme induction.

Ron and Rosenberg (2002) observed that the production of BS is related to the utilization of available hydrophobic substrates by the producing bacteria from their natural habitat, presumably by increasing the surface area of substrates and increasing their apparent solubility. According to Miller and Bartha (1989) low molecular weight BSs like lipopeptides having CMCs increase the apparent solubility of hydrocarbons by incorporating them into the hydrophobic cavities off micelles. Bharali and Konwar (2011) observed that BS of *P. aeruginosa* strains MTCC7815 and MTCC7812 having the concentration of 0.5 $mg.ml^{-1}$ exhibited 41 and 26 $\mu g.\ ml^{-1}$ solubilization of pyrene. Subsequently, the apparent solubility of pyrene was enhanced by factors 5–7, resulting in its higher uptake and metabolism as compared to non-solubilized pyrene. BSs secreted by MTCC7815 and MTCC8163 were lipopeptide containing higher amounts, whereas BSs secreted by MTCC7812, MTCC8165, and MTCC7814 were complex mixtures of lipopeptides and glycoproteins. Moreover, higher pyrene solubilization effect of BS of MTCC7815 as compared to MTCC7812 reinforced the hypothesis that variation in BS isoforms between these two isolates might result in a large variation of the emulsification property and specificity of BSs.

4.6 ROLE OF PH ON BIOSURFACTANTS (BSS)

The efficacy of bacterial BSs is independent of pH. There was little reduction in surface tension of acidic and alkaline conditions as compared to the neutral pH range. Bordoloi and Konwar (2009) found that the dilution of BSs by 10 times (CMD^{-1}) retained the efficacy in reducing surface tension of culture medium similar to that of normal concentration. In 100 times dilution (CMD^{-2}), a similar trend of surface tension reduction in acidic conditions only was exhibited by the all-bacterial isolates. However, the reduction in surface tension at CMD^{-2} was much less in almost all pH levels. This phenomenon might be related to the presence of the higher amount of surfactin isomers in the crude lipopeptide secreted by the isolates due to their preference of higher pH (Morikawa et al., 2000).

4.7 THERMOSTABILITY OF BIOSURFACTANTS (BSS)

Banat reported a thermo-tolerant *Bacillus sp.* growing at 50°C on hydrocarbon-containing medium. BS containing fermentative broth of the bacterium could release 95% oil from the sand pack column suggested a potential application in MEOR and in oil-sludge clean up. Bacterial BSs could be thermostable. The activity of the BSs was retained on heating at 100°C for 60 min with a moderate deviation in the surface tension at CMD^{-1} and CMD^{-2}. This was in good agreement with the earlier reports demonstrating thermostable nature of BSs from *B. subtilis* strains (Cameotra and Makkar, 1998) and *Pseudomonas* strains (Johnson and Boese-Marrazzo, 1980; Turkovskaya et al., 2001). Bharali and Konwar (2011); and Bordoloi and Konwar (2009) observed fall of surface activity by 11.5, 8.3, 6.3, 5.0 and 4.3% in the BSs isolated from *P. aeruginosa strains* MTCC7815, MTCC7812, MTCC7814, MTCC8165, and MTTCC8163, respectively when exposed to 100°C for 60 min.

KEYWORDS

- **biosurfactant**
- **lipopeptides**
- **microbial enhanced oil recovery**
- ***P. aeruginosa***
- **phospholipids**
- **polycyclic aromatic hydrocarbons**

REFERENCES

Bafghi, M. K., & Fazaelipoor, M. H., (2012). Application of rhamnolipid in the formulation of a detergent. *J. Surface. Deterg., 15*, 679–684.

Benincasa, M., (2002). Rhamnolipid production by *P. aeruginosa* LB1 growing on soapstock as the sole carbon source. *J. Food Eng., 54*, 283–288.

Bharali, P., & Konwar, B. K., (2011). Production and physicochemical characterization of a biosurfactant produced by *Pseudomonas aeruginosa* OBP1 isolated from petroleum sludge. *Appl. Biochem. Biotechnol., 164*(8), 1444–1460.

Bharali, P., et al., (2011). Crude biosurfactant from thermophilic *Alcaligenes faecalis*: Feasibility in Petro-spill bioremediation. *Int. Biodeterioration and Biodegradation, 65*(5), 682–690.

Bodour, A. A., & Maier, R. M., (2002). Biosurfactants: Types, screening methods and applications. In: Bitton, G., (ed.), *Encyclopedia of Environmental Microbiology* (pp. 750–770). Wiley, New York.

Bodour, A. A., & Miller-Maier, R. M., (1998). Application of a modified drop collapse technique for surfactant quantitation and screening of biosurfactant-producing microorganisms. *J. Microbiol. Method., 32*, 273–280.

Bordoloi, N. K., & Konwar, B. K., (2007). Microbial surfactant-enhanced mineral oil recovery under laboratory conditions. *Colloids Surf. B. Biointerfaces, 63*, 73–82.

Cameotra, S. S., & Makkar, R. S., (1998). Synthesis of biosurfactants in extreme conditions. *Appl. Microbiol. Biotechnol., 50*, 520–529.

Cameotra, S. S., & Makkar, R. S., (2004). Recent application of biosurfactants as biological and immunological molecules. *Curr. Opin. Microbiol.*, 7, 262–266.

Cameotra, S. S., & Makkar, R. S., (2010). Biosurfactant-enhanced bioremediation of hydrophobic pollutants. *Pure Appl. Chem., 82*, 97–116.

Cooper, D. G., & Paddock, D. A., (1984). Production of biosurfactants from *Torulopsis bombicola, Appl. Environ. Microbiol, 47*, 173–176.

Cooper, D. G., et al., (1981). Enhanced production of surfactin from *B. subtilis* by continuous product removal and metal cation additions. *Appl. Environ. Microbiol., 42*, 408–412.

Costa, S. G. V. A. O., et al., (2009). Cassava wastewater as·a substrate for the simultaneous production of rhamnolipids and, polyhydroxyalkanoates by·*Pseudomonas aeruginosa. J. Ind. Microbiol. Biotechnol., 36*, 1063–1072.

Davey, M. E., et al., (2003). Rhamnolipid surfactant production affects biofilm architecture m *Pseudomonas aeruginosa* PA01. *J. Bacteriol., 185*, 1027–1036.

Desai, J. D., & Desai, A. J., (1993). Production of biosurfactants. In: Kosaric, N., (ed.), *Biosurfactants, Production, Properties, Applications* (pp. 65–97). Marcel Dekker, New York.

Deshpande, M., & Daniels, L., (1995). Evaluation of sophorolipid biostirfactant production by *Candida bombicola* using animal fat. *Bioresour. Technol., 54*, 143–150.

Deziel, E., et al., (2003). rhlA is required for the production of a novel biosurfactant promoting swarming motility in *Pseudomonas aeruginosa:* 3-(3- hydroxyalkanoyloxy) alkanoic acids (HAAs), the precursors of rhamnolipids. *Microbiol., 149*, 2005–2013.

Edward, J. R., & Hayashi, J. A., (1965). Structure of a rhamnolipid from *Pseudomonas aeruginosa. Arch. Biochem. Biophys., 111*, 415–421.
Foght, J. M., & Westlake, D. W. S., (1988). Degradation of polycyclic aromatic hydrocarbons and aromatic heterocycles by a *Pseudomonas species. Canadian J. Microbiol, 34*(10), 1109–1115.
George, S., & Jayachandran, K., (2008). Analysis of rhamnolipid· biosurfactants produced through submerged fermentation using orange fruit peelings as sole carbon source. *Appl. Biochem. Biotechnol., 58*, 428–434.
Gerard, J., et al., (1997). Massetolides A-H, antimycobacterial cyclic depsipeptides produced by two pseudomonads isolated from marine habitats. *J. Nat. Prod., 60*, 223–229.
Haba, E., et al., (2003). Use of liquid chromatography-mass spectrometry for studying the composition and properties of rhamnolipids produced by different strains of *Pseudomonas aeruginosa. Surf. Deterg., 6*, 155–161.
Harshada, K., (2014). Biosurfactant: A potent antimicrobial agent. *J. Microbiol. Expl.*, 1–5.
Hisatsuka, K., et al., (1971). Formation of rhamnolipid by *Pseudomonas aeruginosa*: Its function in hydrocarbon fermentations. *Agric. Biol. Chem., 35*, 686–692.
Holakoo, L., & Mulligan, C. N., (2002). On the capability of rhamnolipids for oil· spill control of surface water. *Proc. Ann. Confe. Can. Soc. Civil. Engg.*, 5–8.
Hommel, R., et al., (1987). Production of water-soluble surface-active exolipids by *Torulopsis apicola. Appl. Microbiol. Biotechnol., 26*, 199–205.
Itoh, S., et al., (1971). Rhamnolipids produced by *Pseudomonas aeruginosa* grown on n-paraffin (mixture of C_{12}, C_{13} and C_{14} fractions). *J. Antibiot., 24*, 855–859.
Jacques, P., Hbid, C., Destain, J., Razafindralambo, H., Paquot, M., De Pauw, E., & Thonart, P., (1999). Optimization of biosurfactant lipopeptide production from *Bacillus subtilis* S499 by Plackett-Burman design. In: *Twentieth Symposium on Biotechnology for Fuels and Chemicals* (pp. 223–233). Humana Press, Totowa, NJ.
Jain, D., et al., (1991). A drop-collapsing test for screening -surfactant-producing microorganisms. *J. Microbiol. Methods, 13*, 271–279.
Jain, R. M., et al., (2012). Isolation and structural characterization of biosurfactant produced by an alkaliphilic bacterium *Cronobacter sakazakii* isolated from oil-contaminated wastewater. *Carbohydrate Polym., 87*, 2320–2326.
Johnson, M. K., & Boese-Marrazzo, D., (1980). Production and properties of heat-stable extracellular hemolysin from *Pseudomonas aeruginosa. Infect. Immun., 29*, 1028–1033.
Joshi, S., et al., (2008). Biosurfactant production using molasses and whey under thermophilic conditions. *Bioresour. Technol., 99*, 195–199.
Kitamoto, D., et al., (2001). Remarkable antiagglomeration effect of yeast biosurfactant, diacylmannosylerythritol, on ice-water slurry for cold thermal storage. *Biotechnol. Prog., 17*, 362–365.
Kluge, B., et al., (1989). Studies on the biosynthesis of surfactin, a lipopeptide antibiotic from *Bacillus subtilis* ATCC-21332. *FEBS Lett., 231*, 107–110.
Lang, S., & Wagner, F., (1987). Structure and properties of biosurfactants. In: Kosaricet, N., et al., (eds.), *Biosurfactants and Biotechnology* (pp. 21–47). Dekker, New York.
Lima, D. C. J. B., et al., (2009). Biosurfactant production by *Pseudomonas aeruginosa* grown in residual soybean oil. *Appl. Biochem. Biotechnol., 152*, 156–168.

Lourith, N., & Kanlayavattanakul, M., (2009). Natural surfactants used in cosmetics: Glycolipids. *Int. J. Cosmet. Sci., 31*, 255–261.
Makkar, R. S., & Cameotra, S. S., (2002). An update on the use of unconventional substrates for biosurfactant production and their new applications. *Appl. Microbiol. Biotechnol., 58*, 428–434.
Maneerat, S., (2005). Production of biosurfactants using substrates from renewable resources. *Songklanakarin J. Sci. Technol., 27*, 675–683.
Mata-Sandoval, J. C., et al., (1999). High-performance liquid chromatography method for the characterization of rhamnolipid mixtures produced by *Pseudomonas aeruginosa* UG2 on com oil. *Chromatogr., 864*, 211–220.
Miller, R. M., & Bartha, R., (1989). Evidence from liposome encapsulation for transport limited microbial metabolism of solid alkanes. *Appl. Environ. Microbiol., 55*, 269–274.
Morikawa, M., et al., (2000). A study on the structure-function relationship of lipopeptide biosurfactants. *Biochimie. Biophys. Acta, 1488*, 211–218.
Mulligan, C. N., et al., (1999). Metal removal from contaminated soil and sediments by the biosurfactant surfactin. *Environ. Sci. Technol., 33*, 3812–3820.
Nakano, M. M., et al., (1991). srfA is an operon required for surfactin production, competence development, and efficient sporulation in *Bacillus subtilis*. *J. Bacteriol., 173*, 1770–1778.
Neidleman, S. L., & Geigert, J., (1984). Biotechnology and oleochemicals: Changing patterns. *J. Am. Oil Chem. Soc., 61*, 290–297.
Nitschke, M., & Pastore, G., (2006). Production and properties of a surfactant obtained from *Bacillus subtilis* grown on cassava wastewater. *Bioresour. Technol., 97*, 336–341.
Noah, K. S., (2005). Surfactin production from potato process effluent by *Bacillus subtilis* in a chemostat. *Appl. Biochem. Biotechnol., 122*, 465–474.
Palejwala, S., & Desai, J. D., (1989). Production of an extracellular emulsifier by a gram-negative bacterium. *Biotechnology Letters, 11*(2), 115–118.
Pornsunthorntawee, O., et al., (2009). Solution properties and vesicle formation of rhamnolipid biosurfactants produced by *Pseudomonas aeruginosa* SP4. *Colloids Surf. B. Biointerfaces, 72*, 6–15.
Read, R. C., et al., (1992). The effect of *Pseudomonas aeruginosa* rhamnolipids on guineapig tracheal mucociliary transport and ciliary beating. *J. Appl. Physiol., 72*, 2271–2277.
Rodrigues, L. R., et al., (2006). Biosurfactants: Potential applications in medicine. *J. Antimicrob. Chemother., 57*, 609–618.
Roongsawang, N., et al., (2002). Isolation and characterization of a halotolerant *Bacillus subtilis* BBK-1 which produces three kinds of lipopeptides: Bacillomycin L, plipastatin, and surfactin. *Extremophiles, 6*(6), 499–506.
Rosenberg, E., & Ron, E. Z., (1996). Bioremediation of petroleum contamination, In: Ronald, L. C., & Don, L. C., (eds.), *Bioremediation: Principles and Applications* (pp. 100–124). Cambridge University Press, UK.
Shaw, A., (1994). Lipid composition as a guide to the classification of bacteria. *Adv. Appl. Microbiol., 17*, 63–108.
Sim, L., et al., (1997). Production and characterization of a biosurfactant isolate from *Pseudomonas aeruginosa* UW-1. *J. Ind. Microbiol. Biotechnol., 19*, 232–238.

Stipcevic, T., et al., (2005). Di-rhamnolipid from *Pseudomonas aeruginosa* displays differential effects on human keratinocyte and fibroblast cultures. *Dermat. Sci., 40*, 141–143.

Suzuki, T., et al., (1974). Sucrose lipids of *Arthrobacteria*, *Corynebacteria*, and *Nocardia* grown on sucrose. *Agric. Biol. Chem., 38*, 557–563.

Syldatk, C., & Wagner, F., (1987). Production of biosurfactants. In: Kosaric, N., et al., (eds.), *Biosurfactants and Biotechnology* (pp. 89–120). Marcel Dekker, New York.

Trzesicka-Mlynarz, D., & Ward, O. P., (1995). Degradation of polycyclic aromatic hydrocarbons (PAHs) by a mixed culture and its component pure cultures, obtained from PAH-contaminated soil. *Canadian J. Microbiol., 41*(6), 470–476.

Turkovskaya, O. V., et al., (2001). A biosurfactant-producing *Pseudomonas aeruginosa* strain. *Appl. Biochem. Microbiol., 37*(1), 71–75.

Ullrich, C., et al., (1991). Cell-free biosynthesis of surfactin, a cyclic lipopeptide produced by *Bacillus subtilis. Biochem., 30*, 6503–6508.

Vatsa, P., et al., (2010). Rhamnolipid biosurfactants as new players in animal and plant defense against microbes. *Inter. J. Mol. Sci., 11*, 5095–5108.

Wicken, A. J., & Knox, K. W., (1970). Studies on the group F antigen of lactobacilli: Isolation of a teichoic acid-lipid complex from *Lactobacillus fermenti* NCTC 6991. *J. Gen. Microbiol., 60*, 303–313.

Yoo, D. S., et al., (2005). Characteristics of microbial biosurfactant as an antifungal agent against plant pathogenic fungus. *J. Microbiol. Biotechnol., 15*, 1164–1169.

Zajic, J. E., et al., (1977). Properties and biodegradation of a bioemulsifier from *Corynebacterium* hydrocarboclastus. *Biotechn. Bioengg., 19*(9), 1303–1320.

FURTHER READING

Das, S., Kalita, S. J., Bharali, P., Konwar, B. K., Das, B., & Thakur, A. J., (2013). Organic reactions in "green surfactant": An avenue to bisuracil derivative. *ACS Sustainable Chem. Eng.,* (p. 301). doi: 10.1021/sc4002774, Publication Date (Web).

Pranjal, B., (2015). *Bioremediation of Crude Oil Contaminated Soil.* (Thesis: Supervisor B K Konwar), Dept. of Mol. Biol. and Biotechnology, Tezpur University (Central), Napaam – 784028, Assam, India.

CHAPTER 5

Biosurfactant Genetics

Peypoux et al. (199) showed biosurfactant (BS) production in bacteria-induced by molecular signals with quorum sensing. Burger et al. (1963) proposed that RL synthesis proceeds by two sequential glycosyl-transfer reactions, each catalyzed by rhamnosyl-transferase. According to Ochsner et al. (1995), mono-rhamnolipid (rhamnolipid 1) synthesis is catalyzed by the enzyme rhamnosyl-transferase I, encoded by the *rhlAB* and present in a single operon. The second rhamnosyl-transferase II is responsible for the synthesis of di-rhamnolipid (rhamnolipid 2), encoded by *rhlC* with the co-coordinated regulation by *rhlAB* having the same quorum sensing. The *rhlR* and *rhlI* genes are arranged sequentially and regulate *rhlAB* genes expression.

The biosynthesis of surfactin is non-ribosomally catalyzed by a large multienzyme peptide synthetase complex called the surfactin synthetase, consisting of three protein subunits-SrfA, SrfB, and SrfC. The peptide synthetase required for amino acid moiety of surfactin and is encoded by four open reading frames (ORFs) in the *srfA* operon namely *SrfAA, SrfAB, SrfAC*, and *SrfAD* or *SrfA-TE*. This operon also contains *comS* gene lying within and out-of-frame of *srfB*. The other three ORFs are essential as compared to *SrfAD* for the biosynthesis of surfactin. The gene *sfp* encodes for phosphopantetheinyl transferase and is essential for the activation of surfactin synthetase by posttranslational modification. When the cell density is high, ComX, a signal peptide after being modified by the gene product of comQ, accumulates in the growth medium (Menkhaus et al., 1993). The histidine protein kinase ComP donates a phosphate to the response regulator ComA and interacts with ComX, which in turn activates the transcription of the *srf* operon (Das et al., 2008).

The other BSs with known molecular mechanism include arthrofactin, iturin, lichenysin, mannosylerythritol lipids (MELs), and emulsan. Arthrofactin is a cyclic lipopeptide-type BS produced by *Pseudomonas* sp. MIS

38. Three genes *arfA*, *arfB,* and *arfC* form the arthrofactin synthetase gene cluster and encodes for ArfA, ArfB, and ArfC proteins, respectively, which assemble to form a unique structure for catalyzing the biosynthetic reactions. Lichenysin is another type of lipopeptide synthesized by non-ribosomally by a multienzyme peptide synthetase complex.

The *lic* operon of *B. licheniformis* is 26.6 kb long and consists of genes *licA*, *licB* (each with three modules), and *licC* (one module). The domain structures of these seven modules resemble the surfactin synthetases SrfA-C. Iturin A is a type of lipopeptide BS produced by *B. subtilis* RB14, composed of four ORFs, contains *ituD*, *ituB, ituC*, and *ituA* genes coding for malonyl coenzyme A transacylase, peptide synthetase consisting of four amino acid adenylation domains and peptide synthetase, respectively while the fourth gene *ituA* encodes for ItuA, having three functional domains homologous to β-ketoacyl synthetase, aminotransferase, and amino acid adenylation. During the stationary phase of growth, *A. lwoffii* RAG-1 secrete a potent bioemulsifier on the cell surface known as emulsan.

5.1 GENETIC ENGINEERING

Chakrabarty (1974) reported that alkanes and simple aromatic hydrocarbon-degrading genes present in the plasmid DNA. Multiple plasmid transfer was accomplished between *Pseudomonas* species to construct a strain that can degrade several hydrocarbons. The genetically engineered plasmid production from conjugation of *Pseudomonas* strains with OCT-CAM and plasmid display Oct^+ and Nah^+ degradation phenotype. By manipulating the exchange of hydrocarbon-degrading genetic material, it is possible to develop strains with extended degradative capability. The ability of microorganisms to degrade hydrocarbon has been shown to be encoded on extrachromosomal-DNA. They also reported that the development in recombinant DNA technology permits the engineering of genes that code for desired enzymatic capabilities. Kolenc et al. (1988) reported that successful transfer of TOL plasmid from a mesophilic *P. putida* to a psychotropic one by conjugation in which the toluene biodegradation genes were expressed at 0°C. Ochser et al. (1995) reported transformation of Pu098 plasmid from *P. auroginosa.* One plasmid was isolated from *P. auroginosa* strain P3, P6, P18, P27, and P31 having the molecular weight 280 ± 15 megadalton. Strains P2, P27, and P31 were found to poses plasmid easily isolated using the alkaline bypass method. The large catabolic plasmids OCT^+ and Nah^+ were isolated from psychotropic *Pseudomonas sp* strains B17, B18, and B19. The isolated plasmid DNA was digested with *Eco* RI and *Bam* H1, compared the data of Oct-plasmid. Two plasmids were isolated from *P. putida* strain HS121 and the second one the strain HS 124. The chromosomal and plasmid DNAs from a mutant strain PCL 1436 of *P. putida* was isolated by *Eco* RI to identify the biosurfactant (BS) producing gene. Dubey and Juwarkar (2004) isolated a high molecular weight plasmid (32.08×10^6 da) from *P. auroginosa* strain BS2 producing BS.

5.2 PLASMID CURING AND ANTIBIOTIC SENSITIVITY

Plasmid curing was carried out and the antibiotic-sensitive bacterial isolates were recovered with the presence of the antibiotic-resistant gene in the plasmid DNA. The curing experiments performed in the bacterial isolates *P. aeruginosa* strains MTCC7815, MTCC7812, MTCC8163, MTCC8165, and MTCC7814 revealed 50–55% curing with a concentration of 500 $mg.ml^{-1}$ of acridine orange. LB medium supplemented with 300 $\mu g.ml^{-1}$ of acridine orange cured the plasmid from aniline-assimilating bacteria. During the log phase of *B. subtilis* culture in nutrient broth at pH 7.6 containing acridine orange (20 $\mu g.ml^{-1}$) with a cell density of $10^4.ml^{-1}$ cured the plasmid-linked polyglutamate gene. The multiple antibiotic resistant gene(s) were linked with R-plasmid of *P. aeruginosa*, and this plasmid was cured with acridine orange in the concentration of 400, 600, 800, 1000, 1200, and 1500 $\mu g.ml^{-1}$. Curing was verified by Bharali et al. (2011) in the strains MTCC7815 and MTCC8163 through culture and plasmid isolation. Following plasmid curing, the isolates failed to grow in hydrocarbon supplemented LB medium. The observations confirmed the presence of the gene(s) responsible for the biosynthesis of BSs and biodegradation of hydrocarbons on the plasmid DNA. These two cured strains failed to grow on plates containing ampicillin, tetracycline, and chloramphenicol as well confirmed the presence of the genes responsible for the resistance to the aforesaid antibiotics. Subsequent non-recovery of plasmid DNA from the isolates indicated its removal following curing. It might therefore be inferred that the genes for antibiotic resistance are in the plasmid as was predicted before.

5.2.1 INCREASE IN SURFACE AREA OF HYDROPHOBIC WATER-INSOLUBLE SUBSTRATES

During the growth of microorganism, certain growth stimulating compounds are produced which tend to emulsify the hydrophobic substrates extending the interfacial area between the microorganisms and the substrate which further facilitates mass transfer of the substrate to the surface of microorganisms (Kitamoto et al., 2002). Work of Zhang and Miller (1994) confirmed the effect of BSs on hydrocarbon biodegradation with the increase of microbial accessibility to insoluble substrates and thus enhanced their biodegradation. Chang et al. (2008) reported that BSs increase the apparent solubility of hydrophobic organic compounds at concentrations above the critical micelle concentration (CMC), which enhances their availability for microbial uptake. Whang et al. (2005) showed that two different types of BSs such as rhamnolipid (RL) and surfactin produced by several sp. of *P. aeruginosa* and *B. subtilis,* respectively, increases the solubility and bioavailability of a petrochemical mixture and stimulate indigenous microorganisms for enhanced biodegradation of diesel contaminated soil. BS-negative mutants of *P. aeruginosa* KY-4025 (Itoh et al., 1971) and *P. aeruginosa* PG-201 (Koch et al., 1991) exhibited poor growth compared to the parent strains on paraffin and hexadecane, respectively, and addition of RL externally to the medium restored growth of the microorganism in the respective hydrocarbon. Franzetti et al. (2008, 2009) reported that the rate of biodegradation is dependent on the chemicophysical properties of the BSs and not by the effects on microbial metabolism. Reid et al. (2000) and Stokes et al. (2005) in their review highlighted that biodegradation assays depends on soil slurries and solubilization of target contaminants, which gives an estimation of bioaccessibility rather than bioavailability. Burgos-Díaz et al. (2013) reported that exogenously added BS can increase the apparent water solubility of organic compounds and alter its bioavailability by mediating interactions between the hydrophobicity of the cell surface and the substrate surface.

5.2.2 INCREASE OF BIOAVAILABILITY OF HYDROPHOBIC WATER-INSOLUBLE SUBSTRATES

The interaction of microorganisms with hydrophobic organic chemicals and the role of BSs in their bioavailability have been reviewed extensively by Salihu et al. (2009); Ron and Rosenberg (2002). Volkering et al. (1993) reviewed the probable modes of how microorganisms interact with hydrophobic organic compounds and suggested that the microorganisms may access the hydrophobic substrate via direct contact or by contact with pseudosolubilized substrate in surfactant micelles or emulsion droplets. In the case of direct contact, the hydrophobicity of both the cell surface and the substrate surface will determine the interaction and BSs may play a role in mediating such interactions (Van Hamme et al., 2006).

Rosenberg (1981) found that emulsan, an extracellular polymeric hetero-polysaccharide capsule, is used by *Acinetobacter calcoaceticus* to facilitate detachment from crude oil droplets exhausted of substrate. Once the utilizable substrates have been consumed the emulsan coat is shed off and changing the hydrophobic oil surface to a hydrophilic one. In *Acinetobacter radioresistens* KA53 the bioemulsifiers, alasan, was found to increase the solubility of PAHs by 6 to 27-fold (Rosenberg et al., 1988). For pseudosolubilization, addition of exogenous BS or surfactant can enhance a noticeable aqueous solubility of organic compounds and modify the bioavailability (Zhao et al., 2005; Boonchan et al., 1998). Miller and Bartha (1989) showed that micelles or other aggregates are formed that partition hydrophobic substrates and may enhance biodegradation by allowing for closer cell-substrate interactions or may fuse directly with microbial membranes resulting in direct substrate delivery. RL has been found to remove LPS in a dose-dependent manner from *P. aeruginosa*, resulting in increased cell surface hydrophobicity and enhanced uptake of hydrophobic substrates (Noordman and Janssen, 2002). Some hydrocarbon-degrading microbes respond to these insoluble carbon sources by producing surface-active compounds, as well as by changing cell surface properties such as cell surface hydrophobicity (Zhang and Miller, 1987; Al-Tahhan et al., 2000; Beal and Betts, 2000).

5.2.3 BINDING OF THE HEAVY METALS

Since many contaminated sites are also co-contaminated with metals, BSs have also been explored for metal chelation (Van Hamme et al., 2006). Mulligan et al. (1984) have recently evaluated remediation technologies for metal-contaminated soils. RLs can remove metals, ions, and forms stable complexes with metals in the following order: Al^{3+}> Cu^{2+}>

Pb^{2+}> Cd^{2+}> Zn^{2+}> Fe^{3+}> Hg^{2+}> Ca^{2+}> Co^{2+}> Ni^{2+}> Mn^{2+}> Mg^{2+}> K^{+} (Osterreicher-Ravid et al., 2000). Tan et al. (1984) studied the effect of mono-rhamnolipid produced by *P. aeruginosa* ATCC 9027 on the formation of metal complexes and reported that surfactant-metal interactions are rapid and stable. The mechanism of reduced toxicity was apparently via RL complexation of cadmium as well as by RL-induced lipopolysaccharide (LPS) removal from the cell surface. Mulligan et al. (1984) have been reported to use surfactin from *Bacillus subtilis* to treat soil and sediments contaminated with Zn, Cu, Cd, oil, and grease. Mulligan and Yong (1997) used BSs from *Bacillus subtilis* ATCC 21332, *P. aeruginosa* ATCC 9027 and *Torulopsis bombicola* ATCC 22214 to examine the removal of metals from oil-contaminated soil. Surfactin, RLs, and sophorolipids produced by the microorganisms were extracted using methods described in Mulligan et al. (2001); and Mulligan and Gibbs (1989). Sandrin et al. (2000) reported that exogenously added RL reduces cadmium toxicity for *Burkholderia* sp. growing on either naphthalene or glucose as sole carbon source. In sorption, metal-ligand complexation, complexation with soil constituents and cation exchange processes are involved (Sposito, 1998) affecting access of the metal to the microorganisms.

5.2.4 PATHOGENESIS

BS noticeably influence the physiological behavior of microbes such as their role in plant and animal pathogenesis and the same have been extensively reviewed by Cameotra and Makkar (1998) and Peypoux et al. (1999). BS is regarded to function as a "dispersing agent" in pathogenic microorganisms effecting plants or animals as a "wetting agent" for the surface of the host cell (Kitamoto et al., 2002). *Pseudomonas syringae* produces two necrosis-inducing lipopeptide toxins, syringopeptin, and syringomycin, types of pore-forming cytotoxins that form ion channels permeable to divalent cations during plant pathogenesis (Hutchison and Gross, 1997). Plant pathogenic bacterium, *Pseudomonas fluorescens* produces CLPs (viscosin), which reduce the surface tension on plant epidermis and thus accelerating wetting of the surface, dispersion of the bacteria and invasion and subsequent decay of the difficult-to-wet, waxy surface, etc. (Hildebrand et al., 1998). *Serratia marcescens*, an opportunistic pathogenic bacterium produces serrawettin a type of nonionic

CLP that helps in wetting of the host cell surface and dispersion of the bacteria (Matsuyama and Nakagawa, 1996). Several sp. of *P. aeruginosa* are reported to be pathogenic and causes serious infections in immuno-compromised patients and individuals suffering from cystic fibrosis (CF) (Abdel-Mawgoud et al., 2010). *P. aeruginosa* has been reported to produce a heat-stable extracellular glycolipid called hemolysin that has hemolytic activity (Johnson and Boese-Marrazzo, 1980). The di-rhamnolipid type BS from *Burkholderia pseudomallei* is similarly hemolytic for erythrocytes of various sp. also cytotoxic at high concentrations for non-phagocytic and phagocytic cell lines (Hauler et al., 1998). Zulianello et al. (2006) showed shown that *P. aeruginosa* requires the production of RLs to invade respiratory epithelia reconstituted with primary human respiratory cells.

5.2.5 ANTIMICROBIAL ACTIVITY

With the interest in developing novel antimicrobials for therapeutic and agricultural applications, several BSs with antibiotic properties have been described. A *Pseudomonas* sp. derived from marine alga produces eight types of Massetolides A-H, novel CLPs. These eight Massetolides A-H were and was found to exhibit *in vitro* antimicrobial activity against *Mycobacterium tuberculosis* and *Mycobacterium avium-intracellulare* (Gerard et al., 1997). Sotirova et al. (2009) showed that RLs from *Pseudomonas* sp. PS-17 interact with *P. aeruginosa* causing a reduction in LPS content, changing the outer membrane proteins and had a direct impact on bacterial cell surface morphology. The antibacterial property of BS and their application in the field of medicine is extensively reviewed by Rodrigues et al. (2006). Kim et al. (1990) studied the effects of RL B on a range of plant pathogenic fungi including *Phytophthora capsici* and *Colletotrichum orbiculare* and observed to cause zoospore lysis, inhibition of zoospore and spore germination, and hyphal growth inhibition. Andersen et al. (2003) reported to isolate a *Pseudomonas* sp. DSS73 strain from the rhizoplane of sugar beet seedlings that showed antagonism towards the root-pathogenic microfungi *Pythium ultimum* and *Rhizoctonia solani*. Yoo et al. (2005) investigated RLs as alternative antifungal agents against typical plant pathogenic oomycetes, including *Phytophthora* sp. and *Pythium* sp. Viscosinamide, a new antibiotic isolated from *Pseudomonas fluorescens*, with BS properties and was found to have antifungal properties (Nielsen et al., 1999). An antimicrobial lipopeptide was isolated from a strain of *B. subtilis* and demonstrated a broad spectrum of activity against Gram-negative bacteria, lesser activity against gram-positive organisms and was active against one of the two fungi assayed. The chelating properties of lichenysin, a cyclic lipopeptide produced by *Bacillus licheniformis*, which might explain the membrane disrupting effect of lipopeptides. Carrillo et al. (2003) studied the molecular mechanism of antibiotic and other important biological actions of surfactin produced by *Bacillus subtilis*. Nielsen and Sorensen (2003) found three cyclic lipopeptides (viscosinamide, tensin, and amphisin) produced by *P. fluorescens* in the rhizosphere of germinating sugar beet seeds and considered that such lipopeptides confer a competitive advantage to the organism during colonization. Vatsa et al. (2010) reviewed the zoosporicidal activity of RLs against various fungal phytopathogens. In the literature, properties of RLs against the algae *Heterosigma akashiwo*, viruses, amoeba-like *Dictyostelium discoideum* and mycoplasma have also been reported but don't have significant effect on yeasts.

5.2.6 BIOSURFACTANT (BS) IN ATTACHMENT OF MICROORGANISM TO SURFACES

One of the most essential survival strategies of microorganisms is their ability to establish themselves in an ecological niche where they can propagate. In such a strategy, the key element is the structure of the cell-surface, responsible for the attachment of the microbes to the proper surface (Rosenberg and Ron, 1999). Neu and Poralla (1995) have reviewed how surfactants can affect the interaction between bacteria and interfaces. Zhang and Miller (1994) observed that cell surface hydrophobicity of *P. aeruginosa* was greatly increased by the presence of cell-bound RL. In the case of *Acinetobacter* strains, the cell-surface hydrophobicity was reduced in the presence of its cell-bound emulsifier. Such observations suggest that microorganisms can use their BSs to regulate their cell-surface properties to attach or detach from surfaces according to the need. Rosenberg (1988) found that emulsan, an extracellular polymeric heteropolysaccharide capsule, is used by *Acinetobacter calcoaceticus* to facilitate detachment from crude oil droplets with the exhaustion of carbon source.

5.2.7 BIOSURFACTANT (BS) PRODUCTION AND QUORUM SENSING

Being a virulence factor, the production of bioemulsifier produced by the pathogens initiates localized attacks on the host (Kim et al., 1990). Ron and Rosenberg (2002) reported that bacteria growing at the oil-water interface starts producing emulsifier when the cellular density becomes higher, resulting in the increase in the surface area of oil droplets and this allows more bacteria to attach on the extended surface area of the oil drops. On the other hand, when these usable fractions of the hydrocarbon present in the oil are consumed, the production of emulsifiers allows the bacteria to get detached from the "used" droplets and find a new one.

5.2.8 ROLE OF BIOSURFACTANT (BS) IN BIOFILMS

RLs are reported to be involved in biofilm development. Exogenous RLs induces a release of LPSs and consequently enhance the cell surface hydrophobicity, which might favor the primary adhesion of planktonic cells (Zhang and Miller, 1994; Al-Tahhan et al., 2000). Alasan, an exocellular polymeric emulsifier produced by *Acinetobacter radioresistens* KA53 strain was reported to bind on the surface of *Sphingomonas paucimobilis* EPA505 and *A. calcoaceticus* RAG-1 and modify their surface properties. Moreover, when the alasan-producing *A. radioresistens* KA53 was co-culture with *A. calcoaceticus* RAG-1, alasan was released from the producing strain and bound to the recipient RAG-1 cells (Osterreicher-Ravid et al., 2000). Such horizontal transfer of bioemulsifier between the bacterial sp. has considerable implications in natural microbial communities, co-aggregation, and biofilm formation. RLs assist the surface-associated migration of bacteria in the biofilm, and therefore, the initial microcolony formation and differentiation of the biofilm structure takes place (Pamp and Tolker, 2007).

RL production appears as a regulator for determining cell-surface hydrophobicity and modification of adhesive interactions, especially when there are changes in nutritional conditions (Vater et al., 2002; Dubey and Juwarkar, 2004). RLs are reported to be involved in biofilm development. Exogenous RLs induces a release of LPSs and consequently enhance the cell surface hydrophobicity, which might favor the primary adhesion of planktonic cells (Gudiana et al., 2010; Daniel et al., 1998). Alasan, an exocellular polymeric emulsifier produced by *Acinetobacter radioresistens* was KA53 reported to bind on the surface of *Sphingomonas paucimobilis* EPA505 and *Acinetobacter calcoaceticus* RAG-1 and modify their surface properties. Moreover, when the alasan-producing *Acinetobacter radioresistens* KA53 was co-culture with *Acinetobacter calcoaceticus* RAG-1, alasan was released from the producing strain and bound to the recipient RAG-1 cells (Fujii et al., 1997; Hara and Ueda, 1982).

5.3 GENETIC REGULATION OF BIOSURFACTANT (BS) SYNTHESIS

The regulation of BS production has been investigated at the molecular level for the glycolipids of *P. aeruginosa* and a few lipopeptides of *Bacillus sp*. The production of BS in bacteria was induced by molecular signal involved in quorum sensing (Pascual et al., 2012). However, whether quorum sensing is the environmental clue to BS production in general is still not known. Burger et al. (1963) proposed that RL synthesis proceeds by two sequential glycosyl transfer reactions, each catalyzed by a different rhamnosyltransferase. Ochsner et al. (1995) made a significant contribution to the genetics of RL biosynthesis. Mono-rhamnolipid (rhamnolipid 1) synthesis is catalyzed by the enzyme rhamnosyltransferase 1, encoded by the *rhlAB* and is present in a single operon.

The second rhamnosyltransferase 2 responsible for the synthesis of di-rhamnolipid (rhamnolipid 2), encoded by *rhlC*, had been characterized and its expression had been shown to be co-coordinately regulated with *rhlAB* by the same quorum sensing system (Reid et al., 2000). The *rhlR* and *rhlI* genes are arranged sequentially and regulate *rhlAB* genes expression. *RhlI* protein forms Nacylhomoserine lactones, which act as autoinducers and influence *RhlR* regulator protein. Induction of *rhlAB* depends on quorum-sensing transcription activator *RhlR* complexes with the autoinducer N-butyryl-homoserine lactone (C4-HSL). The biosynthesis of surfactin is catalyzed non-ribosomally by a large multienzyme peptide synthetase complex called the surfactin synthetase, consisting of three protein subunits-SrfA, ComA (earlier known as SrfB) and SrfC (Bordoloi and Kowar, 2008). The peptide synthetase required for amino acid moiety of surfactin and is encoded by four open reading frames (ORFs) in the *srfA* operon namely s*SrfAA*, *SrfAB, SrfAC*, and *SrfAD* or *SrfA-TE*. This operon also contains *comS* gene lying within and out-of-frame with the *srfB*. The other three ORFs are essential as compared to *SrfAD* for the biosynthesis of surfactin. The gene *sfp* encodes for phosphopantetheinyl transferase and is necessary for the activation of surfactin synthetase by posttranslational modification. When the cell density is high, ComX, a signal peptide after being modified by the gene product of comQ, accumulates in the growth medium (Stokes et al., 2005). Quorum sensing controls *srfA* expression by ComX. The histidine protein kinase ComP donates a phosphate to the response regulator ComA and interacts with ComX, which in turn activates the transcription of the *srf* operon (Pascual et al., 2012). Other types of BSs whose molecular genetics have been decoded in the recent years include arthrofactin, iturin, lichenysin, mannosylerythritol lipids (MEL) and emulsan. Arthrofactin is a cyclic lipopeptide-type BS produced by *Pseudomonas* sp. MIS38. Three genes designated as *arfA*, *arfB,* and *arfC* form the arthrofactin synthetase gene cluster and encodes for ArfA, ArfB, and ArfC proteins, respectively, which assemble to form a unique structure for catalyzing the biosynthetic reactions. Lichenysin is another type of lipopeptide synthesized by non-ribosomally by a multienzyme peptide synthetase complex. The *lic* operon of *B. licheniformis* is 26.6 kb long and consists of genes *licA* (three modules), *licB* (three modules) and *licC* (one module). The domain structures of these seven modules resemble that of surfactin synthetases SrfA-C (Pascual et al., 2012).

Iturin A is a type of lipopeptide BS produced by *B. subtilis* RB14, composed of four ORFs contains *ituD*, *ituB, ituC*, and *ituA* genes which encodes for putative malonyl coenzyme A transacylase, peptide synthetase consisting of four amino acid adenylation domains and peptide synthetase, respectively, while the fourth gene *ituA* encodes for ItuA, having three functional domains homologous to β-ketoacyl synthetase, aminotransferase, and amino acid adenylation. *Ustilago maydis* produces two kinds of glycolipid type BSs, MEL referred to as ustilipids (Burgos-Diaz et al., 2013) and ustilagic acid that are cellobiose lipids. Two genes, *emt1* and *cyp1* are involved in the synthesis of MEL and ustilagic acid, respectively. It is assumed that Cyp1 protein is associated with the terminal and/or sub-terminal hydroxylation of an unusual fatty acid present in cellobiose lipids (Pascual et al., 2012). During the stationary phase of growth, *Acinetobacter lwoffii* RAG-1 secrete a potent bioemulsifier on the cell surface known as emulsan (Boonchan et al., 1998). A 27 kb gene cluster termed wee encodes the genes *wza*, *wzb*, *wzc*, *wzx*, and *wzy* required for the biosynthesis of emulsan (Mulligan and Yong, 1997). It was later established that Wzc and Wzb encode a protein tyrosine kinase and protein tyrosine phosphatase, respectively (Sandrin et al., 2000).

5.4 BIOSYNTHETIC PATHWAYS OF BIOSURFACTANT (BS) SYNTHESIS

BSs are synthesized by two primary metabolic pathways viz hydrocarbon and carbohydrate pathways (Kosswig, 2005; Mata-Sandoval et al., 1999). Metabolic pathways involved in the synthesis of the precursors of hydrophilic and hydrophobic domains of BSs are diverse and utilize a definite set of enzymes. Some of the possible features for the biosynthesis of BSs and their regulation, which includes (i) *de novo* synthesis of hydrophilic and hydrophobic moieties by two independent pathways followed by their linkage to form a complete BS molecule, (ii) *de novo* synthesis of the hydrophilic moiety and the substrate-dependent synthesis of the hydrophobic moiety and its linkages, and (iii) *de novo* synthesis of the hydrophobic moiety and the substrate-dependent synthesis of the hydrophilic moiety followed by its linkage. The biosynthesis of both hydrophobic and hydrophilic moieties depends on the type of substrate used to produce BS (Mata-Sandoval et al., 1999). The biosynthesis of surfactin by *Bacillus subtilis* has been extensively studied by Kluge et al. (1989). The formation of surfactin occurs non-ribosomally and two different mechanisms are involved in the activation of amino acid.

Two components of surfactin synthesizing enzyme complex of *B. subtilis* are homologous to tyrocidine synthase I and gramicidin S synthase. In addition, the biochemical studies confirmed the occurrence of surfactin synthesis via a thio-template mechanism. Enzymatic synthesis of surfactin requires ATP, Mg^{2+}, precursors, and sucrose. The fatty acid component of surfactin is incorporated only as an acetyl-CoA derivative and L-isomer of amino acids are incorporated in the peptide chain (Vatsa et al., 2010). The enzymes involved also catalyze the ATP-P*i* exchange reactions, which are mediated by the amino acid components of surfactin. This pattern was consistent with a peptide-synthesizing system that activates its substrate simultaneously as aminoacyl phosphates (Hildebrand et al., 1998). In case of surface-active compound herbicollin A, both the lipid and peptide domain have been found to be directly synthesized from carbohydrates. Addition of amino acids or fatty acids in the growth medium affected the yield but not the structure of the surfactant (Mata-Sandoval et al., 1999). Research investigations have shown that in Gramicidin-S a type of surface-active antibiotic, lipopeptide, is synthesized non-ribosomally by a multienzyme complex with the involvement of pantetheine cofactor by a thio-template mechanism (Raza et al., 2007). Regarding the biosynthesis of glycolipids, the pathway of the sugar-lipid BS formation depends on the microorganism producing it. An example of glycolipid synthesis is the biosynthesis of anionic RLs by *Pseudomonas* sp.

RL synthesis using enzymology and different radioactively labeled precursors and the proposed biosynthetic pathway has been studied extensively by Banat et al. (1993). Syldatk et al. (1985) reported that the composition of BS produced by *Pseudomonas* sp. is affected by the type of carbon substrate used and the cultivation conditions but the hydrocarbon substrate having different chain length has no effect on the chain length of the fatty acid moiety in glycolipids. Almost the similar results were observed during the production of bioemulsifier by *Acinetobacter sp.* H01-N using alkane as the substrate (Desai et al., 1994). Suzuki et al. (1969) reported the influence of substrate on the sugar moiety of the glycolipid synthesized by *Arthrobacter paraffineus*. The non-ionic trehalose lipid is formed when *A. paraffineus* is grown on n-alkanes, but fructose lipids are produced when fructose is used as the sole carbon source. Other examples of *de novo* synthesis of BS are cellobiose lipid by *Ustilago zeae* (Zawawi, 2005) and sophorolipid by *Torulopsis bombicola* (Bodour et al., 2003) from different hydrophobic substrates. During the synthesis of trehalose mono- and dicorynomycolates in *Rhodococcus erythropolis*, the sugar moiety of the surfactant is *de novo* synthesized and the chain length of the lipid moiety is dependent on hydrocarbon substrate used in the medium (Clarke et al., 2010). A similar pathway has been found to be working in *Rhodococcus erythropolis* for the synthesis of trehalosetetraesters (Sposito, 1989), in *Candida sp.* for the synthesis of MELs (Persson et al., 1990), and in *Nocardia erythropolis* (Ilori et al., 2005) for extracellular glycolipid synthesis.

5.5 REGULATION OF BIOSURFACTANT (BS) SYNTHESIS

The chemical composition, level of production and surface properties of the BS depend not only on the producer strain but also on various factors such as nature and concentration of macro and micronutrients, culture conditions including pH, temperature, agitation, and dilution rate (Lotfabad et al., 2009). The carbon source is the most important factor influencing the BS synthesis either by induction or by repression (Cameotra and Makkar, 1998). The commonly used carbon sources include carbohydrates, hydrocarbons, and vegetable oils. It has been concluded from several studies that different carbon sources can influence the composition of BS formation (Salihu et al., 2009; Raza et al., 2007). *Actinobacter calcoaceticus* and *Arthobacter paraffineus* fail to produce surface-active compounds when grown on organic acids and D-glucose as carbon source, respectively. Previous reports indicate that the addition of water-immiscible substrates result in the induction of BS production. Tulloch et al. (1962) have found the induction of sophorolipid synthesis by the addition of long-chain fatty acids, hydrocarbons, or glycerides to the growth medium of *Torulopsis magnolia* (Desai et al., 1994). *Arthrobacter* produces 75% extracellular BS when grown on acetate or ethanol, but it is totally extracellular when grown on hydrocarbon (Mulligan and Gibbs, 1993).

Hauser and Karnovsky (1958) have demonstrated a severe decrease in the synthesis of RL on addition of glucose, acetate, and tricarboxylic acids during the growth on glycerol. A similar observation was reported for the synthesis of liposan in *Candida lipolytica* (Cirigliano and Carman, 1985). On the other hand, surfactin produced by *Bacillus subtilis* is usually observed with glucose as the carbon source and is inhibited by the addition of hydrocarbons in the medium (Cooper et al., 1981).

Nitrogen is another important factor that plays an important part in the regulation of BS synthesis. It may also contribute to pH control (Zawawi, 2005). Duvnjak et al. (1983) found that urea led to a satisfactory BS production. Moreover, nitrogen limitation also changed the composition of the BS production (Desai and Desai, 1993; Syldat et al., 1985). Among the inorganic salts tested, ammonium salts and urea were preferred for BS production by *Arthobacter paraffineus*, whereas nitrate supported maximum BS production in *P. aeruginosa* (Desai and Banat, 1997). Yeast extract was found to be required for glycolipid production by *Torulopsis bombicola* but was very poor for *P. aeruginosa*. Supplementation of ammonia in the nitrate-containing medium results in the decay production of BS in *Corynebacterium* (Desai and Desai, 1993). Several investigators observed RL in the fermentation broth of *P. aeruginosa* with the exhaustion of nitrogen and the beginning of the stationary phase of growth (Abdel-Mawgoud et al., 2010). Nitrogen limitation not only causes overproduction of BSs but also changes the composition of BSs produced (Desai and Desai, 1993, p. 174). According to Hommel et al. (1994), it is the absolute quantity of nitrogen and not its relative concentration that is important to give an optimum biomass yield while the concentration of hydrophobic carbon source determines the conversion of carbon available to the BS.

Phosphate limitation also influences the metabolism of BS (Zawawi, 2005). Clarke et al. (2010) observed enhanced production of RL by *P. aeruginosa* ATCC 9027 under phosphate limited conditions. The change in activity of several intracellular enzymes dependent on phosphate levels indicated a shift in BS metabolism. Iron limitation is reported to stimulate the production of BSs in *P. fluorescens* (Persson et al., 1990). However, the production of surfactin by *B. subtilis* is reported to be stimulated by the addition of iron and manganese salts to the medium (Cooper et al., 1981). The limitation of multivalent cations also causes overproduction of BSs (Cameotra and Makkar, 1998; Guerra-Santos et al., 1984). Higher yield of RL could be achieved in *P. aeruginosa* DSM 2659 by limiting the concentration of Mg^{2+}, Ca^{2+}, K^{+}, Na^{+} and trace salts (Desai and Desai, 1993). Environmental factors such as temperature, pH, agitation, and oxygen availability also affect the production of BS production through their effect on cellular growth or activity (Salihu et al., 2009; Cameotra and Makkar, 1998; Raza et al., 2007; Ilori et al., 2005). Temperature may cause alteration in the composition of the BS produced by *Pseudomonas* sp. DSM-2874 (Syldatk et al., 1985).

Banat (1993) reported a thermophilic *Bacillus* sp. which could grow and produced BS at temperatures above 40°C. In *Torulopsis bambicola* the pH of the medium plays an important role in sophorolipid production (Gobbert et al., 1984). The penta and disaccharide lipid production by *Nocardia corynbacteroides* was unaffected in the pH range of 6.5 to 8.0. Mulligan and Gibbs (1989) reported that an increase in agitation speed caused the shear effect which reduced the production of BS by *Nocardia erythropolis*. Conversely, production of BS by yeast increased when the agitation and aeration rates increased (Patel and Desai, 1997). The role of abiotic factors on the production of RL by *P. aeruginosa* was extensively reviewed by various authors (Lang and Wagner, 1987; Abdel-Mawgoud et al., 2010; Henkel et al., 2012).

KEYWORDS

- **agitation**
- **critical micelle concentration**
- **cystic fibrosis**
- **lipopolysaccharides**
- **mannosylerythritol lipid**
- **open reading frames**

REFERENCES

Abdel-Mawgoud, A. M., et al., (2010). Rhamnolipids: Diversity of structures, microbial origins and roles. *Appl. Microbiol. Biotechnol., 86*, 1323–1336.

Al-Tahhan, R. A., et al., (2000). Rhamnolipid induced removal of lipopolysaccharide from *Pseudomonas aeruginosa*: Effect on cell surface properties and interaction with hydrophobic substrates. *Appl. Environ. Microbiol., 66*, 3262–3268.

Andersen, J. B., et al., (2003). Surface motility in *Pseudomonas sp*. DSS73 is required for efficient biological containment of the root-pathogenic micro fungi *Rhizoctonia solani* and *Pythium ultimum. Microbiol., 149*, 37–46.

Banat I. M., (1995). Biosurfactants production and possible uses in microbial enhanced oil recovery and oil pollution remediation: A review. *Bioresour. Technol., 51*, 1–12.

Banat, I. M., et al., (1991). Biosurfactant production and use in oil tank clean-up. *World J. Microbiol. Biotechnol., 7*, 80–88.

Banat, I. M., et al., (2010). Microbial biosurfactants production applications and future potential. *Appl. Microbiol. Biotechnol., 87*, 427–444.

Beal, R., & Betts, W. B., (2000). Role of rhamnolipid biosurfactants in the uptake and mineralization of hexadecane in *Pseudomonas aeruginosa. J. Appl. Microbiol., 89*, 158–168.

Bechard, J., et al., (1998). Isolation and partial chemical characterization of an antimicrobial peptide produced by a strain of *Bacillus subtilis. J. Agric. Food. Chem., 46*, 5355–5361.

Bharali, P., & Konwar, B. K., (2011). Production and physicochemical characterization of a biosurfactant produced by *Pseudomonas aeruginosa* OBP1 isolated from petroleum sludge. *Appl. Biochem. Biotechnol., 164*(8), 1444–1460.

Bodour, A. A., & Maier, R. M., (2002). Biosurfactants: Types, screening methods and applications. In: Bitton, G., (eds.), *Encyclopedia of Environmental Microbiology* (pp. 750–770). Wiley, New York.

Boonchan, S., et al., (1998). Surfactant-enhanced biodegradation of high molecular weight polycyclic aromatic hydrocarbons by *Stenotrophomonas maltophilia. Biotechnol. Bioengg., 59*,482–494.

Bordoloi, N. K., & Konwar, B. K., (2008). *Colloids and Surfaces B: Biointerfaces, 63*, 73–82.

Burger, M. M., et al., (1963). The enzymatic synthesis of rhamnose-containing glycolipid by extracts of *Pseudomonas aeruginosa*. *J. Biol. Chem., 238*, 2595–2602.

Burgos-Diaz, C., et al., (2013). The production and physicochemical properties of a biosurfactant mixture obtained from *Sphingobacterium detergens*. *J. Colloid Interface Sci., 394*, 368–79.

Cameotra, S. S., & Makkar, R. S., (1998). Synthesis of biosurfactants in extreme conditions. *Appl. Microbiol. Biotechnol., 50*, 520–529.

Carrillo, C., et al., (2003). Molecular mechanism of membrane permeabilization by the peptide antibiotic surfactin. *Biochim. Biophys. Acta, 1611*, 91–97.

Chakrabarty, A. M., (1974). Dissociation of a degradative plasmid aggregate in *Pseudomonas*. *J. Bacteriology, 118*(3), 815–820.

Chang, M. W., et al., (2008). Molecular characterization of surfactant-driven microbial community changes in anaerobic phenanthrene degrading cultures under methanogenic conditions. *Biotech. Lett., 30*, 1595–1601.

Cirigliano, M. C., & Carman, G. M., (1985). Purification and characterization of liposan, a bioemulsifier from *Candida lipolytica*. *Appl. Environ. Microbiol., 50*, 846–850.

Clarke, K., et al., (2010). Enhanced rhamnolipid production by *Pseudomonas aeruginosa* under phosphate limitation. *World J. Microbiol. Biotechnol., 26*, 2179–2184.

Cooper, D. G., et al., (1981). Enhanced production of surfactin from *B. subtilis* by continuous product removal and metal cation additions. *Appl. Environ. Microbiol., 42*, 408–412.

Daniel, H. J., et al., (1998). Production of sophorolipids in high concentration form deproteinized whey and rapeseed oil in a two-stage fed-batch process using *Candida bombicola* ATCC 22214 and *Cryptococcus curvatus* ATCCC 20509. *Biotechnol. Lett., 20*, 1153–1156.

Das, P., et al., (2008). Genetic regulations of the biosynthesis of microbial surfactants: An overview. *Biotechnol. Genetic Eng. Rev., 5*, 165–186.

Das, P., et al., (2008). Improved bioavailability and biodegradation of a model polyaromatic hydrocarbon by a biosurfactant producing bacterium of marine origin. *Chemosphere, 72*, 1229–1234.

Desai, A. J., et al., (1994). Advances in production of biosurfactants and their commercial applications. *J. Sci. Ind. Res., 53*, 619–629.

Desai, J. D., & Banat, I. M., (1997). Microbial production of surfactants and their commercial potential. *Microbiol. Mol. Bio. Rev., 61*, 47–64.

Desai, J. D., & Desai, A. J., (1993). Production of biosurfactants. In: Kosaric, N., (eds.), *Biosurfactants, Production, Properties, Applications* (pp. 65–97). Marcel Dekker, New York.

Dubey, K., & Juwarkar, A., (2004). Determination of Genetic Basis for Biosurfactant Production in Distillery and Curd Whey Wastes Utilizing Pseudomonas aeruginosa Strain BS2. *Indian J. Biotech 3*, 74–81.

Duvnjak, Z., et al., (1983). Effect of nitrogen source on surfactant production by *Arthrobacter paraffines* ATCC 19558. In: Zajic, J. E., et al., (eds.), *Microbial. Enhanced Oil Recovery* (pp. 66–72). Pennwell Books, Tulsa, Okla.

Franzetti, A., et al., (2008). Surface-active compounds and their role in bacterial access to hydrocarbons in *Gordonia* strains. *FEMS Microbiol. Ecol., 63*, 238–248.

Fujii, T., et al., (1997). Plasmid-encoded genes specifying aniline oxidation from *Acinetobacter sp*. strain YAA. *Microbiology, 143*(1), 93–99.

Gerard, J., et al., (1997). Massetolides A-H, antimycobacterial cyclic depsipeptides produced by two pseudomonads isolated from marine habitats. *J. Nat. Prod., 60*, 223–229.

Gobbert, et al. (984). Soforose lipid formation by resting cells of *Torulopsis bombicola. Biotech. Letters, 6*, 225–230.

Grangemard, I., et al., (2001). Lichenysin: A more efficient cation chelator than surfactin. *Appl. Biochem. Biotechnol., 90*, 199–210.

Gudiana, E. J., et al., (2010). Isolation and functional characterization of a biosurfactant produced by *Lactobacillus paracasei. Colloids Surf. B. Biointerfcaes, 16*, 298–304.

Guerra-Santos, L., et al., (1984). *Pseudomonas aeruginosa* biosurfactant production in continuous culture with glucose as carbon source. *Appl. Environ. Microbiol., 48*, 30–305.

Hara, T., & Ueda, S., (1982). Regulation of polyglutamate production in *Bacillus subtilis* (natto): Transformation of high PGA productivity. *Agri. Biol Chem., 46*(9), 2275–2281.

Hauler, S., et al., (2003). Structural and functional cellular changes induced by *Burkholderia pseudomallei* rhamnolipid. *Infect. Immun., 71*, 2970–2975.

Hauser, G., & Karnovsky, M. L., (1958). Studies on the biosynthesis of L-rhamnose. *J. Biol. Chem., 233*, 287–291.

Henkel, M., et al., (2012). Rhamnolipids as biosurfactants from renewable resources: Concepts for next-generation rhamnolipid production. *Process Biochem., 47*, 1207–1219.

Hildebrand, P. D., et al., (1998). Role of the biosurfactant viscosin in broccoli head rot caused by a pectolytic strain of *Pseudomonas fiuorescens. Can. J. Plant Pathol., 20*, 296–303.

Hommel, R., et al., (1987). Production of water-soluble surface-active exolipids by *Torulopsis apicola. Appl. Microbiol. Biotechnol., 26*, 199–205.

Hutchison, M. L., & Gross, D. C., (1997). Lipopeptide phytotoxins produced by *Pseudomonas syringae* pv. *syringae*: Comparison of the biosurfactant and ion channel-forming activities of syringopeptin and syringomycin. *Mol. Plant Microb. Interact., 10*, 347–54.

Ilori, M. O., et al., (2005). Factors affecting biosurfactant production by oil-degrading *Aeromonas spp.* isolated from a tropical environment. *Chemosphere, 61*, 985–992.

Itoh, S., et al., (1971). Rhamnolipids produced by *Pseudomonas aeruginosa* grown on n-paraffin (mixture of C_{12}, C_{13} and C_{14} fractions). *J. Antibiot., 24*, 855–859.

Johnson, M. K., & Boese-Marrazzo, D., (1980). Production and properties of heat-stable extracellular hemolysin from *Pseudomonas aeruginosa. Infect. Immun., 29*, 1028–1033.

Kamp, P. F., & Chakrabarty, A. M., (1979). Plasmids specifying p-chlorobiphenyl degradation in enteric bacteria. *Plasmids of Medical, Environmental and Commercial Importance* (pp. 257–285). Elsevier/North-Holland Biomedical Press, Amsterdam.

Kim, B. S., et al., (2000). *In vivo* control and *in vitro* antifungal activity of rhamnolipid B, a glycolipid antibiotic, against *Phytophthora capsici* and *Colletotrichum orbiculare. Pest Manage. Sci., 56*, 1029–1035.

Kitamoto, D., et al., (2001). Remarkable antiagglomeration effect of yeast biosurfactant, diacylmannosylerythritol, on ice-water slurry for cold thermal storage. *Biotechnol. Prog., 17*, 362–365.

Kitamoto, D., et al., (2002). Functions and potential applications of glycolipid biosurfactants-from energy-saving materials to gene delivery carriers. *J. Biosci. Bioeng., 94*, 187–201.

Kluge, B., et al., (1989). Studies on the biosynthesis of surfactin, a lipopeptide antibiotic from *Bacillus subtilis* ATCC-21332. *FEBS Lett., 231*, 107–110.

Koch, A. K., et al., (1991). Hydrocarbon assimilation and biosurfactant production in *Pseudomonas aeruginosa* mutants. *J. Bacteriol., 173*, 4212–4219.

Kolenc, R. J., et al., (1988). Transfer and expression of mesophilic plasmid-mediated degradative capacity in a psychrotrophic bacterium. *Appl. Envit. Microbiol., 54*(3), 638–641.

Kosswig, K., (2005). *Surfactants, in Ullmann's Encyclopedia of Industrial Chemistry.* Wiley-VCH, Weinheim.

Kuiper, I., et al., (2004). Characterization of two *Pseudomonas putida* lipopeptide biosurfactants, putisolvin I and II, which inhibit biofilm formation and break down existing biofilms. *Molecular Microbiology, 51*(1), 97–113.

Lang, S., & Wagner, F., (1987). Structure and properties of biosurfactants. In: Kosaricet, N., et al., (eds.), *Biosurfactants and Biotechnology* (pp. 21–47). Dekker, New York.

Lotfabad, T. B., et al., (2009). An efficient biosurfactant-producing bacterium *Pseudomonas aeruginosa* MR01, isolated from oil excavation areas in south of Iran. *Colloids Surf. B. Biointerfaces, 69*, 183–193.

Mata-Sandoval, J. C., et al., (1999). High-performance liquid chromatography method for the characterization of rhamnolipid mixtures produced by *Pseudomonas aeruginosa* UG2 on com oil. *Chromatogr., 864*, 211–220.

Matsuyama, T., & Nakagawa, Y., (2009). Surface-active exolipids: Analysis of absolute chemical structures and biological functions. *J. Microbiol. Methods, 5*, 165–175.

Menkhaus, M., et al., (1993). Structural and functional organization of the surfactin synthetase multienzyme system. *J. Biol. Chem., 268*, 7678–768.

Miller, R. M., & Bartha, R., (1989). Evidence from liposome encapsulation for transport limited microbial metabolism of solid alkanes. *Appl. Environ. Microbiol., 55*, 269–274.

Mulligan, C. N., & Yong, R. N., (1997). *Contaminated Ground: Fate of Pollutants and Remediation* (pp. 461–466). Thomas Telford Publishers, London.

Mulligan, C. N., et al., (1999). Metal removal from contaminated soil and sediments by the biosurfactant surfactin. *Environ. Sci. Technol., 33*, 3812–3820.

Nakano, M. M., et al., (1991). srfA is an operon required for surfactin production, competence development, and efficient sporulation in *Bacillus ·subtilis. J. Bacteriol., 173*, 1770–1778.

Neu, T. R., & Poralla, K., (1990). Emulsifying agent from bacteria isolated during screening for cells with hydrophobic surfaces. *Appl. Microbiol. Biotechnol., 32*, 521–525.

Nielsen, T. H., & Sørensen, J., (2003). Production of cyclic lipopeptides by *Pseudomonas fluorescens* strains in bulk soil and in the sugar beet rhizosphere. *Appl. Environ. Microbiol., 69*, 861–868.

Nielsen, T. H., et al., (1999). Viscosinamide, a new cyclic depsipeptide with surfactant and antifungal properties produced by *Pseudomonas fluorescens* DR54. *J. Appl. Microbiol., 87*, 80–90.

Noordman, W. H., & Janssen, D. B., (2002). Rhamnolipid stimulates uptake of hydrophobic compounds by *Pseudomonas aeruginosa. Appl. Environ. Microbiol., 68*, 4502–4508.

Ochsner, U. A., et al., (1994). Isolation, characterization, and expression in *Escherichia coli* of the *Pseudomonas aeruginosa* rhlAB genes encoding a rhamnosyltransferase involved in rhamnolipid biosurfactant synthesis. *J. Biol. Chern., 269*, 19787–19795.

Ochsner, U. A., et al., (1995). Production of *Pseudomonas aeruginosa* rhamnolipid biosurfactants in heterogonous hosts. *Appl. Environ. Microbiol., 61*, 3503–3506.

Osterreicher-Ravid, D., et al., (2000). Horizontal transfer of an exopolymer complex from one bacterial species to another. *Environ. Microbiol., 2*, 366–372.

Pamp, S. J., & Tolker, N. T., (2007). Multiple roles of biosurfactants in structural biofilm development by *Pseudomonas aeruginosa. J. Bacteriol., 189*, 2531–2539.

Park, S. Y., & Kim, Y., (2009). Surfactin inhibits the immunostimulatory function of macrophages through blocking NK-KB, MAPK and Akt pathway. *Inter. Immunopharmacol., 9*, 886–893.

Pascual, J., et al., (2012). *Pseudomonas litoralis* sp. Nov., isolated from Mediterranean seawater. *Int. J. Syst. Evol. Microbiol., 62*(2), 438–444.

Patel, R. M., & Desai, A. J., (1997). Biosurfactant production by *Pseudomonas aeruginosa* GS3 from molasses. *Lett. Appl. Microbiol., 25*, 91–94.

Persson, A., et al., (1990). Physiological and morphological changes induced by nutrient limitation of *Pseudomonas fluorescens* 378 in continuous culture. *Appl. Environ. Microbiol., 56*, 686–692.

Peypoux, F., et al., (1999). Recent trends in the biochemistry of surfactin. *Appl. Microbiol. Biotechnol., 51*, 553–563.

Raza, Z. A., et al., (2007). Improved production of biosurfactant by a *Pseudomonas aeruginosa* mutant using vegetable oil refinery wastes. *Biodegr., 18*, 115–121.

Reid, B. J., et al., (2000). Bioavailability of persistent organic pollutants in soils and sediments-a perspective on mechanisms, consequences and assessment. *Environ. Pollu., 108*, 103–112.

Rodrigues, L. R., et al., (2006). Biosurfactants: Potential applications in medicine. *J. Antimicrob. Chemother., 7*, 609–618.

Rosenberg, E., & Ron, E. Z., (1996). Bioremediation of petroleum contamination. In: Ronald, L. C., & Don, L. C., (eds.), *Bioremediation: Principles and Applications* (pp. 100–124). Cambridge University Press, UK.

Rosenberg, E., et al., (1988). Purification and chemical properties of *Acinetobacter calcoaceticus* A2 biodispersan. *Appl. Environ. Microbiol., 54*, 323–326.

Rosenberg, M., (1981). Bacterial adherence to polystyrene-a replica method of screening for bacterial hydrophobicity. *Appl. Environ. Microbiol., 42*, 375–377.

Salihu, A., et al., (2009). An investigation for potential development on biosurfactants. *Biotechnol. Mol. Biol. Rev., 3*, 111–117.

Sandrin, T. R., et al., (2000). A rhamnolipid biosurfactant reduces cadmium toxicity during naphthalene biodegradation. *Appl Environ. Microbiol., 66*, 4585–4588.

Sotirova, A., et al., (2009). Effects of rhamnolipid-biosurfactant on cell surface of *Pseudomonas aeruginosa. Microbiol. Res., 164*, 297–303.

Sposito, G., (1989). *The Chemistry of Soils* (p. 277). Oxford University Press, Oxford.

Stokes, J. D., et al., (2005). Prediction of polycyclic aromatic hydrocarbon biodegradation in contaminated soils using an aqueous hydroxypropyl-beta-cyclodextrin extraction technique. *Environ. Toxicol. Chem., 24*, 1325–1330.

Suzuki, T., et al., (1974). Sucrose lipids of *Arthrobacteria*, *Corynebacteria*, and *Nocardia* grown on sucrose. *Agric. Biol. Chem., 38*, 557–563.

Syldatk, C., & Wagner, F., (1987). Production of biosurfactants. In: Kosaric, N., et al., (eds.), *Biosurfactants and Biotechnology* (pp. 89–120). Marcel Dekker, New York.

Syldatk, C., et al., (1985). Chemical and physical characterization of four interfacial-active rhamnolipids. From *Pseudomonas. sp.* DSM 2874 grown on n-alkanes. *Z. Naturforsch. C., 40*, 51–60.

Tan, H., et al., (1994). Complexation of cadmium by a rhamnolipid biosurfactant, *Environ. Sci Technol., 28*, 2402–2406.

Thimon, L., et al., (1992). Interactions of surfactin, a biosurfactant from *Bacillus subtilis* with inorganic cations. *Biotechnol. Lett., 14*, 713–718.

Tiquia, S. M., et al., (1996). Effects of composting on phytotoxicity of spent pig-manure sawdust litter. *Environ. Pollut., 93*, 249–256.

Tulloch, A. P., et al., (1962). The fermentation of long-chain compounds by *Torulopsis magnoliae.* Structures of the hydroxyl fatty acids obtained by fermentation of fatty acids and hydrocarbons. *Can. J. Chem., 40*, 1326–1338.

Van, H. J. D., et al., (2006). Physiological aspects: Part 1 in a series of papers devoted to surfactants in microbiology and biotechnology. *Biotechnol. Adv., 24*, 604–620.

Vater P. J., (1986). Lipopeptides in food applications. In: Kosaric, N., (eds.), *Biosurfactants-Production, Properties and Applications* (pp. 419–446). Dekker, New York.

Vatsa, P., et al., (2010). Rhamnolipid biosurfactants as new players in animal and plant defense against microbes. *Inter. J. Mol. Sci., 11*, 5095–5108.

Volkering, F., et al., (1998). Microbiological aspects of surfactant use for biological soil remediation. *Biodegr., 8*, 401–417.

Wang, X., et al., (2005). Algicidal activity of rhamnolipid biosurfactants produced by *Pseudomonas aeruginosa. Harmful Algae, 4*, 433–443.

Wyrwas, B., et al., (2012). Utilization of triton X-I00 and polyethylene glycols during surfactant mediated biodegradation of diesel fuel. *Hazard. Mater., 197*, 97–103.

Yoo, D. S., et al., (2005). Characteristics of microbial biosurfactant as an antifungal agent against plant pathogenic fungus. *J. Microbiol. Biotechnol., 15*, 1164–1169.

Zawawi, R. B. M., (2005). *Production of Biosurfactant by Locally Isolated Bacteria from Petrochemical Waste.* MSc Thesis, Faculty of Science, Universiti-Teknologi Malaysia, Malaysia.

Zhang, Y., & Miller, R. M., (1992). Enhanced octadecane dispersion and biodegradation by a *Pseudomonas rhamnolipid* surfactant (biosurfactant). *Appl. Environ. Microbiol., 58*, 3276–3282.

Zhang, Y., & Miller, R. M., (1994). Effect of a *Pseudomonas* rhamnolipid biosurfactant on cell hydrophobicity and biodegradation of octadecane. *Appl. Environ. Microbiol., 60*, 2101–2106.

Zhao, B., et al., (2005). Solubilization and·biodegradation of phenanthrene in mixed anionic-nonionic surfactant solutions. *Chemosphere, 58*, 33–40.

Zulianello, L., et al., (2006). Rhamnolipids are virulence factors that promote early infiltration of primary human airway epithelia by *Pseudomonas aeruginosa. Infect. Immun., 74*, 3134–3147.

FURTHER READING

Bharali, P., Das, S., Konwar, B. K., & Thakur, A. J., (2011). Crude biosurfactant from thermophilic *Alcaligenes faecalis*: Feasibility in Petro-spill bioremediation. *Inter. Biodeter. Biodegr., 65*, 682–690.

Das, S., Kalita, S. J., Bharali, P., Konwar, B. K., Das, B., & Thakur, A. J., (2013). Organic reactions in "green surfactant": An avenue to bisuracil derivative. *ACS Sustainable Chem. Eng.* (p. 301). doi: 10.1021/sc4002774, Publication Date (Web).

Pranjal, B., (2015). *Bioremediation of Crude Oil Contaminated Soil.* (Thesis: Supervisor B K Konwar), Dept. of Mol. Biol. and Biotechnology, Tezpur University (Central), Napaam – 784028, Assam, India.

CHAPTER 6

Screening for Biosurfactant Producing Microorganisms

Numerous methods have been tried for the high-throughput screening (HTS). These procedures are reliable and significantly accelerate the screening process towards high biosurfactant (BS) producer strains (Muller et al., 2012). The Du-Nouy-Ring assay using a tensiometer is most widely applied for screening of BS-producing microbes (Henkel et al., 2012). This method was based on measuring the force required to detach a ring or loop of wire from an interface or surface. Jain et al. (2012) developed the rapid drop-collapsing test, a simple method of detecting BS production. In this technique, a drop of a cell suspension is placed on an oil-coated surface, and the drops containing BS collapse within a few seconds, whereas non-surfactant containing drops remain stable.

Persson and Molin (1987) described a similar assay using a glass surface instead of the oil-coated surface. Vaux and Cottingham developed and patented a method called microplate assay. This assay was based on the change in the optical distortion, which was caused by surfactants in an aqueous solution. Maczek et al. (2007) developed a qualitative technique suitable for HTS, known as penetration assay. This assay was based on the contact of two insoluble phases, which leads to a change in color. Another method of detection of BS production is the oil spreading technique. This technique measures the diameter of clear zones caused when a drop of BS-containing solution is placed on an oil-water surface (Morikawa et al., 2000). Another most widely used assay which was based on the emulsification capacity of BSs was developed by Cooper and Goldenberg. Emulsification capacity was expressed as emulsification index (E_{24}), where E_{24} is the emulsification percentage obtained by vigorously shaking of culture supernatant with kerosene. This method is most suitable for emulsifying BSs. Rosenberg et al. developed the bacterial adhesion to hydrocarbons (BATH) method, a simple photometrical assay for measuring the hydrophobicity of bacteria. Cell surface hydrophobicity is an important characteristic associated with the adherence of bacterial cells to various liquid hydrocarbons (Van der Mei et al., 1987). A simple replica plate assay for the identification and isolation of hydrophobic microbes was developed by Rosenberg. The basis of this assay is the adherence of bacterial strains to hydrophobic polystyrene surface which correlates to cell surface hydrophobicity. Siegmund and Wagner (1991) developed a semi-quantitative CTAB agar plate method for the detection of extracellular glycolipids or specifically for anionic type BSs.

The interaction between the anionic BS secreted by the microbes with the cationic surfactant CTAB (cetyltrimethylammonium bromide) and methylene blue results in the formation of insoluble ion pair. The resulting productive colonies are surrounded by dark blue halos (Walter et al., 2010). Another method of detecting BS-producing microbes is by their ability to cause hemolysis of RBC on solid media plates and was developed by Mulligan et al. (2001).

Blood agar lysis has been used to quantify surfactin (Moran et al., 2002) and rhamnolipids (RLs) (Johnson and Boese-Marrazzo, 1980). Schenk et al. (1995) developed a high-performance liquid chromatographic method of detection of RL produced by *P. aeruginosa*. A similar method of detection of BS production in the cell-free fermentative broth of *Bacillus subtilis* ATCC 21332 was proposed by Lin et al. (1998). Among all the known techniques reported till date, drop collapse assay, microplate assay and penetration assay are considered as HTS methods because these techniques are rapid and reliable (Walter et al., 2010).

6.1 PRODUCTION OF BIOSURFACTANT (BS) FROM ALTERNATIVE CARBON SOURCES

The success of commercial level BS production depends on the development of cheaper processes and the use of low-cost raw materials, accounting for 50% of the final product cost (Cameotra and Makkar, 1998). Agro-industrial wastes are obtained at low cost from the respective processing industries and are as potent as low-cost substrates for industrial level BS production. Potato process effluents generated from potato processing industries were reported to be used to produce BS by *B. subtilis* (Noah, 2005). Cassava wastewater generated during the preparation of cassava flour is a potential substrate for the production of surfactin and rhamnolipid (RL) by *B. subtilis* (Nitschke and Pastore, 2006) and *P. aeruginosa* (Costa et al., 2009), respectively. George and Jayachandran (2008) analyzed the RL BSs produced through submerged fermentation using orange fruit peelings as sole carbon source.

Maria et al. reported the utilization of cashew apple juice supplemented with peptone and nutritive broth for the cultivation of *P. aeruginosa* to obtain BSs. Patel and Desai reported the use of molasses and corn-steep liquor as the primary carbon and nitrogen source to produce RL using *P. aeruginosa* GS3. Dubey and Juwarkar studied the production of BS using industrial waste from distillery using *P. aeruginosa* BS2. Kitchen waste oils generated from domestic uses, vegetable oil refineries or the soap industries have been reported to be suitable to produce BS through microbial fermentation (Nitschke and Pastore, 2003; Benincasa, 2009).

Lima et al. (2009) reported the use of residual waste of soybean oils for producing BS by submerged fermentation in stirred tank reactors using *P. aeruginosa* PACL. Soap stock, an industrial waste by-product has been used to produce emulsan, bio-dispersant, and RL by *Acinetobacter calcoaceticus A* (Zhang et al., 2005) and *P. aeruginosa* LBI (Zhang et al., 2009) respectively through batch fermentation. Anastasia et al. (2010) reported the use of sunflower seed oil and oleic acid to produce RLs by *Thermus thermophilus* HB8. Palm oil was reported to be used for the simultaneous production of polyhydroxyalkanoates and RLs by *P. aeruginosa* (Marsudi et al., 2008). Hazra et al. (2013) reported the utilization of de-oiled cakes of mahua (*Madhuca indica*), Karanja (*Pongamia pinnata*), jatropha (*Jatropha curcus*) and neem (*Azadirachta indica*) in production of RL using *P. aeruginosa* AB4. Pratap et al. (2011) reported the use of non-edible traditional oils such as neem oil, jatropha oil and Karanja oil in the production of RL using *P. aeruginosa* ATCC 10145. *P. aeruginosa* 47T2 was reported to produce RL when grown on olive oil wastewater or in waste frying oils obtained from olive/sunflower (50:50; v/v) (Zhang et al., 2005, 2009; Haba et al., 2003). Daniel et al. (1998) used dairy wastes as carbon substrates and achieved production of high concentrations of sophorolipids using two-stage cultivation process for the *yeast Cryptococcus curvatus* ATCC 20509. Deshpande and Daniels used animal fat to produce sophorolipids BS using the yeast, *Candida bombicola*.

KEYWORDS

- **bacterial adhesion to hydrocarbons**
- **cetyltrimethylammonium bromide**
- **emulsification**
- **high-throughput screening**
- **microplate assay**
- **sophorolipids**

REFERENCES

Benincasa, M., & Accorsini, F. R., (2008). *Pseudomonas aeruginosa* LBI production as an integrated process using the wastes from sunflower-oil refining as a substrate. *Bioresour. Technol., 99*, 3843–3849.

Cameotra, S. S., & Makkar, R. S., (1998). Synthesis of biosurfactants in extreme conditions. *Appl. Microbiol. Biotechnol., 50*, 520–529.

Cooper, D., & Goldenberg, B., (1987). Surface-active agents from 2 *Bacillus* species. *Appl. Environ. Microbiol., 53*, 224–229.

Costa, S. G. V. A. O., et al., (2009). Cassava wastewater as substrate for the simultaneous production of rhamnolipids and, polyhydroxyalkanoates by *Pseudomonas aeruginosa. J. Ind. Microbiol. Biotechnol., 36*, 1063–1072.

Deshpande, M., & Daniels, L., (1995). Evaluation of sophorolipid biostirfactant production by *Candida bombicola* using animal fat. *Bioresour. Technol., 54*, 143–150.

Dubey, K., & Juwarkar, A., (2004). *Determination of Genetic Basis for Biosurfactant Production in Distillery and Curd Whey Wastes Utilizing Pseudomonas aeruginosa Strain BS2, 82*, 1012–1020.

Haba, E., et al., (2000). Screening and production of rhamnolipids by *Pseudomonas aeruginosa* 47T2 NCIB 40044 from waste frying oils. *J. Appl. Microbiol., 88*, 379–387.

Hazra, C., et al., (2010). Screening and identification of *Pseudomonas aeruginosa* AB4 for improved production, characterization and application of glycolipid biosurfactant using low-cost agro-based raw materials. *J. Chem. Technol. Biotechnol., 86*,185–198.

Henkel, M., et al., (2012). Rhamnolipids as biosurfactants from renewable resources: Concepts for next-generation rhamnolipid production. *Process Biochem., 47*, 1207–1219.

Johnson, M. K., & Boese-Marrazzo, D., (1980). Production and properties of heat-stable extracellular hemolysin from *Pseudomonas aeruginosa. Infect. Immun., 29*, 1028–1033.

Lima, D. C. J. B., et al., (2009). Biosurfactant production by *Pseudomonas aeruginosa* grown in residual soybean oil. *Appl. Biochem. Biotechnol., 152*, 156–168.

Lin, S. C., et al., (1998). General approach for the development of high-performance liquid chromatography methods for biosurfactants analysis and purification. *J. Chromatogr., A 825*, 149–159.

Maria, V. P., et al., (2007). *Production of Biosurfactant by Pseudomonas aeruginosa Grown on Cashew Apple Juice* (pp. 136–140). Humana Press Inc.

Marsudi, S., et al., (2008). Palm oil utilization for the simultaneous-production of Polyhydroxyalkanoates and rhamnolipids by *Pseudomonas aeruginosa*. *Appl. Microbiol. Biotechnol., 78*, 955–961.

Moran, A. C., et al., (2002). Quantification of surfactin in culture supernatant by haemolytic activity. *Biotechnol. Lett., 24*, 177–180.

Morikawa, M., et al., (2000). A study on the structure-function relationship of lipopeptide biosurfactants. *Biochimie. Biophys. Acta, 1488*, 211–218.

Muller, M. M., et al., (2012). Rhamnolipids-next generation surfactants. *J. Biotechnol., 162*, 366–380.

Mulligan, C. N., et al., (1999). Metal removal from contaminated soil and sediments by the Biosurfactant surfactin. *Environ. Sci. Technol., 33*, 3812–3820.

Nitschke, M., & Pastore, G., (2003). Cassava flour wastewater as substrate for biosurfactant production. *Appl. Biochem. Biotechnol., 108*, 295–301.

Noah, K. S., (2005). Surfactin production from potato process effluent by *Bacillus subtilis* in a chemostat. *Appl. Biochem. Biotechnol., 122*, 465–474.

Pratap, A., et al., (2011). Non-traditional oils as newer feedstock for rhamnolipid production by *Pseudomonas aeruginosa* (ATCC 10145). *J. Am. Oil Chem. Soc., 88*, 1935–1943.

Rosenberg, E., (1988). Production of biodispersan by *Acientobacter calcoaceticus* A2. *Appl. Environ. Microbiol., 54*, 317–322.

Rosenberg, E., et al., (1988). Purification and chemical properties of *Acinetobacter calcoaceticus* A2 biodispersan. *Appl. Environ. Microbiol., 54*, 323–326.

Schenk, T., et al., (1995). High-performance liquid chromatographic determination of the rhamnolipids produced by *Pseudomonas aeruginosa*. *J. Chromatogr.*, A 693, 7–13.

Van, D. M. H. C., et al., (1987). A comparison of various methods to determine hydrophobic properties of *Streptococcal* cell surfaces. *J. Microbiol. Methods, 6*, 277–287.

Walter, V., Christoph, S., & Hausmann, R., (2010). Screening Concepts for the isolation of biosurfactant producing microorganisms. In: Ramkrishna, S., (ed.), *Biosurfactants*. Landes Bioscience and Springer Science Business Media.

Zhang, G. L., et al., (2005). Biodegradation of crude oil by *Pseudomonas aeruginosa* in the presence of rhamnolipids. *J. Zhejiang University Sci. Biol., 6*, 725–730.

Zhang, H., et al., (2009). Enhanced treatment of waste frying oil in an activated sludge system by addition of crude rhamnolipid solution. *J. Hazard. Mater., 167*, 217–223.

CHAPTER 7

Application of Biosurfactants

Biosurfactants (BSs) are beginning to attain a status of potential effective substance in various fields (Rahman and Gakpe, 2008). Various applications of BSs have been extensively reviewed (Banat et al., 2000). At the production level, along with the utilization of cheap renewable substrates and organic wastes, the cost of the BSs has become competitive with that of the cost of the synthetic chemical surfactants.

7.1 IN PRODUCTION OF SPECIFIC COMPOUNDS

Biosurfactants (BSs) are being considered as an alternative to the high-value synthetic chemical whose use may have toxic environmental impacts (Makkar and Cameotra, 2002). The pyrenacylester of rhamnolipids (RLs) are reported to be synthesized for its use in monitoring the polarity and fluidity of solid surfaces and also used in determining the impact of coatings on the surface properties (Ishigami et al., 1996). RL from *P. aeruginosa* is a superior source of rhamnose as it is excreted in late log and stationary phases of growth (Aparna et al., 2012). RLs have been a source of stereospecific L-rhamnose, which is used in the production of high-quality flavoring compounds and as starting material for the synthesis of some organic compounds (Linhardt et al., 1989).

7.2 IN LAUNDRY AND OTHER SECTORS

RL were applied in the formulation of laundry detergents and examined their effectiveness in removing sunflower oil, chocolate, and albumen stains from cotton fabrics (Bafghi and Fazaelipoor, 2012). Some other commercial applications of BSs are in the pulp and paper, the paint, textiles, and ceramics industries (Makkar and Cameotra, 2002).

7.3 IN COSMETIC INDUSTRY

The cosmetic and health care industries use large amounts of BSs in several different formulations. Products like insect repellents, antacids, acne pads, anti-dandruff products, contact lens solutions, deodorants, nail care products, anti-wrinkle, and anti-aging products and toothpastes require surfactants that have high surface and emulsifying activities. These characteristics of surfactants play a vital role in maintaining the texture consistency of these products (Maier and Soberon-Chavez, 2000; Lourith and Kanlayavattanakul, 2009; Haba et al., 2003). BSs also acquires the position in the market of personal care products due to its low toxicity, excellent moisturizing properties and skin compatibility (Brown, 1991). Currently, there are patents for the use of RLs to make liposomes and emulsions (Lourith and Kanlayavattanakul, 2009).

7.4 IN FOOD INDUSTRY

BSs are routinely used in the food industry as emulsifiers in the processing of raw materials. Other applications of BSs are in bakery and meat products where they influence the rheological characteristics of flour or to emulsify the partially broken fat tissue (Vater, 1986). Lecithin and its derivatives are currently used as emulsifiers in the food industry worldwide (Bloomberg, 1991). BSs produced by thermophilic dairy *Streptococci sp.* used for fouling control of heat-exchanger plates in pasteurizers as they hindered the colonization of *Staphylococcus thermophilus* responsible for fouling (Busscher et al., 1996). There could be a possible application of BSs immunonutrition (Bengmark, 1998). Diacylmannosylerythritol, a glycolipid type BS produced by *Candida Antarctica* as an effective anti-agglomeration agent in the slurry system (Kitamoto et al., 2001). Such type of BSs exhibited a remarkable effect on the slurry, attaining a high ice-packing factor (35%) for 8 h at a BS concentration of 10 $mg.l^{-1}$.

In the food industry, BSs provide multiple functions and act as emulsifying/foaming agents, stabilizers, antioxidant agents, and anti-adhesives (Nitschke and Costa, 2007). The addition of polymeric surfactants forms very stable emulsions which improve the texture and creaminess of low-fat dairy products such as soft cheese and ice creams (Rosenberg and Ron, 1999). Shepherd et al. reported the successful use of extracellular carbohydrate-rich compound from *Candida utilis* as an emulsifying agent in salad dressing formulations. BSs have been reported to control the agglomeration of fat globules, stabilize aerated systems, improve texture and shelf-life of starch-containing products, modify rheological properties of wheat dough and improve consistency and texture of fat-based products (Kachholz and Schlingmann, 1987). In bakery and ice cream formulations, BSs act by controlling consistency, retarding staling, and solubilizing flavor oils; they are also utilized as a fat stabilizer and anti-spattering agents during cooking of oil and fats (Kosaric, 2001). Nitschkea and Costa (2007) suggested the use of RLs to improve the properties of buttercream, croissants, and frozen confectionery products. Iyer et al. (2006) reported the isolation of a bioemulsifier from a marine strain of *Enterobacter cloac*ae and described it as a potential viscosity-enhancing agent because it imparts high-quality viscosity at acidic pH, allowing its use in food products containing citric or ascorbic acid. BSs have been reported to be used in pre-conditioning of material surfaces (stainless steel, polystyrene, and polytetrafluoroethylene) found in food processing to prevent the adhesion of food-borne pathogens to such solid surfaces (Kim et al., 2006; Nitschke et al., 2009).

7.5 IN METALLURGY PROCESSES

BSs are used in the dispersion of inorganic minerals in mining and various manufacturing processes. Rosenberg et al. (1988) reported the production of an anionic polysaccharide called biodispersan by *Acinetobacter calcoaceticus* A2 that prevents the flocculation and effect dispersion of limestone 10% in water. The use of BS isolated from *Nocardia amarae* was reported to be used for the removal and recovery of non-ionic organics from aqueous solutions (Sutton, 1992). Polman et al. (1994) reported partial solubilization of North Dakota Beulah Zap lignite coal with the use of crude BS isolated from *Candida bombicola*. Surfactin, RLs, and sophorolipid were used in batch washing experiments to remove heavy metals from sediments (Mulligan et al., 2001). Other commercial applications of BSs involve the processing of uranium ore and mechanical dewatering of peat (Makkar and Cameotra, 2002; Ron and Rosenberg, 2002). The foaming ability of BSs isolated from *Pseudomonas aeruginosa* was investigated for its possible application in coal and mineral flotation as a frother and co-frother (Abbasi et al., 2012).

7.6 IN PETROLEUM INDUSTRY

Over the recent years, many studies have shown the capability of BSs and BS-producing bacterial strains to enhance availability and biodegradation rates of organic contaminants (Zhang and Miller, 1994; Deziel et al., 1996; Rahman et al., 2003). Research investigations conducted by Zhang and Miller (1992) confirmed that BS affects the degradation of hydrocarbon by increasing microbial accessibility to insoluble substrates and thus enhance their biodegradation. Glycolipid BSs have also been shown to enhance the hydrocarbon removal (80–95%) from soil; furthermore, the BS was reported to increase hydrocarbon mineralization by two-fold and shorten the adaptation time of microbial populations to fewer hours. Holakoo and Mulligan (2002) reported the usefulness of BSs for oil spills remediation and for dispersing oil slicks into fine droplets and converting mousse oil into oil-in-water emulsion. Barkay et al. (1999) used the bioemulsifier Alasan produced by *Acinetobacter radioresistens* KA53 to enhance PAH solubility, and degradation results showed 6.6, 25.7 and 19.8-fold increases in the solubilities of phenanthrene, fluoranthene, and pyrene respectively. Similarly, the solubilization of PAH has been reported with the RLs produced by *Pseudomonas aeruginosa* and other pseudomonads (Bordoloi and Konwar, 2009; Polman et al., 1994).

Balachandran et al. (2012) reported the degradation of petroleum and polyaromatic hydrocarbons (PHA) and metabolism of naphthalene by *Streptomyces* sp. isolated from oil-contaminated soil. Mata-Sandoval et al. (2001) reported that biodegradation of chlorinated and polychlorinated biphenyl hydrocarbons can be enhanced by addition of glycolipids to the medium. Several sp of *P. aeruginosa* and *B. subtilis* produces RL and surfactin, respectively; these two BSs have been shown by Whang et al. (2005) to increase solubility and bioavailability of a petrochemical mixture and also stimulate indigenous microorganisms for enhanced biodegradation of diesel contaminated soil. Bordoloi and Konwar (2009) reported that pyrene was solubilized more by the BS of *P. aeruginosa* (MTCC7815) and *P. aeruginosa* (MTCC7812); phenanthrene by *P. aeruginosa* (MTCC8165); fluorene by *P. aeruginosa* (MTCC7812) and *P. aeruginosa* (MTCC8163); crude oil by the BS of *P. aeruginosa* (MTCC8165). Addition of RLs above their CMC enhanced the apparent aqueous solubility of hexadecane, enhanced biodegradation of hexadecane, octadecane, n-paraffins, creosotes, and other hydrocarbon mixtures in soil and promoted bioremediation of petroleum sludges (Bordoloi and Konwar,

2007; Noordman and Janssen, 2002; Rahman et al., 2003). *Gordonia* sp. BS29 growing on aliphatic hydrocarbons as sole carbon source has been found to produce bioemulsan, which effectively degrades crude oil, PAHs, and other recalcitrant branched hydrocarbons from the contaminated soils (Franzetti et al., 2009). Reddy et al. (2009) reported 93.92% degradation of phenanthrene by a BS producing *Brevibacterium* sp. PDM-3 strain and reported the ability of the bacterial strain to degrade other PHA such as anthracene and fluorene.

7.7 IN MICROBIAL ENHANCED OIL RECOVERY (MEOR)

BSs have been shown in many cases to have emulsification properties equivalent to that of the industrially available emulsifying agents, and the most desirable character is their biodegradable nature. There is a possible use of BSs in mobilizing heavy crude oil, transporting petroleum in pipelines, viscosity control, managing oil spills, oil-pollution control, cleaning oil sludge from oil storage facilities, soil/sand bioremediation and microbial enhanced oil recovery (MEOR) (Banat, 1995). MEOR is a less expensive process as compared to CEOR because microorganisms can synthesize useful products by fermenting low-cost substrates or raw materials. Furthermore, microbial products are biodegradable and have low toxicity (Lazar et al., 2007; Suthur et al., 2008; Banat et al., 2010; Gudina et al., 2012). Single microorganism or consortium could be used to degrade heavy oil fractions, as a result, the oil viscosity decreases, and it becomes more fluid, lighter, and more valuable (Jinfeng et al., 2005). BSs were produced by culturing the necessary microbes in the basal salt medium containing 2% w/v glucose and oleic acid together as carbon source and was used as a substitute for the chemical surfactants in a test carried out on an oil storage tank belonging to Kuwait Oil Company (Banat, 1995). Clark et al. (2010) estimated that about 27% of oil reservoirs in the USA are amenable to microbial growth and MEOR based on computational survey.

Biosurfactant-mediated MEOR represents one of the most promising methods to recover a substantial proportion of the residual oil from mature oil fields (Banat et al., 2010). The efficiency of MEOR has been proven in field studies in Czech Republic, Romania, Hungary, Poland, U.S., and Holland, with a significant increase in oil recovery observed in all cases (Kosswig, 2005; Lang and Wagner, 1987). Jinfeng et al. (2005) conducted microbial enhanced water-flooding experiment in a Guan 69 Unit in

Dagang Oilfield in China by injection of a mixture of *Arthrobacter* sp. (A02), *Pseudomonas* sp. (P15) and *Bacillus* sp. (B24) strain suspension and the nutrient solution through injection wells in an ongoing waterflood reservoir where the temperature reached 73°C. They observed that the oil production steadily increased after microbial water-flooding. In recent years, physical stimulation test such as sand pack column experiment has been served as an excellent laboratory instrument to investigate and understand the mechanism and performance of BS flooding in enhanced oil recovery (Xia et al., 2011; Banat, 1995). Bordoloi and Konwar (2009) treated crude oil-saturated sand pack column with cell-free culture broths containing BS of four different *P. aeruginosa* strains (MTCC7815, MTCC7814, MTCC7812, and MTCC8165) and were reported to release about 15% more crude oil at 90°C than at room temperature (RT) and 10% more than at 70°C under laboratory conditions. Pornsunthorntawee et al. compared the oil recovery from the sand-packed column saturated with motor oil using the BSs (2009) produced by *B. subtilis* PT2 and *P. aeruginosa* SP4 with three synthetic surfactants. The results showed that the BSs produced by *B. subtilis* PT2 and *P. aeruginosa* SP4 were more efficient in oil recovery, removing about 62% and 57%, respectively, of the tested oil while synthetic surfactants were able to release approximately 53–55%.

Suthur et al. (2008) compared the oil recovery upon application of bioemulsifier and BS to a sand pack column designed to stimulate an oil reservoir. They reported that crude bioemulsifier produced by *B. licheniformis* K125 gave better oil recovery than the BS by *B. mojavensis* JF2 and *B. licheniformis* TT42. Oil recovery experiments in physical simulation showed 7.2–14.3% recovery of residual oil after water flooding when the BS of three strains of. *aeruginosa*, *B. subtilis* and *R. erythropolis* was added (Xia et al., 2011). Gudiña et al. (2012) studied the efficiency of BS produced by the strains of *B. subtilis*, isolated from the crude oil samples of Brazilian Oil field and suggested their usefulness for MEOR applications due to their unique properties including thermo- and salt-tolerance; stable surface activity and hydrocarbon degradation.

7.8 IN ENVIRONMENTAL APPLICATIONS

BSs are extensively studied and applied directly in the presence and absence of microorganisms for bioremediation of organic pollutants and heavy metals on the laboratory scale (Mulligan, 2005; Lebron-Paler, 2008).

BSs are used in various industrial and environmental applications, which frequently involve exposure to extreme environmental conditions. As a result, researchers have focused on isolating and screening strains that are able to produce BS under extreme environments, especially for MEOR and bioremediation purposes (Darvishi et al., 2011). The effectiveness of BS in separating crude oil was equivalent to those of synthetic surfactant and much higher than that of natural plant-derived surfactant-saponin (Urum et al., 2006) and synthetic Tween 60 (Kuyukina et al., 2005). In the case of removing hexadecane from the contaminated sand, BS was found to be much more efficient than SDS and Tween 80 (Bai et al., 1997). The addition of BS produced by *Candida Antarctica* to the fermentation process of n-undecane improved degradation rate of petroleum hydrocarbons, while application of synthetic surfactant Tween 40 and Span 80 didn't show any improvement (Hua et al., 2003). There are several microorganisms such as *Pseudomonas aeruginosa* known to degrade hydrocarbons by using as carbon sources (Wentzel et al., 2007) and produce BSs (Tang et al., 2007). BSs are very effective in enhancing oil biodegradation either by enhancing the uptake of hydrocarbon or by specific adhesion/desorption mechanisms (Rosenberg and Ron, 1996). In the recent time, focus is given to the possible application of BS in its attempt to recover residual oil from oil sludge and in enhanced biodegradation of oil sludge process (Joseph and Joseph, 2009; Helmy et al., 2010). BSs form complexes preferably with toxic heavy metal cations which include Cd^{2+}, Pb^{2+}, Zn^{2+}, Ar^{2+}, etc., (Christofi and Ivshina, 2002; Herman et al., 1995) than with other non-toxic metals such as Ca^{2+} and Mg^{2+} for which they have much lesser affinity (Singh and Cameotra, 2004). Due to the anionic nature of RLs, they are able to take out metal ions from the soil such as arsenic, cadmium, copper, lanthanum, lead, and zinc due to their complexation ability (Christofi and Ivshina, 2002; Herman et al., 1995). The order of RL stability constant for the complexation with metals, tested at pH 6.9 was in the order $Al^{2+} > Cu^{2+} > Pb^{2+} > Cd^{2+} > Zn^{2+} > Fe^{3+} > Hg^{2+} > Ca^{2+} > Co^{2+} > Ni^{2+} > Mn^{2+} > Mg^{2+} > K^{2+}$ (Francisco et al., 2001)

Soil washing technology is characterized by chemico-physical properties of the BS and not by their effect on metabolic activities or changes in cell-surface properties of bacteria (Banat et al., 2010). However, the processes may enhance the bioavailability for bioremediation. Such cleanup process is highly desirable as it is economically rewarding and environmentally friendly (Lillienberg et al., 1992). Abu-Ruwaida et al. (1991)

showed that cell-free broth of *Rhodococcus sp.* containing BS removed about 86% of adsorbed crude oil from the contaminated sand. Urum et al. (2006) investigated the efficiency of different surfactant solutions in removing crude oil from contaminated soil using a soil washing process. They demonstrated higher crude oil elimination by synthetic surfactant-sodium dodecyl sulfate (SDS) and RL BSs (46% and 44%, respectively) than natural surfactant saponins (27%). Franzetti et al. (2009) showed that the BS29 bioemulsans from *Gordonia sp.* are promising washing agents for remediation of hydrocarbon-contaminated soils. The mean of the crude oil removal for bioemulsans was 33%. The BS29 bioemulsans were also able to remove metals (Cu, Cd, Pb, Zn, Ni) but their potential in the process was lower than RLs. Costa et al. reported that increasing the RL concentration produced by *P. aeruginosa* L2-1 on cassava wastewater enhanced the removal of crude oil from the contaminated sand from 69% at CMC to 84% at CMC + 5% (w/w). Aparna et al. (2012) reported that the *Pseudomonas* sp. 2B BS solution at 0.01% and 0.05% BS concentrations was capable to remove 89% and 92% of the oil absorbed in the sand, respectively, while the distilled water and SDS removed 48% and 63% of the contaminated oil, respectively.

Thimon et al. reported that glutamate residues of surfactin can bind metals such as Mg, Mn, Ca, Ba, Li, and rubidium. Soil washing with 0.25% surfactin removed 70% of the Cu and 22% of the Zn. Using micellar-enhanced ultrafiltration, 85–100% removal of cadmium, copper, and zinc by surfactin from contaminated water was achieved (Mulligan et al., 1999). RLs have been reported to be used for the removal of heavy metals such as Ni and Cd from soils due to their anionic nature, with efficiencies of 80–100% in the lab and 20–80% in the field samples (Neilson et al., 2003). Fermentative broth of the bacterial strain (Pet 106) containing BS produced by using 2% (w/v) glucose followed by 2% (v/v) oleic acid as carbon source in basal salt medium was used as a substitute for chemical surfactants in a test carried out on an oil storage tank belonging to Kuwait Oil Company, Kuwait. Joseph and Joseph separated the residual oil from the petroleum sludge generated from the crude oil refinery by directly inoculating the strains of *Bacillus sp.* and by addition of the cell-free culture supernatant of the bacteria. The removal efficiency of the bacterial strains was in the range between 91.67 and 97.46%. The application of BS of *Azotobacter vinelandii* AV01 increased oil recovery by 15% from the oil sludge.

7.9 IN AGRICULTURE

BSs have been evaluated for their potential role in controlling various plant pathogens. The proposed mechanism for their action is that it intercalates into and disrupts the plasma membrane (Makkar and Cameotra, 2002). The RLs isolated from *P. aeruginosa* was efficient in biological control of zoosporic plant pathogens at very low concentrations and reported to be successful in controlling the disease in a hydrophonic recirculating cultural system (Stanghellini and Miller, 1997). Surfactin and a similar lipopeptide, iturin A, produced by *Bacillus subtilis* RB14 were reported to suppress the damping-off disease of tomato seedlings caused by *Rhizoctonia solani* (Haferburg et al., 1987). RLs and other BSs were reported to have zoosporicidal activity against species of *Pythium*, *Phytophthora*, and *Plasmophora* at concentrations ranging from 5 to 30 μg ml^{-1} (Makkar and Cameotra, 2002). RL at a rate of 1% emulsion was successfully used for the treatment of *Nicotiana glutinosa* leaves infected with tobacco mosaic virus and for the control of potato virus X disease (Haferburg et al., 1987).

RLs have a direct biocide action on various plant disease-causing bacteria and fungi (Vatsa et al., 2010). They are reported to increase the susceptibility of certain gram-positive bacteria to specific antibiotics (Helmy et al., 1995). Nielsen et al. (2003) demonstrated that viscosinamide, a new cyclic depsipeptide (CLP) produced by *P. fluorescens* DR54 exhibited strong BS properties, and some had antibiotic properties towards root-pathogenic microfungi. Andersen et al. (2003) isolated a cyclic lipopeptide amphibian from *Pseudomonas sp.* DSS73 from the rhizoplane of sugar beet seedlings that exhibited antagonism towards the root-pathogenic microfungi *Pythium ultimum* and *Rhizoctonia solani*. RLs were reported to exert high zoosporicidal activity, probably through zoospore lysis, against various zoosporic phytopathogens, including sp. from the *Pythium, Phytophthora*, and *Plasmopara* genera (Stanghellini and Miller, 1997). Interestingly, fluorescent *Pseudomonads* are effective in the biological control of plant pathogens.

Furthermore, antiviral, algicidal, mycoplasmicidal, and antiamoebal properties of RLs have also been reported (Itoh et al., 1971; Wang et al., 2005). The efficacy of RL has also been demonstrated in the near commercial, hydroponics, recirculating cultural system (Stanghellini and Miller, 1997). Apart from antimicrobial properties, surface-active compounds are used in the agricultural sector for hydrophilization of heavy soil, which results in soil improvement. Pattel and Gopinathan (1986) reported that glycolipopeptide produced by *Bacillus* strains were able to form a stable emulsion in the presence of the organophosphorus pesticide fenthion and helps in spontaneous distribution in water. Banat et al. (2010) reported the biodegradation of around 40% of chlorinated pesticide α- and β endosulfan by the BS produced by *B. subtilis* MTCC2423.

7.10 IN MEDICINES AND THERAPEUTICS

Currently, BSs are being investigated and exploited for medical purposes (Yoo et al., 2005; Rodrigues et al., 2006). BSs have a range of therapeutic applications; such as RLs produced by *P. aeruginosa*, lipopeptides produced by *B subtilis* and *B licheniformis* as biocidic agents (Makkar and Cameotra, 2002; Yoo et al., 2005). The possible applications of BSs as emulsifying agent for transporting drugs to the site of infection, for supplementing pulmonary surfactant, and as adjuvant for vaccines were assessed (Kosaric, 1996). The surfactin produced by *Lactobacillus acidophilus* RC 14 has been investigated for its possible application as anti-adhesive biological coatings for catheter materials (Velraeds-Martin et al., 1997). They are also known to have the potential use as major antimycoplasmic, antiviral, anti-tumor agent, inhibitor of fibrin clot formation, immunomodulatory, hypocholesterolemic, antiadhesive, and most recently as dispersants for nanoparticles (Kim et al., 1990; Mulligan et al., 2001).

RLs have permeabilizing effects on Gram-positive and gram-negative human bacterial strains reinforcing their potential in biomedicine (Sotirova et al., 2008). Lipopeptides such as pumilacidin and surfactin have been reported to act as antiviral agents (Itokawa et al., 1994). The loss of membrane integrity as opposed to other vital physiological processes makes surfactin and other lipopolypeptides potentially important as the next generation of antibiotics (Cameotra and Makkar, 2004). Gan et al. (2002) reported that both *Lactobacillus fermentum* RC-14 and its secreted BS significantly inhibited *Staphylococcus aureus* infection and bacterial adherence to surgical implants. In current biofilm preventive strategies, various research investigations suggested the use of BSs as antiadhesion with antimicrobial biological coatings for catheter materials or other medical surfaces (Rodrigues et al., 2006). Sophorolipids are reported to have activity against human immunodeficiency virus (Shah et al., 2005). Similarly, RL, and its complex with alginate, both produced by a

Pseudomonas sp. strain showed significant antiviral activity against herpes simplex virus types 1 and 2 (Remichkova et al., 2008). Research Cao et al. (2010) indicated that surfactin has potent immune-suppressive capabilities which suggested important therapeutic implications for transplantation and autoimmune diseases including allergy, arthritis, and diabetes. Clinical trials using RLs for the treatment of psoriasis, lichen planus, neurodermatitis, and human burn wound healing have confirmed excellent ameliorative effects of RLs when compared to conventional therapy using corticosteroids (Stipcevic et al., 2005, 2006). RLs also display differential effects on human keratinocyte and fibroblast cultures (Hauler et al., 1998).

7.11 PROSPECT OF BIOSURFACTANT (BS) IN NANOTECHNOLOGY

Considering the requirement of greener bioprocesses and novel enhancers for the synthesis using microbial processes, BSs, and/or BS producing microbes are emerging as an alternate source for nanomaterial synthesis / functionalization and its subsequent applications (Xie et al., 2006; Kiran et al., 2011). These nanomaterial biomolecule multifunctional systems could be used to mimic the behavior of biomolecules in cells and, therefore, could be helpful in explaining the mechanisms of complex biological processes with several potential applications (Christof, 2001). Recent developments show that BSs are multifunctional smart molecules which would become a part of diverse biotechnological applications, including biocontrol, drug delivery vehicle, and bioremediation. BS-mediated nanomaterial synthesis and/or stabilization is a recent development in the field of nanotechnology. BSs, especially RLs have been used as green capping agents for nanoparticle synthesis. At present, focus on BS-mediated processes is steeply increasing due to their potential implication in the synthesis of various metal nanoparticles such as NiO, ZnS, Ag, Au, etc., (Kumar et al., 2010; Bharali et al., 2013).

BS-mediated nanomaterial synthesis and/or stabilization is a recent development in the field of nanotechnology. The BS-mediated process and microbial synthesis of nanoparticles are now emerging as clean, non-toxic, and environmentally acceptable "green chemistry" procedures (Reddy et al., 2009; Kiran et al., 2011). The focus on the BS-mediated processes is steeply increasing due to their potential implication on the synthesis of silver nanoparticles (SNP) (Palanisamy and Raichur, 2009; Reddy et al., 2009). Palanisamy (2008) reported the synthesis of stable NiO nano-rods by a water-in-oil (w/o) microemulsion technique using RL BS. Literature related to NiO nano-rod synthesis using RLs as BSs revealed that particle morphology can be tuned by altering the pH (Palanisamy and Raichur, 2009). Recently BSs have been shown to be promising candidates for the "green" stabilizing agent of nanoparticles. Mulligan et al. (1999) reported the use of RL BSs as dispersants for nanoparticles. Biswas and Raichur (2008) evaluated the efficiency of RLs for the synthesis and stabilization of nano zirconia particles. Reddy et al. (2010) reported the use of surfactin as an environmentally friendly stabilizing agent in the synthesized SNP. Reddy et al. (2010) synthesized, for the first time, surfactin-mediated gold nanoparticles, opening the way to a new and fascinating application of BSs in the biomedical field. Kiran et al. (2009) reported the application of brevifactin, a novel lipopeptide BS produced by the marine actinobacterium *Brevibacterium casei* MSA19 for the synthesis and stabilization of SNP. The use of gold nanoparticles is currently undergoing a dramatic expansion in the field of drug and gene delivery, targeted therapy and imaging (Pissuwan et al., 2009). Hazra et al. (2013) reported the biomimetic fabrication of biocompatible and biodegradable core-shell polystyrene/biosurfactant bionanocomposites for protein drug release.

KEYWORDS

- **biosurfactants**
- **cyclic depsipeptide**
- **microbial enhanced oil recovery**
- **polystyrene**
- **rhamnolipid**
- **sodium dodecyl sulfate**

REFERENCES

Abbasi, H., et al., (2012). Biosurfactant-producing bacterium, *Pseudomonas aeruginosa* MAO1 isolated from spoiled apples: Physicochemical and structural characteristics of isolated biosurfactant. *J. Biosci. Bioengg.*, *113*, 211–219.

Abu-Ruwalda, A. S., et al., (1991). Isolation of biosurfactant producing bacteria product characterization and evaluation. *Acta Biotechnol., 11*, 315–24.

Aparna, A., et al., (2012). Production and characterization of biosurfactant produced by a novel *Pseudomonas* sp. 2B. *Colloids Surf. B. Biointerfcaes, 95*, 2–2 9.

Bafghi, M. K., & Fazaelipoor, M. H., (2012). Application of rhamnolipid in the formulation of a detergent. *J. Surfact. Deterg., 15*, 679–684.

Bai, G., et al., (1997). Biosurfactant enhanced removal of residual hydrocarbon from soil. *J. Contaminant. Hydrol., 25*, 157–170.

Banat, I. M., (1995). Biosurfactants production and possible uses in microbial enhanced oil recovery and oil pollution remediation: A review. *Bioresour. Technol., 51*, 1–12.

Banat, I. M., et al., (1991). Biosurfactant production and use in oil tank clean-up. *World J. Microbiol. Biotechnol., 7*, 80–88.

Banat, I. M., et al., (2000). Potential commercial applications of microbial surfactants. *Appl. Microbiol. Biotechnol., 53*, 495–508.

Banat, I. M., et al., (2010). Microbial biosurfactants production applications and future potential. *Appl. Microbiol. Biotechnol., 87*, 427–444.

Beal, R., & Betts, W. B., (2000). Role of rhamnolipid biosurfactants in the uptake and mineralization of hexadecane in *Pseudomonas aeruginosa. J. Appl. Microbiol., 89*, 158–168.

Bengmark, S., (1998). Immunonutrition: Role of biosurfactants, fiber and probiotic bacteria. *Nutrition, 14*, 585–594.

Bharali, P., et al., (2011). Crude biosurfactant from thermophilic *Alcaligenes faecalis*: Feasibility in petro-spill bioremediation. *Int. Biodeter. Biodegr., 65*, 682–690.

Bloomberg, G., (1991). Designing proteins as emulsifiers. *Lebensmittel Technologie, 24*, 130–131.

Bordoloi, N. K., & Konwar, B. K., (2007). Microbial surfactant-enhanced mineral oil recovery under laboratory conditions. *Colloids Surf. B. Biointerfaces, 63*, 73–82.

Bordoloi, N. K., & Konwar, B. K., (2009). Bacterial biosurfactant in enhancing solubility and metabolism of petroleum hydrocarbons. *J. Hazardous Materials, 170*(1), 495–505.

Brown, M. J., (1991). Biosurfactants for cosmetic applications. *Int. Cosmet. Sci., 13*, 61–64.

Busscher, H. J., et al., (1996). Biosurfactants from thermophilic dairy streptococci and their potential role in the fouling control of heat exchanger plates. *Ind. Microbiol., 16*, 15–21.

Cameotra, S. S., & Makkar, R. S., (2004). Recent application of biosurfactants as biological and immunological molecules. *Curr. Opin. Microbiol., 7*, 262–266.

Christof, M. N., (2001). Nanoparticles, proteins and nucleic acids: Biotechnology meets materials science. *Angewandte Chemie., 40*, 4128–4158.

Christofi, N., & Ivshina, I. B., (2002). Microbial surfactants and their use in field studies of soil remediation. *J. Appl. Microbiol., 93*, 915–929.

Darvishi, P., et al., (2011). Biosurfactant production under extreme environmental conditions by an efficient microbial consortium ERCPPI-2. *Colloids Surf B. Biointeraces, 84*, 292–300.

Deziel, E., et al., (1996). Biosurfactant production by a soil pseudomonas strains growing on polycyclic aromatic hydrocarbons. *Appl. Environ. Microbiol., 62*, 1908–1912.

Franzetti, A., et al., (2009). Potential, applications of surface-active compounds by *Gordonia* sp. strain BS29 in soil remediation technologies. *Chemosphere, 75*, 801–807.

Gudina, E. J., et al., (2012). Isolation and study of microorganisms from oil samples for application in microbial enhanced oil recovery. *Inter. Biodeter. Biodegrad., 68*, 56–64.
Haba, E., et al., (2003). Physicochemical characterization and antimicrobial properties of rhamnolipids produced by *Pseudomonas aeruginosa* 47T2 NCBIM 40044. *Biotechnol. Bioengg., 81*, 316–322.
Haba, E., et al., (2003). Use of liquid chromatography-mass spectrometry for studying the composition and properties of rhamnolipids produced by different strains of *Pseudomonas aeruginosa. Surf. Deterg., 6*, 155–161.
Haferburg, D., et al., (1987). Antiviral activity of rhamnolipids from *Pseudomonas aeruginosa. Acta Biotech., 1*(7), 353–356.
Hauler, S., et al., (1998). Purification and characterization of a cytotoxic exolipid of *Burkholderia pseudomallei. Infect. Immun., 66*, 1588–1593.
Helmy, Q., et al., (2010). Application of biosurfactant produced by *Azotobacter vinelandii* AVOI for enhanced oil recovery and biodegradation of oil sludge. *Inter. J. Civil Environ. Eng., 10*, 7–14.
Herman, D. C., et al., (1995). Removal of cadmium lead and zinc from soil by a rhamnolipid biosurfactant. *Environ. Sci. Technol., 29*, 2280–2285.
Hua, Z., et al., (2003). Influence of biosurfactants produced by Candida Antarctica on surface properties of microorganism and biodegradation of n-alkanes. *Wat. Res., 37*, 4143–4150.
Itoh, S., et al., (1971). Rhamnolipids produced by *Pseudomonas aeruginosa* grown on n-paraffin (mixture of C_{12}, C_{13} and C_{14} fractions). *J. Antibiot., 24*, 855–859.
Itokawa, H., et al., (1994). Structural and conformational studies of [Ile7] and [Leu7] surfactants from *Bacillus subtilis. Chem. Pharm. Bull (Tokyo), 42*, 604–607.
Jinfeng, L., et al., (2005). The field pilot of microbial enhanced oil recovery in a high temperature petroleum reservoir. *J. Petroleum Sci. Eng., 48*, 265–271.
Joseph, P. J., & Joseph, A., (2009). Microbial enhanced separation of oil from petroleum refinery sludge. *J. Hazard. Mater., 161*, 522–525.
Kachholz, T., & Schlingmann, M., (1987). Possible food and agricultural applications of microbial surfactants: An assessment. In: Kosaric, N., et. al., (eds.), *Biosurfactants and Biotechnology* (pp. 183–210). Marcel, Dekker, New York.
Kim, H., et al., (2006). Attachment and biofilm formation by *Enterobacter sakazakii* on stainless steel and enteral feeding tubes. *Appl. Environ: Microbiol., 72*, 5846–5856.
Kim, J. S., et al., (1990). Microbial glycolipid production under nitrogen limitation and resting cell conditions. *J. Biotechnol., 13*, 257–266.
Kiran, G. S., et al., (2011). Biosurfactants as green stabilizers for the biological synthesis of nanoparticles. *Critical-Rev. Biotechnol., 31*, 354–364.
Kitamoto, D., et al., (2001). Remarkable anti-agglomeration effect of yeast biosurfactant, diacylmannosylerythritol, on ice-water slurry for cold thermal storage. *Biotechnol. Prog., 17*, 362–365.
Kosaric, N., (2001). Biosurfactants and their application for soil bioremediation. *Food-Technol. Biotechnol., 39*, 295–304.
Kosaric, N., et al., (1984). The role of nitrogen in multiorganism strategies for biosurfactant production. *J. Am. Oil. Chem. Soc., 61*, 1735–1743.
Kosswig, K., (2005). Surfactants. In: *Ullmann's Encyclopedia of Industrial Chemistry*. Wiley-VCH, Weinheim.

Kumar, C. G., et al., (2010). Synthesis of biosurfactant-based silver nanoparticles with purified rhamnolipids isolated from *Pseudomonas aeruginosa* BS-161R. *J. Microbiol. Biotechnol., 20*, 1061–1068.

Kuyukina, M. S., et al., (2005). Effect of biosurfactants on crude oil desorption and mobilization in a soil system. *Env. Inter., 31*, 155–161.

Lang, S., & Wagner, F., (1987). Structure and properties of biosurfactants. In: Kosaricet, N., et al., (eds.), *Biosurfactants and Biotechnology* (pp. 21–47). Dekker, New York.

Lazar, I., et al., (2007). Microbial enhanced oil recovery (MEOR). *Petroleum Sci. Technol., 25*, 1353–1366.

Lebron-Paler, A., (2008). *Solution and Interfacial Characterization of Rhamnolipid Biosurfactant from P. aeruginosa ATCC·9027*. PhD Thesis, University of Arizona Graduate College, Arizona, USA.

Lillienberg, L., et al., (1992). Health-effects of tank cleaners. *Am. Ind. Hygiene Assoc. J., 53*, 95–102.

Linhardt, R. J., et al., (1989). Microbially produced rhamnolipid as a source of rhamnose. *Biotechnol. Bioeng., 33*, 365–368.

Lourith, N., & Kanlayavattanakul, M., (2009). Natural surfactants used in cosmetics: Glycolipids. *Int. J. Cosmet. Sci., 31*, 255–261.

Maier, R. M., & Soberon-Chavez, G., (2000). *Pseudomonas aeruginosa* rhamnolipids: biosynthesis and potential applications. *Appl. Microbiol. Biotechnol., 54*, 625–633.

Makkar, R. S., & Cameotra, S. S., (2002). An update on the use of unconventional substrates for biosurfactant production and their new applications. *Appl. Microbiol. Biotechnol., 58*, 428–434.

Mulligan, C. N., (2005). Environmental applications for biosurfactants. *Environ. Pollut., 133*, 183–198.

Mulligan, C. N., et al., (1999). Metal removal from contaminated soil and sediments by the biosurfactant surfactin. *Environ. Sci. Technol., 33*, 3812–3820.

Mulligan, C. N., et al., (2001). Heavy metal removal from sediments by biosurfactants. *J. Hazard. Mater., 85*, 111–125.

Nitschke, M., & Costa, S. G. Y. A. O., (2007). Biosurfactants in food industry. *Trends Food Sci. Technol., 18*, 252–259.

Nitschke, M., et al., (2009). Surfactin reduces the adhesion of food-borne pathogenic bacteria to solid surfaces. *Lett. Appl. Microbiol., 49*, 241–247.

Noordman, W. H., & Janssen, D. B., (2002). Rhamnolipid stimulates uptake of hydrophobic compounds by *Pseudomonas aeruginosa*. *Appl. Environ. Microbiol., 68*, 4502–4508.

Palanisamy, P., & Raichur, A. M., (2009). Synthesis of spherical NiO nanoparticles through a novel biosurfactant mediated emulsion technique. *Mater. Sci. Eng., C 29*, 199–204.

Pissuwan, D., et al., (2009). The forthcoming applications of gold nanoparticles in drug and gene delivery systems. *J. Contr. Release, 149*, 65–71.

Polman, J. K., et al., (1994). Solubilization of bituminous and lignite coals by chemically and biologically synthesized surfactants. *J. Chem. Tech: Biotechnol., 61*, 11–17.

Rahman, K. S. M., et al., (2003). Enhanced bioremediation of n-alkane petroleum sludge using bacterial consortium amended with rhamnolipid and micronutrients., *Bioresour. Technol., 90*, 159–168.

Rahman, P. K. S. M., & Gakpe, E., (2008). Production, characterization and application of biosurfactants-review. *Biotechnol., 7*, 360–370.

Reddy, S. A., et al., (2009). Synthesis of silver nanoparticles using surfactin: A biosurfactant as stabilizing agent. *Mat. Lett., 63*, 1227–1230.
Remichkova, M., et al., (2008). Anti-herpesvirus activities of Pseudomonas sp. S-17 rhamnolipid and its complex with alginate. *Z. Naturforsch. C., 63*, 75–81.
Rodrigues, L. R., et al., (2006). Biosurfactants: Potential applications in medicine. *J. Antimicrob. Chemother., 57*, 609–618.
Ron, E., & Rosenberg, E., (2002). Biosurfactants and oil bioremediation. *Curr. Opinion Biotechnol., 13*, 249–252.
Rosenberg, E., & Ron, E. Z., (1996). Bioremediation of petroleum contamination. In: Ronald, L. C., & Don, L. C., (eds.), *Bioremediation: Principles and Applications* (pp. 100–124). Cambridge University Press, UK.
Rosenberg, E., & Ron, E. Z., (1999). High-and low-molecular-mass microbial surfactants. *Appl. Microbiol. Biotech., 52*(2), 154–162.
Shah, V., et al., (2005). Sophorolipids microbial glycolipids with anti-human immunodeficiency virus and sperm immobilizing activities. *Antimicrobial. Agents Chemother., 49*, 4093–4100.
Singh, P., & Cameotra, S., (2004). Potential application of microbial surfactants in biomedical sciences. *Trends Biotechnol., 22*, 142–146.
Sotirova, A. V., et al., (2008). Rhamnolipid-biosurfactant permeabilizing effects on gram-positive and gram-negative bacterial strains. *Curr. Microbiol., 56*, 639–6441.
Stanghellini, M. E., & Miller, R. M., (1997). Biosurfactants: Their identity and potential efficacy in the biological control of zoosporic plant pathogens. *Plant Dis., 81*, 4–12.
Stipcevic, T., et al., (2005). Di-rhamnolipid from *Pseudomonas aeruginosa* displays differential effects on human keratinocyte and fibroblast cultures. *Dermat. Sci., 40*, 141–143.
Stipcevic, T., et al., (2006). Enhanced healing of full-thickness burn wounds using dirhamnolipid. *Bums, 32*, 24–34.
Suthur, H., et al., (2008). Evaluation of bioemulsifier mediated microbial enhanced oil recovery using sand pack column. *Microbiol. Methods, 75*, 225–230.
Sutton, R., (1992). Use of biosurfactants produced by *Nocardia amarae* for removal and recovery of non-ionic organics from aqueous solutions. *Water Sci. Technol., 26*, 9–11.
Tang, H., et al., (2007). Enhanced crude oil biodegradability of *Pseudomonas aeruginosa* ZJU after preservation in crude oil-containing medium. *World Microbiol. Biotechnol., 23*, 7–14.
Urum, K., et al., (2006). A comparison of the efficiency of different surfactants for removal of crude oil from contaminated soils. *Chemosphere, 62*, 403–1410.
Vater P. J., (1986). Lipopeptides in food applications: In: Kosaric, N., (ed.), *Biosurfactants-Production, Properties and Applications* (pp. 419–446). Dekker, New York.
Vatsa, P., et al., (2010). Rhamnolipid biosurfactants as new players in animal and plant defense against microbes. *Inter. J. Mol. Sci., 11*, 5095–5108.
Velraeds-Martin, M. C., et al., (1997). Inhibition of initial adhesion of uropathogenic *Enterococcus faecalis* to solid substrate by an absorbed biosurfactant layer from *Lactobacillus acidophilus*. *Urol., 49*; 790–794.
Wang, X., et al., (2005). Algicidal activity of rhamnolipid biosurfactants produced by *Pseudomonas aeruginosa*. *Harmful Algae, 4*, 433–443.

Wentzel, A., et al., (2007). Bacterial metabolism of long-chain n-alkanes. *Appl. Microbiol. Biotechnol., 76*, 1209–1221.

Xia, W., et al., (2011). Comparative study of biosurfactant produced by microorganisms isolated from formation water of petroleum reservoir. *Colloids Surf. A Physicochem. Eng. Aspects, 392*, 124–130.

Xie, Y., et al., (2006). Synthesis of silver nanoparticles in reverse micelles stabilized by natural biosurfactant. *Colloids Surf. A Physicochem. Eng. Aspects, 279*, 175–178.

Yoo, D. S., et al., (2005). Characteristics of microbial biosurfactant as an antifungal agent against plant pathogenic fungus. *J. Microbiol. Biotechnol., 15*, 1164–1169.

Zhang, Y., & Miller, R. M., (1994). Effect of a *Pseudomonas* rhamnolipid biosurfactant on cell hydrophobicity and biodegradation of octadecane. *Appl. Environ. Microbiol., 60*, 2101–2106.

FURTHER READING

Bharali, P., & Konwar, B. K., (2011). Production and physicochemical characterization of a biosurfactant produced by *Pseudomonas aeruginosa* OBP1 isolated from petroleum sludge. *Appl. Biochem. Biotechnol., 164*, 1444–1460.

Das, S., Kalita, S. J., Bharali, P., Konwar, B. K., Das, B., & Thakur, A. J., (2013). Organic reactions in "green surfactant": An avenue to bisuracil derivative. *ACS Sustainable Chem. Eng.* (p. 301). doi: 10.1021/sc4002774, Publication Date (Web).

Pranjal, B., (2015). *Bioremediation of Crude Oil Contaminated Soil.* (Thesis: Supervisor B K Konwar), Dept. of Mol. Biol. and Biotechnology, Tezpur University (Central), Napaam – 784028, Assam, India.

CHAPTER 8

Role of Biosurfactant in Producing Bacteria

Biosurfactants (BS) possess diverse properties and physiological functions such as an increase in the surface area and bioavailability of hydrophobic water-insoluble substrates, heavy metal binding, bacterial pathogenesis, quorum sensing, and biofilm formation (Singh and Cameotra, 2004). However, it is impossible to make any generalization or identify one or more common roles to all microbial surfactants (Ron and Rosenberg, 2002). Most of the bacteria isolated from sites having a history of contaminations by hydrocarbons and derivatives are gram-negative, and it might be a characteristic that contributes to the survival of the populations in such harsh environments (Maria et al., 2007; Yalcin and Ergene, 2009). Most of them produce BSs, amphiphilic molecules of diverse chemical nature, which improve the ability of microbial cells to utilize hydrophobic compounds as growth substrates (Desai and Banat, 1997; Rosenberg and Ron, 1999). BSs are either produced on microbial cell surfaces or excreted extracellularly (Kosswig, 2005; Bodour and Maier, 2002; Maier, 2003). The capacity of bacteria to produce BS, specifically with antimicrobial property, could be a survival strategy allowing them to flourish ahead of other organisms in the competitive environments (Ilori et al., 2008). The mode of action of BSs is the modification of the cell surface hydrophobicity and/or in promoting emulsification and/or solubilization of substrates (Hommel, 1994).

8.1 BIOSURFACTANTS (BSS) WITH SPECIAL REFERENCE TO RHAMNOLIPIDS (RLS)

Biosurfactants (BSs) were grouped on the basis of molecular weights, chemical properties and cellular localization. The low molecular weight

BSs such as glycolipids, lipopeptides, flavolipids, corynomycolic acids and phospholipids lowers the surface and interfacial tensions (IFTs) at the air/water interfaces. The high molecular weight BSs are called bioemulsans, such as emulsan, alasan, liposan, polysaccharides, and protein complexes. These BSs are efficient emulsifiers at low concentrations and exhibit considerable substrate specificity in stabilizing oil-in-water emulsions (Franzetti et al., 2008; Salihu et al., 2009). However, general classification based on parent chemical structure and classified as glycolipids, lipopeptides, lipoproteins, lipopolysaccharides (LPSs), phospholipids, fatty acids, and polymeric lipids (Eddouaouda et al., 2011).

Rhamnolipids (RLs) are known to be produced by the different species of *Pseudomonas* bacteria, especially by *P. aeruginosa*, a leading commercial BS which is suitable for considerable applications (Abdel-Mawgoud et al., 2009; Lourith and Kanlayavattanakul, 2009). RLs, the glycolipid-type BSs produced by *P. aeruginosa* in the late log and stationary phases of growth (Maier and Soberon-Chavez, 2000) are among the most effective BSs and have been applied in various industries and bioremediations (Maier and Soberon-Chavez, 2000; Wei et al., 2005; Abdel-Mawgoud et al., 2009). The production of RLs is known to be crucial for *P. aeruginosa* to survive and thrive under specific conditions (Lebron-Paler, 2008). RLs basically are glycolipid in nature, composed of a hydrophilic head formed by one or two rhamnose molecules, known respectively as mono-rhamnolipid and di-rhamnolipid, and a hydrophobic tail which contains up to three molecules of hydroxyl fatty acids of varying chain length from 8 to 14 of which β-hydroxydecanoic acid is predominant (Banat et al., 2000; Monterio et al., 2007; Silva et al., 2010). Thus, distribution of RL congeners always exists. Although, there are several types of RL species reported in the literature, all of them possess similar chemical structure and have an average molecular weight of 577 (Wei et al., 2005; Abalos et al., 2001; Dos Santos et al., 2010). The crude BSs extracted from the liquid culture of *Pseudomonas* strains were found to reduce the surface tension of water from 72 to 30 mNm^{-1} (Abalos et al., 2001) with a critical micelle concentration (CMC) of 5–200 $mg.l^{-1}$ and exhibited an emulsification index of above 70% (Nitschke et al., 2005; Monterio et al., 2007; Silva et al., 2010; Abdel-Mawgoud et al., 2009; Bodour and Miller-Maier, 1998). With respect to their production, they show a higher yield as compared to other known BSs. Another advantage of RLs over the other bio-surfactants is the ease at which they can be isolated from the culture as they are extracellularly produced (Makkar et al., 2011; Wei et al., 2005). Being of microbial origin, RLs with high biodegradability possess an additional advantage over the synthetic surfactants in soil washing and bioremediation processes (Aparna et al., 2012; Mata-Sandoval et al., 1999). RLs reveal the potential to improve the microbial degradation of chlorinated hydrocarbons, PAHs, heavy metals and petroleum hydrocarbons from contaminated soil and water (Mulligan, 2005). They are not only efficient surfactants but also exhibit excellent antimicrobial activity against several other microorganisms (Rodrigues et al., 2006; Lourith and Kanlayavattanakul, 2009; Benincasa et al., 2004) and disrupt host defenses during infections (Read et al., 1992).

RLs have been investigated in several applications, which include bacteriocide (Haba et al., 2003), fungicide (Yoo et al., 2005), wound healing (Stipcevic et al., 2005) and others (Lourith and Kanlayavattanakul, 2009). They also assist in cell surface motility (Deziel et al., 2003) and influence the architecture of biofilms, especially in the formation and maintenance of fluid channels within the exo-polymeric matrix after bacteria adhere irreversibly on a substratum (Davey et al., 2003). The production of RL is known to be regulated by quorum sensing mechanism and it depends on various environmental and nutritional factors, which include pH, temperature, phosphates, and iron content, as well as the nature of the carbon source (Lebron-Paler, 2008).

8.2 PETROLEUM CONTAMINATIONS AS SOURCE FOR BIOSURFACTANT (BS) PRODUCING BACTERIA

The Indian petroleum industry is one of the six core industries of the country and contributes over 15% to GDP. India is the 6th largest consumer of oil in the world and the ninth-largest crude oil importer. Oil continued to remain the top item in the country's import list during 2012–2013. In India, particularly the state Assam has a huge reserve of oil and natural gas. It has 1.3 billion tons of proven crude oil. Assam is the first state in the country where for the first time, oil well was dug mechanically at Digboi in Tinsukia district in the year 1889. The state has the oldest refinery in the country, established in the year 1901, which started to produce 500 barrels of crude oil per day and established to refine the crude oil in Digboi itself. Subsequently, various petroleum industries were set up to exploit the crude oil and natural gas of Assam, which includes Oil India Limited

(OIL), Oil and Natural Gas Corporation (ONGC), Indian Oil Corporation Limited (IOCL), consisting of two major refineries, Guwahati refinery and Bongaigaon refinery, Assam Gas Company Limited (AGCL) and Numaligarh Refinery Limited (NRL).

Industrial wastes from petroleum-based industries are identified as one of the major sources of pollution. The oil-enriched geographical regions of Assam are very fertile and under the traditional agricultural practices since time immemorial. The people of the state mostly depend on agriculture and the major crops of the region are rice and tea. Exploration activities in the oil fields often cause spillages of crude oil from the oil wells into the nearby forests, tea cultivations and agricultural fields. Such problems are raising with the increase in the scale of oil exploration activities. The incidences of crude oil contamination are more often in the Upper Assam area resulting in the destruction of soil quality that consequently damages crop cultivations. Moreover, there have been instances where pipelines transporting crude oil from the site of drilling to the refinery are damaged and large quantities of oil discharged into open fields. Apart from exploration activities, refining, and transportation of crude and refined products do contribute towards pollution of soil and water affecting agriculture, aquaculture, and human health (Bordoloi, 2008). Therefore, the problem of crude oil pollution in Upper Assam needs effective remediation solutions for the security and the sustenance of crop cultivations and biodiversity of the region.

Soil which is accidentally contaminated with petroleum hydrocarbons can be remediated by physical, chemical or biological methods. Among all, *in situ* bioremediation is environmentally friendly because it restores the soil structure, requires less energy input, and involves the complete destruction or immobilization of the contaminations rather than their transfer from one environmental compartment to another which mainly occur in physical or chemical treatment processes (Sabate et al., 2004). Although most of the hydrocarbons are biodegradable, but the rate of biodegradation in the environment is limited due to their hydrophobicity or less accessibility to microbes and low aqueous solubility (Desai and Banat, 1997). One of the approaches to enhance biodegradation of crude oil contamination is the use of BS (Rahman and Gakpe, 2008), which could increase the solubility of hydrophobic substrates/oils in an aqueous medium to enhance the bioavailability of the hydrophobic substrates leading to higher oil degradation. The treatment of crude oil contaminated soil and environments through the indigenous BS-producing bacteria having the capacity to degrade the petroleum hydrocarbons seems to be advantageous (Christofi and Ivshina, 2002). As most of the bacteria dwelling in the crude oil contaminated environment have the capacity to utilize the components of crude oil as the source of carbon and energy for their growth and tend to secrete BSs, which in turn help in the degradation of crude oil components (Rosenberg and Ron, 1999; Ilori et al., 2005). Moreover, the production of BS at the site of treatment with the producer bacteria doesn't require rigorous testing like that of the chemical surfactants because of their environmentally compatible nature (Wei et al., 2005). Therefore, the application of BSs in bioremediation might be more acceptable from the social point of view. Thus, there is a need for increased production of BSs and their characterization.

8.3 ISOLATION OF MICROORGANISMS BY ENRICHMENT CULTURE

The bacterial strains were isolated from the crude oil contaminated soil and water as well as petroleum sludge samples using enrichment culture technique. A sample weighing 5.0 g was added to 100 ml of mineral salt medium (MSM) formulated according to Bordoloi and Konwar (2009) in a 500 ml Erlenmeyer flask supplemented with 1% (v/v) of membrane filtered (0.22 μm) n-hexadecane as the sole source of carbon and incubated at 37°C on an orbital incubator shaker for 7 days at 180 rpm. After 10 cycles of enrichment, 1 ml of the saturated culture was diluted 10^4 times and an aliquot of 100 μl was spread on MSM agar plates having 0.1% (v/v) hexadecane as the carbon source. The culture plates were incubated at 37°C for 48 h, morphologically distinct bacterial colonies thus obtained were further purified in MSAM plates with or without 0.1% (v/v) n-hexadecane to eliminate autotrophs and agar-utilizing bacteria. The procedure was repeated and isolates showing pronounced growth on n-hexadecane were selected and preserved for further characterization (Sood and Lal, 2008; Mehdi and Giti, 2008).

8.4 ISOLATION OF MICROORGANISMS BY DIRECT CULTURE TECHNIQUE

Crude oil contaminated soil and water and petroleum sludge samples were serially diluted with sterile saline water (0.9% NaCl, w/v) using the standard dilution technique. The total viable bacterial populations were determined by spread-plating each sample after appropriate dilution (10^{-5} to 10^{-6} fold) onto nutrient agar. The morphologically different bacterial colonies thus obtained were further purified on nutrient agar. Bacterial isolates were further grown on Bushnell-Hass medium to ensure their hydrocarbon utilizing capacity. The medium was supplemented with 1% (v/v) n-hexadecane and was used for the screening of potential hydrocarbon-degrading and BS-producing bacterial isolates (Casellas et al., 1997).

8.5 MAINTENANCE AND PRESERVATION OF ISOLATED MICROORGANISMS

The bacterial pure cultures were preserved at 4°C in both nutrient agar plates and slants. Cultures were sub-cultured at an interval of 30 days. For long time storage, frozen stock cultures were prepared in 15% (v/v) glycerol and stored at –70°C.

8.5.1 SCREENING OF BIOSURFACTANT (BS) PRODUCING BACTERIA FOR SURFACE ACTIVITY

The broth culture of each bacterial isolate was centrifuged at 8,000 rpm at 4°C in a Remi 22R centrifuge for 15 min after 96 h of culture. The culture supernatant was collected, and its surface tension was measured using a digitalized tensiometer (Krűss Tensiometer K9 ET/25). Before using the platinum ring, it was thoroughly washed three times with distilled water followed by acetone and then allowed to dry at room temperature (RT) (Bodour and Miller-Maier, 1998).

8.5.2 DROP COLLAPSE TEST

Screening for BS production was done using the qualitative drop collapse test described by Bodour and Maier (2002). For the experiment, 2 μl of crude oil was dropped at the center of each well on a 96-well microtiter plate lid and allowed to stand for 24 h for equilibration. An aliquot of 5 μl of culture medium of 48 h duration, before, and after centrifugation (10^3

rpm for 15 min at 4°C), was dropped on the oil-coated wells and then the drop size was observed after 1 min with the help of a magnifying glass. If the drop diameter was found to be at least 1 mm larger than the one produced by the de-ionized water drop, the result was positive for the BS production.

8.5.3 OIL-DISPLACEMENT TEST

The oil displacement test was done by adding 50 ml of distilled water to a Petri dish with a diameter of 45 cm. An aliquot of 500 µl of crude oil was dropped onto the surface of the water; this was followed by the addition of 100 µl of cell-free culture supernatant of the bacterial culture, onto the surface of the oil. The diameter of the clear zone was then measured. Each experiment was repeated thrice to determine an averaged value of the diameter of the appeared clear zone (Rodrigues et al., 2006).

KEYWORDS

- **Assam Gas Company Limited**
- **Indian Oil Corporation Limited**
- **mineral salt medium**
- **Numaligarh Refinery Limited**
- **Oil and Natural Gas Corporation**
- **Oil India Limited**

REFERENCES

Abalos, A., et al., (2001). Physicochemical and antimicrobial properties of new rhamnolipids produced by *Pseudomonas aeruginosa* AT10 from soybean oil refinery wastes. *Langmuir, 17*, 1367–1371.

Abdel-Mawgoud, A. M., et al., (2009). Characterization of rhamnolipid produced by *Pseudomonas aeruginosa* isolate BS20. *Appl. Biochem. Biotechnol., 2*, 329–345.

Aparna, A., et al., (2012). Production and characterization of biosurfactant produced by a novel *Pseudomonas* sp. 2B. *Colloids Surf. B. Biointerfcaes, 95*, 2–2 9.

Bafghi, M. K., & Fazaelipoor, M. H., (2012). Application of rhamnolipid in the formulation of a detergent. *J. Surfact. Deterg., 15*, 679–684.

Banat, I. M., et al., (1991). Biosurfactant production and use in oil tank clean-up. *World J. Microbiol. Biotechnol., 7*, 80–88.

Benincasa, M., & Accorsini, F. R., (2008). *Pseudomonas aeruginosa* LBI production as an integrated process using the wastes from sunflower-oil refining as a substrate. *Bioresour. Technol., 99*, 3843–3849.

Bodour, A. A., & Maier, R. M., (2002). Biosurfactants: Types, screening methods and applications. In: Bitton, G., (ed.), *Encyclopedia of Environmental Microbiology* (pp. 750–770). Wiley, New York.

Bodour, A. A., & Miller-Maier, R. M., (1998). Application of a modified drop collapse technique for surfactant quantitation and screening of biosurfactant-producing microorganisms. *J. Microbiol. Method., 32*, 273–280.

Bordoloi, N. K., & Konwar, B. K., (2009). Bacterial biosurfactant in enhancing solubility and metabolism of petroleum hydrocarbons. *J. Hazardous Materials, 170*(1), 495–505.

Bordoloi, N. K., (2008). *Biochemical and Molecular Characterization of Certain Bacteria for Application in Bioremediation of Petroleum Contamination.* PhD: Thesis, Tezpur University at Tezpur, Assam, India.

Casellas, M., et al., (1997). New metabolites in the degradation of fluorene by *Arthrobacter* sp., strain F101. *Appl. Environ. Microbiol., 63*, 819–26.

Christof, M. N., (2001). Nanoparticles, proteins and nucleic acids: Biotechnology meets materials science. *Angewandte Chemie., 40*, 4128–4158.

Davey, M. E., et al., (2003). Rhamnolipid surfactant production affects biofilm architecture m *Pseudomonas aeruginosa* PA01. *J. Bacteriol., 185*, 1027–1036.

Desai, J. D., & Banat, I. M., (1997). Microbial production of surfactants and their commercial potential. *Microbiol. Mol. Bio. Rev., 61*, 47–64.

Deziel, E., et al., (1996). Biosurfactant production by a soil Pseudomonas strains growing on polycyclic aromatic hydrocarbons. *Appl. Environ. Microbiol., 62*, 1908–1912.

Deziel, E., et al., (2003). rhlA is required for the production of a novel biosurfactant promoting swarming motility in *Pseudomonas aeruginosa:* 3-(3-hydroxyalkanoyloxy) alkanoic acids (HAAs), the precursors of rhamnolipids. *Microbiol., 149*, 2005–2013.

Dos, S. S. C., et al., (2010). Evaluation of substrates from renewable-resources in biosurfactants production by *Pseudomonas* strains. *Afr. 1. Biotechnol., 9*, 5704–5711.

Eddouaouda, K., et al., (2011). Characterization of a novel biosurfactant produced by *Staphylococcus* sp. strain 1E with potential application on hydrocarbon bioremediation. *J. Basic Microbiol., 51*, 1–11.

Franzetti, A., et al., (2009). Potential, applications of surface-active compounds by *Gordonia* sp. strain BS29 in soil remediation technologies. *Chemosphere, 75*, 801–807.

Hommel, R. K., (1994). Formation and function of biosurfactants for degradation of water-insoluble substrates. In: Ratledge, C., (ed.), *Biochemistry of Microbial Biodegradation* (pp. 63–87). Kluwer Academic Publishers, Dordrecht.

Ilori, M. O., et al., (2005). Factors affecting biosurfactant production by oil-degrading *Aeromonas spp.* isolated from a tropical environment. *Chemosphere, 61*, 985–992.

Ilori, M. O., et al., (2008). Isolation and characterization of hydrocarbon-degrading and biosurfactant producing yeast strains obtained from lagoon water. *World. J. Microbiol. Biotechnol., 24*, 2539–2545.

Lebron-Paler, A., (2008). *Solution and Interfacial Characterization of Rhamnolipid Biosurfactant from P. aeruginosa ATCC·9027.* PhD Thesis, University of Arizona Graduate College, Arizona, USA.

Lourith, N., & Kanlayavattanakul, M., (2009). Natural surfactants used in cosmetics: Glycolipids. *Int. J. Cosmet. Sci., 31*, 255–261.

Maier, R. M., & Soberon-Chavez, G., (2000). *Pseudomonas aeruginosa* rhamnolipids: biosynthesis and potential applications. *Appl. Microbiol. Biotechnol., 54*, 625–633.

Maier, R. M., (2003). Biosurfactants: Evolution and diversity in bacteria. *Adv. Appl. Microbiol., 52*, 101–121.

Makkar, R. S., et al., (2011). Advances in utilization of renewable substrates for biosurfactant production. *AMB Express, 1*, 5–10.

Maria, V. P., et al., (2007). *Production of Biosurfactant by Pseudomonas aeruginosa Grown on Cashew Apple Juice* (pp. 136–140) Humana Press Inc.

Mata-Sandoval, J. C., et al., (2001). Influence of rhamnolipids and Triton X-100 on the biodegradation of three pesticides in aqueous and soil slurries. *J. Agric. Food Chem., 49*, 3296–3303.

Mehdi, H., & Giti, E., (2008). Investigation of alkane biodegradation using the microtiter plate method and correlation between biofilm formation, biosurfactant production and crude oil degradation. *Inter. Biodeter. Biodegr., 62* 170–178.

Monterio, S. A., et al., (2007). Molecular and structural characterization of the biosurfactant produced by *Pseudomonas aeruginosa* DAUPE 614. *Chem Phys Lipids, 147*, 1–13.

Nitschke, M., et al., (2009). Surfactin reduces the adhesion of food-borne pathogenic bacteria to solid surfaces. *Lett. Appl. Microbiol., 49*, 241–247.

Rahman, K. S. M., et al., (2003). Enhanced bioremediation of n-alkane petroleum sludge using bacterial consortium amended with rhamnolipid and micronutrients. *Bioresour. Technol., 90*, 159–168.

Rahman, P. K. S. M., & Gakpe, E., (2008). Production, characterization and application of biosurfactants-review. *Biotechnol., 7*, 360–370.

Read, R. C., et al., (1992). The effect of *Pseudomonas aeruginosa* rhamnolipids on guineapig tracheal mucociliary transport and ciliary beating. *J. Appl. Physiol., 72*, 2271–2277.

Rodrigues, L. R., et al., (2006). Biosurfactants: Potential applications in medicine. *J. Antimicrob. Chemother., 57*, 609–618.

Ron, E., & Rosenberg, E., (2002). Biosurfactants and oil bioremediation. *Curr. Opinion Biotechnol., 13*, 249–252.

Rosenberg, E., & Ron, E. Z., (1996). Bioremediation of petroleum contamination. In: Ronald, L. C., & Don, L. C., (eds.), *Bioremediation: Principles and Applications* (pp. 100–124). Cambridge University Press, UK.

Sabate, J., et al., (2004). Laboratory scale bioremediation experiments on hydrocarbon contaminated soils. *Int. Biodeter. Biodegrad., 54*, 19–25.

Salihu, A., et al., (2009). An investigation for potential development on biosurfactants. *Biotechnol. Mol. Biol. Rev., 3*, 111–117.

Silva, S. N. R. L., et al., (2010). Glycerol as substrate for the production of biosurfactant by *Pseudomonas aeruginosa* UCP 0992. *Colloids Surf. B. Biointerfaces, 79*, 174–183.

Singh, P., & Cameotra, S., (2004). Potential application of microbial surfactants in biomedical sciences. *Trends Biotechnol., 22*, 142–146.

Sood, N., & Lal, B., (2008). Isolation and characterization of a potential paraffin-wax degrading thermophilic bacterial strain *Geobacillus·kaustophilus* TERINSM for application in oil wells with ·paraffin deposition problems. *Chemosphere, 70*, 1445–1451.

Stipcevic, T., et al., (2005). Di-rhamnolipid from *Pseudomonas aeruginosa* displays differential effects on human keratinocyte and fibroblast cultures. *Dermat. Sci., 40*, 141–143.

Wei, Y. R., et al., (2005). Rhamnolipid production by indigenous *Pseudomonas aeruginosa* J4 originating from petrochemical wastewater. *Biochem. Eng., 127*, 146–154.

Yalcin, E., & Ergene, A., (2009). Screening antimicrobial activity of biosurfactants produced by microorganisms isolated from refinery wastewaters. *J. Appl. Biol. Sci., 3*, 148–153.

Yoo, D. S., et al., (2005). Characteristics of microbial biosurfactant as an antifungal agent against plant pathogenic fungus. *J. Microbiol. Biotechnol., 15*, 1164–1169.

FURTHER READING

Bharali, P., & Konwar, B. K., (2011). Production and physicochemical characterization of a biosurfactant produced by *Pseudomonas aeruginosa* OBP1 isolated from petroleum sludge. *Appl. Biochem. Biotechnol., 164*, 1444–1460.

Bharali, P., Das, S., Konwar, B. K., & Thakur, A. J., (2011). Crude biosurfactant from thermophilic *Alcaligenes faecalis*: Feasibility in Petro-spill bioremediation. *Inter. Biodeter. Biodegr., 65*, 682–690.

Das, S., Kalita, S. J., Bharali, P., Konwar, B. K., Das, B., & Thakur, A. J., (2013). Organic reactions in "green surfactant": An avenue to bisuracil derivative. *ACS Sustainable Chem. Eng.*, (p. 301). doi: 10.1021/sc4002774, Publication Date (Web).

Pranjal, B., (2015). *Bioremediation of Crude Oil Contaminated Soil.* (Thesis: Supervisor B K Konwar), Dept. of Mol. Biol. and Biotechnology, Tezpur University (Central), Napaam – 784028, Assam, India.

CHAPTER 9

Biosurfactant Producing Bacteria Utilizing Hydrocarbon from the Environment

9.1 USE OF HYDROCARBON FROM THE ENVIRONMENT BY BIOSURFACTANT PRODUCING BACTERIA

The environmental samples consist of crude oil-contaminated soils of oil fields, petroleum sludge, and waste residual crude oil dumping sites from ONGC, Jorhat, and petroleum sludge from the oil fields of ONGC, Sibsagar, Assam; contaminated soil samples from the different oil depots of Tezpur, Assam. The bacterial strains were isolated from the above-mentioned environmental samples using both the enrichment and direct plate culture techniques. The bacterial isolates were grown for 96 h at 37°C with 150 rpm to determine the total dry biomass yield. The bacterial strains were isolated from the above-mentioned environmental samples using both the enrichment and direct plate culture techniques. The isolates with their colony characters are presented in Table 9.1.

A total of 52 culturable isolates were obtained based on the distinct colony morphology. The bacterial isolates were further re-cultured to obtain the pure colonies of each individual strain and maintained in nutrient agar plates and in stab agar cultures at 4°C. The bacterial isolates were sub-cultured at an interval of 30 days in nutrient broth agar. For long-term maintenance, the bacterial isolates were preserved in 15% (w/v) glycerol and stored at –70°C.

TABLE 9.1 Morphological Characters of Bacterial Isolates Obtained from Crude Oil-Contaminated Samples

SL. No.	Bacterial Isolates	Biomass (g.l^{-1})	SL. No.	Bacterial Isolates	Biomass (g.l^{-1})
1.	JB08S11	2.32 ± 0.75	27.	JB10OD13	2.83 ± 0.93
2.	JB08S12	0.45 ± 0.32	28.	JB10OD14	4.86 ± 0.83
3.	JB08S13	1.87 ± 0.43	29.	JB10OD15	4.25 ± 0.53
4.	JB08S14	4.78 ± 0.21	30.	JB10OD21	3.52 ± 0.77
5.	JB08S15	4.23 ± 0.65	31.	JB10OD22	5.02 ± 0.84
6.	JB08S16	2.15 ± 0.31	32.	JB10OD23	3.94 ± 0.53
7.	JB08S17	4.54 ± 0.65	33.	JB10OD24	2.17 ± 0.91
8.	JB08S18	0.20 ± 0.55	34.	JB10OD25	4.94 ± 0.63
9.	JB08S21	4.93 ± 0.83	35.	S10PSS11	5.05 ± 0.82
10.	JB08S22	2.8 ± 0.32	36.	S10PSS12	3.32 ± 0.32
11.	JB08S23	4.46 ± 0.27	37.	S10PSS13	4.93 ± 0.54
12.	JB08S24	2.83 ± 0.11	38.	S10PSS14	4.05 ± 0.82
13.	JB08S25	5.02 ± 0.45	39.	S10PSS15	4.88 ± 0.43
14.	JB08S31	4.93 ± 0.83	40.	S10PSS16	3.62 ± 0.81
15.	JB08S32	2.20 ± 0.29	41.	S10PSS17	5.09 ± 0.80
16.	JB08S33	4.75 ± 0.43	42.	S10PSS21	3.58 ± 0.31
17.	JB08S34	0.32 ± 1.2	43.	S10PSS22	4.95 ± 0.52
18.	JB08PS35	5.10 ± 0.82	44.	S10PSS23	0.07 ± 1.1
19.	JB08PS36	0.71 ± 0.37	45.	S10PSS24	0.27 ± 0.62
20.	JB08PS37	1.65 ± 0.44	46.	S10PSS25	5.02 ± 0.41
21.	JB08PS38	4.77 ± 0.62	47.	T11PS11	4.75 ± 0.53
22.	JB08PS39	2.53 ± 0.83	48.	T11PS12	1.88 ± 0.41
23.	JB08PS41	0.16 ± 0.97	49.	T11PS21	3.91 ± 0.74
24.	JB08PS42	4.92 ± 1.1	50.	T11PS22	5.10 ± 0.18
25.	JB10OD11	5.08 ± 0.41	51.	T11PS23	4.90 ± 0.42
26.	JB10OD12	0.67 ± 0.67	52.	T11PS24	2.96 ± 0.62

The bacterial isolates were further re-cultured to obtain the pure colonies of each individual strain and maintained in nutrient plates and in stab agar cultures at 4°C. The bacterial isolates were sub-cultured at an interval of 30 days in nutrient broth agar. For long-term maintenance, the bacterial isolates were preserved in 15% (w/v) glycerol and stored at –70°C.

After the initial screening at 37°C, all 52 bacterial strains were re-cultured in 100 ml mineral salt medium (MSM) supplemented with 1%

(v/v) of n-hexadecane as the sole source of carbon and incubated at 37°C for 7 days at a constant rpm of 150. After the sixth cycle of subcultures, the bacterial isolates were cultured on Bushnell-Hass medium supplemented with 1% (v/v) n-hexadecane to establish their hydrocarbon utilizing ability. The bacterial isolates were grown for 96 h at 37°C with 150 rpm to determine the total dry biomass yield and the same are presented in Table 9.2.

TABLE 9.2 Biomass Yields in Bushnell-Hass Medium Supplemented with 1% (v/v) n-Hexadecane After 96 h

SL. No.	Bacterial Isolates	Biomass ($g.l^{-1}$)	SL. No.	Bacterial Isolates	Biomass ($g.l^{-1}$)
1.	JB08S11	2.32 ± 0.75	27.	JB10OD13	2.83 ± 0.93
2.	JB08S12	0.45 ± 0.32	28.	JB10OD14	4.86 ± 0.83
3.	JB08S13	1.87 ± 0.43	29.	JB10OD15	4.25 ± 0.53
4.	JB08S14	4.78 ± 0.21	30.	JB10OD21	3.52 ± 0.77
5.	JB08S15	4.23 ± 0.65	31.	JB10OD22	5.02 ± 0.84
6.	JB08S16	2.15 ± 0.31	32.	JB10OD23	3.94 ± 0.53
7.	JB08S17	4.54 ± 0.65	33.	JB10OD24	2.17 ± 0.91
8.	JB08S18	0.20 ± 0.55	34.	JB10OD25	4.94 ± 0.63
9.	JB08S21	4.93 ± 0.83	35.	S10PSS11	5.05 ± 0.82
10.	JB08S22	2.8 ± 0.32	36.	S10PSS12	3.32 ± 0.32
11.	JB08S23	4.46 ± 0.27	37.	S10PSS13	4.93 ± 0.54
12.	JB08S24	2.83 ± 0.11	38.	S10PSS14	4.05 ± 0.82
13.	JB08S25	5.02 ± 0.45	39.	S10PSS15	4.88 ± 0.43
14.	JB08S31	4.93 ± 0.83	40.	S10PSS16	3.62 ± 0.81
15.	JB08S32	2.20 ± 0.29	41.	S10PSS17	5.09 ± 0.80
16.	JB08S33	4.75 ± 0.43	42.	S10PSS21	3.58 ± 0.31
17.	JB08S34	0.32 ± 1.2	43.	S10PSS22	4.95 ± 0.52
18.	JB08PS35	5.10 ± 0.82	44.	S10PSS23	0.07 ± 1.1
19.	JB08PS36	0.71 ± 0.37	45.	S10PSS24	0.27 ± 0.62
20.	JB08PS37	1.65 ± 0.44	46.	S10PSS25	5.02 ± 0.41
21.	JB08PS38	4.77 ± 0.62	47.	T11PS11	4.75 ± 0.53
22.	JB08PS39	2.53 ± 0.83	48.	T11PS12	1.88 ± 0.41
23.	JB08PS41	0.16 ± 0.97	49.	T11PS21	3.91 + 0.74
24.	JB08PS42	4.92 ± 1.1	50.	T11PS22	5.10 ± 0.18
25.	JB10OD11	5.08 ± 0.41	51.	T11PS23	4.90 ± 0.42
26.	JB10OD12	0.67 ± 0.67	52.	T11PS24	2.96 ± 0.62

Results represent mean ± S.D of three individual experiments.

A total of 23 isolates possessed better growth in Bushnell-Hass medium supplemented with 1% (v/v) n-hexadecane based on increased dry biomass yield. The isolates were further screened for their ability to produce BS by culturing on n-hexadecane supplemented media.

To evaluate BS production, the bacterial isolates were grown in MSM supplemented with 1% (v/v) n-hexadecane and incubated at 37°C and 150 rpm for 8 days. After 96 h of incubation, the reduction in the surface tension of the culture medium by the individual bacterial isolates was determined and is presented in Table 9.3.

TABLE 9.3 Surface Activity Exhibited by Bacterial Isolates Grown in Mineral Salt Medium (MSM) supplemented with 1% (v/v) n-Hexadecane After 96 h

SL. No.	Bacterial Isolates	Reduction in Surface Tension (mNm^{-1})	Drop Collapse Test	Oil Displacement Test (cm^2)
1.	JB08S14	43.8 ± 0.42	+	12.3 ± 0.23
2.	JB08S17	32.6 ± 0.85	+++	35.6 ± 0.34
3.	JB08S21	38.2 ± 0.24	+++	33.4 ± 0.42
4.	JB08S23	36.0 ± 0.42	+++	35.3 ± 0.15
5.	JB08S25	47.8 ± 0.62	+	11.6 ± 0.32
6.	JB08S31	39.5 ± 0.29	++	28.5 ± 0.26
7.	JB08S33	38.9 ± 0.73	+++	30.7 ± 0.44
8.	JB08PS35	52.3 ± 0.44	–	–
9.	JB08PS38	39.2 ± 0.82	+++	29.8 ± 0.21
10.	JB08PS42	38.7 ± 0.48	+++	30.2 ± 0.34
11.	JB10OD11	43.6 ± 0.57	+	26.8 ± 0.40
12.	JB10OD14	55.8 ± 0.23	–	–
13.	JB10OD22	46.3 ± 0.55	+	23.9 ± 0.27
14.	JB10OD25	34.3 ± 0.65	+++	36.± 0.33
15.	S10PSS11	39.1 ± 0.63	+++	29.9 ± 0.28
16.	S10PSS13	44.7 ± 1.1	+	19.8 ± 0.32
17.	S10PSS15	56.2 ± 0.45	–	–
18.	S10PSS17	39.8 ± 0.28	+++	31.0 ± 0.41
19.	S10PSS22	40.4 ± 0.73	++	26.8 ± 0.11
20.	S10PSS25	51.1 ± 0.11	–	–
21.	T11PS11	46.8 ± 0.45	+	21.7 ± 0.23
22.	T11PS22	30.6 ± 0.23	+++	35.7 ± 0.34
23.	T11PS23	47.9 ± 0.97	+	16.6 ± 0.27

Results represent mean ± S.D of three experiments.

Abbreviations: – = negative, + = positive, ++ = significant, +++ = Excellent.

Out of 23 bacterial strains, 17 strains which include JB08S17, JB08S21, JB08S23, JB08S31, JB08S33, JB08PS38, JB08PS42, JB10OD11, JB10OD22, JB10OD25, S10PSS11, S10PSS13, S10PSS17, S10PSS22, T11PS11, T11PS22, and T11PS23 exhibited the reduction in the surface tension of the culture medium from 67.5 ± 0.93 mNm^{-1} to a minimum of 32.6 ± 0.55 mNm^{-1}. The flask containing the culture medium alone was taken as control and it exhibited the maintenance of the same surface tension value of 67.8 ± 1.3 after 96 h. Out of 23 bacterial isolates, 19 showed positive value in the drop collapse test. In the case of the drop-collapse test, if the cell-free culture supernatant contains BS, the droplets of the culture broth on the oil-coated wells will collapse, but no change occurs in the shape of the droplets if the broth is without BS. Similarly, in oil displacement test, the addition of cell-free culture supernatant of 19 bacterial isolates caused oil to spread and formed a wide clear zone on the oil-water surface, confirming the presence of BS.

The objective of the present investigation was to identify bacteria capable of producing BS and could utilize the petroleum hydrocarbons efficiently. Out of 23 bacterial isolates, only 17 could exhibit both properties. On further experimentation, only 4 isolates viz. JB08S17, JB08S21, JB08S23, and JB10OD25 were found to be the potential users of hydrocarbons and producers of BSs as determined by their biomass yield and surface properties exhibited while growing on n-hexadecane. Therefore, these four bacterial isolates were selected for taxonomic identification and further studies. The bacterial isolates JB08S17, JB08S21, JB08S23, and JB10OD25 were further designated as OBP1, OBP2, OBP3, and OBP4, respectively for the sake of convenience.

9.2 MORPHOPHYSIOLOGICAL CHARACTERIZATION OF POTENTIAL BIOSURFACTANT (BS) PRODUCING BACTERIA

The biochemical, morphological, and physiological characterization of the bacterial isolates was carried out and data thus obtained are presented in Table 9.4.

TABLE 9.4 Biochemical Characterization of Bacterial Isolates

Biochemical Test	Bacterial Isolate			
	OBP1	**OBP2**	**OBP3**	**OBP4**
1. Gram-staining	Negative	Negative	Negative	Negative
2. Shape of the cell	Straight rod	Straight rod	Straight rod	Straight rod
3. Capsule staining	Positive	Positive	Positive	Positive
4. Endospore staining	Negative	Negative	Negative	Negative
5. Motility test	Positive	Positive	Positive	Positive
6. Acid production from:				
a. Glucose;	Positive	Positive	Positive	Positive
b. Fructose;	Positive	Positive	Positive	Positive
c. Xylose;	Positive	Positive	Positive	Positive
d. Maltose;	Negative	Negative	Negative	Negative
e. Lactose;	Negative	Negative	Negative	Negative
f. Mannitol;	Negative	Negative	Negative	Negative
g. Salicin;	Negative	Negative	Negative	Negative
h. Sucrose;	Negative	Negative	Negative	Negative
i. Oxidase test.	Positive	Positive	Positive	Positive
7. Catalase test	Positive	Positive	Positive	Positive
8. H_2S production	Negative	Negative	Negative	Negative
9. Gelatin hydrolysis test	Positive	Positive	Positive	Positive
10. Starch hydrolysis	Negative	Negative	Negative	Negative
11. Methyl red test	Negative	Negative	Negative	Negative
12. Voges-Proskauer test	Negative	Negative	Negative	Negative
13. Indole test	Negative	Negative	Negative	Negative
14. Citrate utilization test	Positive	Positive	Positive	Positive
15. Nitrate reduction test	Positive	Positive	Positive	Positive
16. Prod. of fluorescence	Positive	Positive	Positive	Positive
17. Prod. of pyocyanin	Negative	Negative	Negative	Negative

The colony morphology of the bacterial isolates is shown in Figure 9.1. The characterization details revealed four selected bacterial isolates belonging to the genus *Pseudomonas* and their phylogenetic relationship is depicted in Figure 9.2.

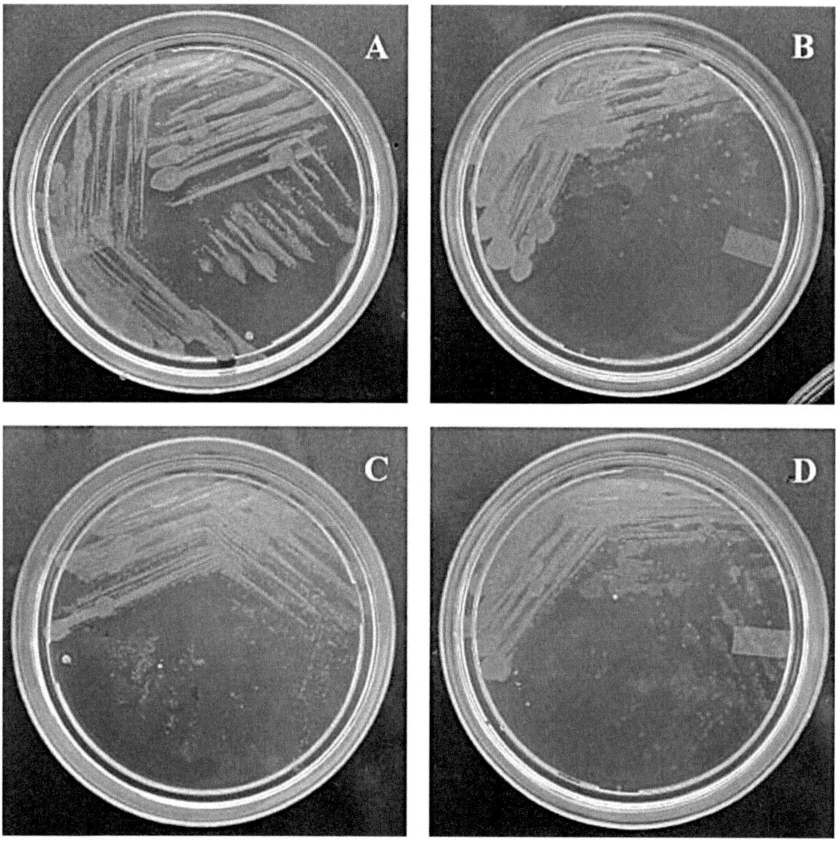

FIGURE 9.1 Colony morphology of pure cultures on nutrient agar (A) *P. aeruginosa* OBP1; (B) *P. aeruginosa* OBP2; (C) *P. aeruginosa* OBP3; and (D) *P. aeruginosa* OBP4.

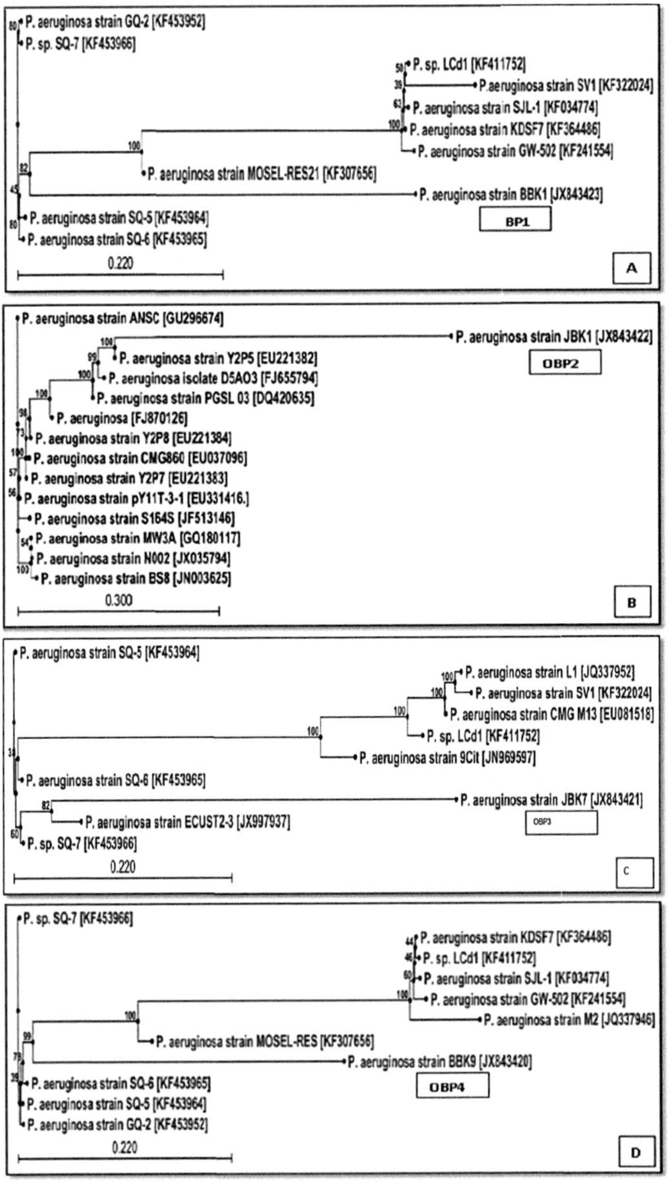

FIGURE 9.2 Phylogenetic tree generated using neighbor-Joining method showing the similarity of selected strains with other 16S rRNA gene sequences of *P. aeruginosa*. Bootstrap values are expressed as percentages of 1000 replications. Bar, 0.01 substitutions per nucleotide position. (A) *P. aeruginosa* OBP1; (B) *P. aeruginosa* OBP2; (C) *P. aeruginosa* OBP3; and (D) *P. aeruginosa* OBP4.

A total of 52 bacterial strains with the ability to produce BS and utilize hydrocarbons, particularly the purified petroleum components such as n-hexadecane were isolated from hydrocarbon-contaminated soil environments using the enrichment technique. Among them four bacterial strains were selected based on their efficiency to utilize n-hexadecane and produce BSs. Several reports suggested that the enrichment culture technique is efficient in isolating BS producing as well as hydrocarbon utilizing bacteria (Batista et al., 2006; Janbandhu and Fulekar, 2011). Biochemical tests further confirmed that all four strains belonged to the species *P. aeruginosa*. Different dominant mesophilic bacteria such as *P. putida*, *P. aeruginosa*, *P. saccharophila*, *Flavobacterium sp.*, *Burkholderia cepacia*, *Rhodococcus sp.*, *Stenotrophomonas sp.* and *Mycobacterium sp.*, etc., were reported to be isolated from the petroleum-contaminated environments (Karanth et al., 1999). *Pseudomonas sp.* is commonly detected using both culture-dependent (Iwabuchi et al., 2011; Das and Mukherjee, 2007) and culture-independent (Chaillan et al., 2004; Alexander, 2000) approaches and bacteria belonging to the species were isolated from the industrial wastewater samples and screened for growth on hydrocarbons and BS production (Whiteley and Bailey, 2000). The genus *Pseudomonas* is the most versatile group due to its inherent ability to utilize a diverse range of substrates as carbon source, particularly those found in petroleum (Vasileva-Tonkova and Galabova, 2003; Vasileva-Tonkova et al., 2006). Superior performance of *P. aeruginosa* is due to the evolution of the alkane oxidation genes, which allow them to grow on alkanes as a sole carbon source. The genes involved in the hydrocarbon degradation were present in the plasmids. Previous studies of Saikia et al. (2012) clearly supported the predominance of bacterial genus *Pseudomonas* in the North-Eastern region of India, especially in the petroleum-contaminated environments of Assam.

9.3 CARBON SOURCE IN OPTIMIZATION OF CULTURE CONDITIONS FOR BIOSURFACTANT (BS) PRODUCTION

Different carbon sources based on their increasing complexity and hydrophobicity, including glucose, glycerol, vegetable oil (soybean oil) and petroleum hydrocarbons (n-hexadecane, octadecane, diesel, and crude oil) were screened to determine their effectiveness in BS production. The data are presented in Tables 9.5 and 9.6, respectively.

TABLE 9.5 Growth ($g.l^{-1}$), Reduction of Surface Tension (mNm^{-1}) and Yield of Biosurfactant *P. aeruginosa* Strains

Carbon Sources	Biomass ($g.l^{-1}$)	Maximum Reduction in Surface Tension (mNm^{-1})	Yield of Crude Biosurfactant ($g.l^{-1}$)
Glucose			
P. aeruginosa OBP1	5.04 ± 0.45	50.7 ± 0.23	0.12 ± 0.42
OBP2	5.06 ± 0.23	51.4 ± 0.16	0.09 ± 0.12
OBP3	4.91 ± 0.63	50.6 ± 0.41	0.10 ± 0.38
OBP4	4.96 ± 0.28	51.8 ± 0.36	0.07 ± 0.18
Glycerol			
P. aeruginosa OBP1	4.96 ± 0.21	49.2 ± 0.27	0.30 ± 0.35
OBP2	5.12 ± 0.40	46.6 ± 0.38	0.39 ± 0.52
OBP3	5.03 ± 0.29	49.5 ± 0.53	0.26 ± 0.11
OBP4	5.08 ± 0.57	48.6 ± 0.45	0.33 ± 0.50
n-Hexadecane			
P. aeruginosa OBP1	4.87 ± 0.63	31.1 ± 0.88	4.57 ± 0.53
OBP2	5.03 ± 0.37	37.6 ± 0.51	2.86 ± 0.28
OBP3	4.73 ± 0.72	35.5 ± 0.38	2.83 ± 0.43
OBP4	5.10 ± 0.21	33.2 ± 0.79	3.17 ± 0.37
Octadecane			
P. aeruginosa OBP1	4.83 ± 0.34	31.9 ± 0.85	4.21 ± 0.34
OBP2	3.82 ± 0.56	39.6 ± 0.24	2.23 ± 0.52
OBP3	4.12 ± 0.74	38.4 ± 0.42	2.37 ± 0.50
OBP4	4.24 ± 0.37	36.8 ± 0.65	2.58 ± 0.44
Diesel			
P. aeruginosa OBP1	4.54 ± 0.93	32.0 ± 0.42	3.04 ± 0.60
OBP2	3.97 ± 0.53	37.5 ± 0.28	2.47 ± 0.38
OBP3	4.91 ± 0.40	36.2 ± 0.68	2.64 ± 0.87
OBP4	5.04 ± 0.62	34.3 ± 1.0	2.96 ± 0.92
Crude Oil			
P. aeruginosa OBP1	3.71 ± 0.44	32.7 ± 0.66	2.53 ± 0.47
OBP2	3.27 ± 0.62	40.4 ± 0.25	1.48 ± 0.52
OBP3	3.86 ± 0.82	39.5 ± 0.41	2.20 ± 0.73
OBP4	4.07 ± 0.53	37.7 ± 0.16	2.46 ± 0.28
Soyabean Oil			
P. aeruginosa OBP1	2.51 ± 0.56	38.3 ± 0.61	1.71 ± 0.22
OBP2	2.05 ± 0.29	42.8 ± 0.52	1.24 ± 0.45
OBP3	1.13 ± 0.37	45.6 ± 0.37	1.05 ± 0.58
OBP4	1.05 ± 0.35	46.5 ± 0.29	0.97 ± 0.33

Results represent mean ± S.D of three experiments. DB, dry biomass; ST, surface tension; BS, yield of biosurfactant.

TABLE 9.6 Growth Bacterial (*P. aeruginosa*) Strains in Liquid Culture Media Supplemented with Various Petroleum Hydrocarbons and Influence on the Surface Tension

Carbon Sources	Properties	*P. aeruginosa* Strains			
		OBP1	OBP2	OBP3	OBP4
Pentane	DB ($g.l^{-1}$)	–	–	–	–
	ST (mNm^{-1})	68.3 ± 0.21	67.5 ± 0.42	68.8 ± 0.43	67.9 ± 0.32
Hexane	DB ($g.l^{-1}$)	–	–	–	–
	ST (mNm^{-1})	67.9 ± 0.32	68.5 ± 0.61	68.1 ± 0.45	67.7 ± 0.32
Heptane	DB ($g.l^{-1}$)	–	–	–	–
	ST (mNm^{-1})	68.4 ± 0.24	67.8 ± 0.57	67.6 ± 0.31	68.3 ± 0.36
Iso-octane	DB ($g.l^{-1}$)	–	–	–	–
	ST (mNm^{-1})	67.9 ± 0.36	68.2 ± 0.51	67.7 ± 0.48	67.9 ± 0.52
Dodecane	DB ($g.l^{-1}$)	3.28 ± 0.65	3.31 ± 0.34	2.55 ± 0.53	2.37 ± 0.65
	ST (mNm^{-1})	36.8 ± 0.21	43.4 ± 0.63	41.6 ± 0.27	38.9 ± 0.52
Triadecane	DB ($g.l^{-1}$)	3.37 ± 0.64	3.48 ± 0.45	3.61 ± 0.44	3.54 ± 0.32
	ST (mNm^{-1})	36.0 ± 0.24	42.5 ± 0.57	39.7 ± 0.31	37.5 ± 0.36
n-Hexadecane	DB ($g.l^{-1}$)	4.87 ± 0.63	5.03 ± 0.37	4.73 ± 0.72	5.10 ± 0.21
	ST (mNm^{-1})	31.1 ± 0.88	37.6 ± 0.51	35.5 ± 0.38	33.2 ± 0.79
Octadecane	DB ($g.l^{-1}$)	4.83 ± 0.34	3.82 ± 0.56	4.12 ± 0.74	4.24 ± 0.37
	ST (mNm^{-1})	31.9 ± 0.85	39.6 ± 0.24	38.4 ± 0.42	36.8 ± 0.65
Eicosane	DB ($g.l^{-1}$)	–	–	–	–
	ST (mNm^{-1})	68.6 ± 0.18	68.4 ± 0.61	67.9 ± 0.28	68.2 ± 0.20
Triacontane	DB ($g.l^{-1}$)	–	–	–	–
	ST (mNm^{-1})	67.7 ± 0.49	68.3 ± 0.61	68.5 ± 0.28	67.9 ± 0.19
Paraffin	DB ($g.l^{-1}$)	1.38 ± 0.22	1.13 ± 0.46	1.52 ± 0.27	1.47 ± 0.51
	ST (mNm^{-1})	45.3 ± 0.29	47.2 ± 0.27	44.5 ± 0.25	45.6 ± 0.55
Phenol	DB ($g.l^{-1}$)	–	–	–	–
	ST (mNm^{-1})	68.3 ± 0.23	68.6 ± 0.23	67.8 ± 0.27	67.4 ± 0.82
Benzene	DB ($g.l^{-1}$)	–	–	–	–
	ST (mNm^{-1})	68.3 ± 0.18	68.6 ± 0.23	67.9 ± 0.41	68.4 ± 0.21
Toluene	DB ($g.l^{-1}$)	0.36 ± 0.13	–	0.78 ± 0.52	1.04 ± 0.22
	ST (mNm^{-1})	54.7 ± 0.39	68.7 ± 0.41	52.2 ± 0.17	52.5 ± 0.48
Xylene	DB ($g.l^{-1}$)	–	–	–	–
	ST (mNm^{-1})	68.2 ± 0.16	68.5 ± 0.42	68.7 ± 0.19	68.5 ± 0.15
Naphthalene	DB ($g.l^{-1}$)	–	–	0.83 ± 0.66	–
	ST (mNm^{-1})	68.7 ± 0.69	68.2 ± 0.36	53.5 ± 0.35	68.7 ± 0.92
Anthracene	DB ($g.l^{-1}$)	0.45 ± 0.37	–	0.66 ± 0.27	0.93 ± 0.41
	ST (mNm^{-1})	54.9 ± 0.83	67.7 ± 0.23	54.2 ± 0.44	53.6 ± 0.58
Phenanthrene	DB ($g.l^{-1}$)	0.61 ± 0.34	–	1.05 ± 0.29	1.18 ± 0.73
	ST (mNm^{-1})	54.2 ± 0.59	68.6 ± 0.15	53.8 ± 0.43	53.2 ± 0.46

TABLE 9.6 *(Continued)*

Carbon Sources	Properties	*P. aeruginosa* Strains			
		OBP1	OBP2	OBP3	OBP4
Pyrene	DB ($g.l^{-1}$)	–	–	–	–
	ST (mNm^{-1})	67.8 ± 0.91	68.7 ± 0.20	67.6 ± 0.82	68.7 ± 0.73
Fluorene	DB ($g.l^{-1}$)	–	–	–	–
	ST (mNm^{-1})	68.8 ± 0.83	67.9 ± 0.93	68.6 ± 0.43	67.9 ± 0.42
Diesel	DB ($g.l^{-1}$)	4.54 ± 0.93	3.97 ± 0.53	4.91 ± 0.40	5.04 ± 0.62
	ST (mNm^{-1})	32.0 ± 0.42	37.5 ± 0.28	36.2 ± 0.69	34.3 ± 1.0
Kerosene	DB ($g.l^{-1}$)	2.73 ± 0.76	2.30 ± 0.56	2.54 ± 3.2	2.78 ± 0.74
	ST (mNm^{-1})	39.6 ± 0.62	42.2 ± 0.29	40.5 ± 0.83	39.2 ± 0.44
Lubricating oil	DB ($g.l^{-1}$)	2.31 ± 0.56	1.67 ± 0.21	2.39 ± 0.82	2.24 ± 0.29
	ST (mNm^{-1})	41.5 ± 0.37	43.8 ± 0.18	40.7 ± 0.64	41.4 ± 1.0
Crude oil	DB* ($g.l^{-1}$)	3.71 ± 0.44	3.27 ± 0.62	3.86 ± 0.82	4.07 ± 0.53
	ST**(mNm^{-1})	32.7 ± 0.66	40.4 ± 0.25	39.5 ± 0.41	37.7 ± 0.16

* DB: Dry biomass, **ST: Surface tension, Mean ± S.D of three individual experiments.

In the case of aromatic hydrocarbons like toluene the bacterial strains OBP4, OBP3, and OBP1 showed slight growth but no growth on benzene, phenol, and xylene supplemented media. The bacterial strains OBP4, OBP3, and OBP1 exhibited minimum growth on PAHs like phenanthrene and anthracene but no growth on pyrene and fluorene. Among the four *P. aeruginosa* strains, only OBP3 exhibited growth on naphthalene supplemented medium. Among the tested hydrocarbons, the bacterial strains showed good performance in n-hexadecane supplemented MSM. The capability of the bacterial strains to utilize a hydrocarbon as the sole source of carbon and energy was different from each other as revealed by their biomass yield and efficiency to reduce the surface tension.

The population of OBP3 bacterial strain got reduced with increase in the concentration of n-hexadecane above 1.5%, whereas the same happened in the case of the other three bacterial strains at a higher concentration (2.0%). A maximum $cfu.ml^{-1}$ of OBP1, OBP2, and OBP4 was detected by the plate count technique in the culture medium supplemented with 2.0% n-hexadecane whereas OBP3 possessed a maximum $cfu.ml^{-1}$ in 1.5% n-hexadecane. As shown in Table 9.7, the number of $cfu.ml^{-1}$ of the bacterial strains OBP1, OBP2, and OBP4 increased from 5.5×10^7 to 7.5×10^{11} with the increase in the concentration of n-hexadecane from 1.0 to 2.0%; however, a similar change from 4.8×10^8 to 5.4×10^9 $cfu.ml^{-1}$ in the case of OBP3 occurred from 1.0 to 1.5% n-hexadecane. The increase in the concentration of n-hexadecane above 2.0% showed a sharp reduction in the cfu.ml^{-1} of the bacterial strains and the same in the case of OBP3 at above 1.5%.

The selected bacterial strains exhibited better growth in the tested petroleum hydrocarbons. The overall preference of hydrocarbons was found to be in the order of n-hexadecane > octadecane > diesel > crude oil > tridecane > dodecane > kerosene > lubricating oil as evident from the increase in the bacterial biomass. The bacterial strains have the potentiality to degrade almost all the tested aliphatic hydrocarbons. The bacterial strain OBP1 followed by OBP3 and OBP4 exhibited the highest growth in dodecane, tridecane, hexadecane, and octadecane-supplemented media as evident from the production of bacterial biomass. However, the bacterial isolate OBP3 exhibited almost similar biomass production in hexadecane and octadecane supplemented media but lower in tridecane and dodecane supplemented media. None of the strains were capable to utilize eicosane and triacontane. There are several reports on the degradation of short (C_8–C_{16}) and long-chain (C44) hydrocarbons (Mehdi and Giti, 2008; Wonga et al., 2004). With the increase in the chain length of alkanes, the hydrophobicity of the molecule increases, which in turn makes the molecules less soluble in water and reduces the bioavailability (Zhang et al., 2005). Several workers reported the capability to use aliphatic chains in the range of C_{12}–C_{24}, C_{12}–C_{34}, C_6–C_{28}, C_{12}–C_{32}, and C_{12}–C_{28} by *P. aeruginosa* strains PAO1, RR1, (A1–A6), *P fluorescens* CHA0 and *Pseudomonas sp*. PUP 6 (Yalcin and Ergene, 2010; Kullen et al., 2001). Among the tested aromatic hydrocarbons such as benzene, toluene, and xylene; the bacterial strains OBP4, OBP3 and OBP1 exhibited growth on toluene but not by the OBP2 strains. However, the growth of the bacterial strains on aromatic hydrocarbons was not significant as compared to that on aliphatic hydrocarbons. The results are in consistent with the observation of Jones and Edington, who reported that only 0.5% of a large group of soil organisms could use aromatic hydrocarbons. In the case of polyaromatic hydrocarbons (PAHs), bacterial strains OBP4, OBP3, and OBP1 exhibited growth on anthracene and phenanthrene. The growth was significant in the case of OBP4. But no growth was observed in naphthalene except for OBP3.

The complete mineralization of high molecular weight PAHs could be achieved by only a limited number of microorganisms. Preference for toluene, anthracene, and phenanthrene as the carbon source could be due to their isolation being from the environments, very frequently contaminated by aromatic and polyaromatic fractions of the crude oil. Bordoloi and Konwar (1998) reported the isolation of several strains of

P. aeruginosa from the petroleum contaminated sites of Assam capable of utilizing various PAHs such as phenanthrene, pyrene, and fluorene as carbon substrates for their growth.

The bacterial strains exhibited better growth in diesel, crude oil and kerosene supplemented media, as evident from the bacterial biomass. A greater portion of diesel, crude oil and kerosene contains aliphatic hydrocarbons and are reported to be more prone to microbial degradation (Mehdi and Giti, 2008; Wonga et al., 2004; Smith et al., 2002).

After the completion of the fermentation, various parameters such as reduction in the surface tension, yield of BS and dry biomass were assessed. Concentrations of the efficient carbon source 1.0, 1.5, 2.0, 2.5 and 3% (w/v) were used to determine the optimum level of BS production by the bacterial strains. After the completion of the fermentation process, various parameters such as reduction in the surface tension, yield of BS and dry biomass were assessed.

All four bacterial strains could grow on MSM supplemented with the substrates. The carbon sources allowed good bacterial growth, whereas the production of BS was quite different. Among the substrates, n-hexadecane was found to be most suitable to produce BS by all four bacterial strains followed by diesel. The substrates octadecane and crude petroleum could also significantly increase the BS production by all four bacterial strains. The strains showed significant difference in their growth and surface properties when grown on the selected vegetable oil. However, OBP1 exhibited better performance in terms of bacterial biomass and BS production in soyabean oil. The substrates glucose and glycerol were proficient in terms of biomass production, but not so in BS production.

The carbon source plays a crucial role in the production of RL (Maier and Soberon-Chavez, 2000; Mata-Sandoval et al., 1999). In the present investigation, different carbon sources were used based on their increasing complexity and hydrophobicity, which included glucose, glycerol, vegetable oils, and petroleum hydrocarbons (n-hexadecane, octadecane, diesel, and crude oil) to determine their effectiveness on RL production and data thus obtained are presented in Table 9.9. The findings confirmed the ability of the bacterial strains in utilizing both the water-soluble and insoluble substrates to produce BSs. The hydrophobic substrates such as n-hexadecane, octadecane, diesel, and crude oil could induce better BS production as compared to the vegetable oils and water-miscible substrates. BS production was reported to be induced by hydrocarbons or other water-insoluble substrates (Deziel et al., 1999). Another phenomenon is the catabolic repression of BS synthesis by glucose and other primary metabolites (Karanth et al., 1999). No surface-active agent could be isolated from the culture of *Arthrobacter paraffineus* when glucose was used as the carbon source instead of hexadecane (Syldatk and Wagner, 1987). Similarly, *P. aeruginosa* S7B1 was reported to produce a protein-like activator for n-alkane oxidation when grown on hydrocarbon supplemented medium but not on glucose, glycerol, or palmitic acid (Manresa et al., 1991). RL production by *P. aeruginosa* on glycerol was reduced sharply on adding glucose, acetate, succinate or citrate to the culture medium.

Hydrocarbons are the excellent carbon sources to produce RL from *P. aeruginosa* strains. The carbon source n-hexadecane was found to be the most efficient for the reduction of surface tension of the culture medium and achieving the highest BS production. It was observed that the bacterial strain OBP1 also showed preference for the vegetable oil, whereas OBP3 and OBP4 preferred mineral oils such as diesel and crude oil. However, the strain OBP2 preferred n-hexadecane. Such anomalous behavior of the bacterial strains clearly suggested that the carbon source preference for RL production entirely depends on the bacterial strain. The effect of the nutrient medium and particularly the carbon source on the synthesis of RLs is still not well understood (Chaerun et al., 2004). The bacterial strains were found to grow easily on mineral oils such as kerosene, diesel, and crude oil. However, the pattern of BS production was different. The mineral oil, especially kerosene was much less efficient in RL production by the bacterial strains. The bacterial strains possessed an efficient alkane utilizing ability. The involvement of *P. aeruginosa* in the production of RL using water-immiscible hydrocarbon like hexadecane as the carbon substrate was reported by many workers. The tested vegetable oils: mustard oil, sunflower oil, soyabean oil, sesame seed oil, castor oil, Nahor oil and jatropha oil, were less efficient in inducing RL production but found to be better than glucose and glycerol. The strain OBP1 produced RL efficiently when cultivated on the vegetable oil, but production was less as compared to n-hexadecane. The probable reason for such behavior might be since most of *P. aeruginosa* species produce lipases which facilitate the assimilation of fatty acids present in the vegetable oil (Maier and Soberon-Chavez, 2000; Haba et al., 2000). The long-chain fatty acids can either be further degraded via β-oxidation to support cell growth or might be transformed into the lipid precursor to promote biosynthesis of RL (Maier and Soberon-Chavez, 2000; Wei et al., 2005). The production of RL by the other three strains in vegetable oils was much lower. The low production of RL in water-soluble substrates like glucose and glycerol could be reasonable as these substrates are quite soluble and hence more readily available to the bacterial cells.

There is no requirement for the bacterial cells to synthesize BS to improve their solubility or availability (Wei et al., 2005). Further, catabolic repression of BS synthesis by glucose or primary metabolites is one of the important regulatory mechanisms found to be operative in the hydrocarbon utilizing microorganisms. The function of BS is related to the hydrocarbon availability or their uptake and therefore, they are synthesized predominantly by hydrocarbon-degrading microorganisms (Desai and Desai, 1993). The preference for carbon substrates among the studied bacterial strains of *P. aeruginosa* is quite different, and the complex hydrocarbons were found to be better than easily available carbon sources in producing RL. This clearly suggests that the preference for carbon sources for RL production entirely depends on the bacterial strains.

The quality and quantity of BS produced by the bacterial strains were reported to be influenced by the nature of the carbon substrate (Henkel et al., 2012; Haba et al., 2000). All four strains of *P. aeruginosa* were able to produce higher quantity of BSs in n-hexadecane (conc. 1.5–2.0%) supplemented medium with maximum yield in the range of 2.83–4.57 g.l^{-1}. The culture medium of the bacterial strains attained a lower surface tension in the range of 31.1–37.6 mNm^{-1}.

The bacterial strains exhibited better growth, as was evident from the increased biomass density in n-hexadecane-supplemented medium. Bacterial strains OBP1, OBP2, and OBP4 formed the highest cfu.ml^{-1} in the range of 5.9 × 109–7.5 × 10^{11} on 2.0% n-hexadecane while OBP3 exhibited the maximum of 5.4 × 10^{9} cfu.ml^{-1} on 1.5% n-hexadecane during incubation at 37°C with 180 rpm. The high concentration of n-hexadecane (>2.0%) reduced both the growth and the production of BSs by all the four strains. Concentrations of n-hexadecane more than the optimum caused a drastic reduction in the bacterial population growth. The exact reason for such behavior is not well understood but may be related to the availability of dissolved oxygen. The inefficient oxygen supply in shake-flask cultures might be responsible for the poor growth of *P. aeruginosa* WatG on petroleum refined products like kerosene and diesel and reported it to be an oxygen-intensive metabolic process. The effects of toxicity, enzyme inhibition, and oxygen limitation were minimized by using relatively low concentrations of hydrocarbons. The high concentration of crude oil reduced the growth rate of the different crude oil-degrading bacterial strains of *Pseudomonas, Rhodococcus, and Bacillus.*

9.4 DETECTION AND QUANTIFICATION OF BIOSURFACTANT (BS)

The cell density of the individual bacterial strains was reduced by serial dilution and then they were spread over the MSM agar plates supplemented with CTAB (0.2 $g.l^{-1}$), methylene blue dye (5 $mg.l^{-1}$) and n-hexadecane (0.1%, v/v) and incubated at 37°C for 48 h. All four bacterial strains could grow on CTAB agar plates forming blue halos around the colonies. The appearance of blue halos around the colonies on the blue agar plates confirmed the production of extracellular anionic BSs by the strains, and the same are shown in Figure 9.3.

The bacterial strains OBP1, OBP2, OBP3, and OBP4 could grow on MSM supplemented with n-hexadecane causing reduction of surface tension of the culture medium from 68.5 mNm^{-1} to 31.1, 37.6, 35.5 and 33.2 mNm^{-1}, respectively between 84 and 96 h of incubation. Growth curve of the bacterial strains revealed the maximum biomass productions between 120 and 144 h of incubation.

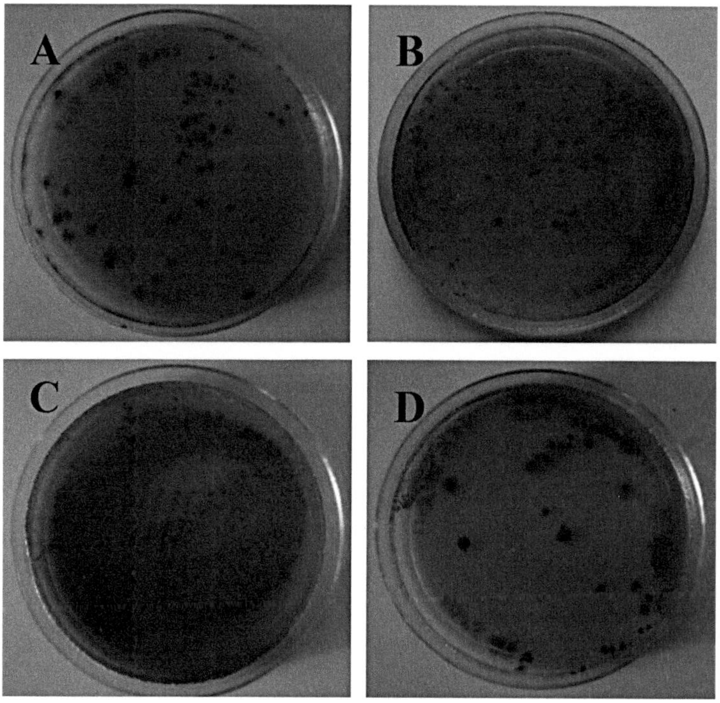

FIGURE 9.3 CTAB agar assay for the detection of glycolipid biosurfactant secreted by the *P. aeruginosa* strains (A) OBP1; (B) OBP2; (C) OBP3; and (D) OBP4.

To detect and quantify the surface-active glycolipids, three independent experiments were performed with blood agar assay, CTAB agar test and orcinol assay. The cell-free culture supernatants of the selected bacterial strains were analyzed for their hemolytic activity on blood agar plates at 37°C overnight. The bacterial strains exhibited distinct zone of hemolysis in blood-agar plates containing 2% (v/v) goat blood and the same is shown in Figure 9.4. The hemolysis assay confirmed the production of BS and the same was considered as the preliminary criterion to produce BS.

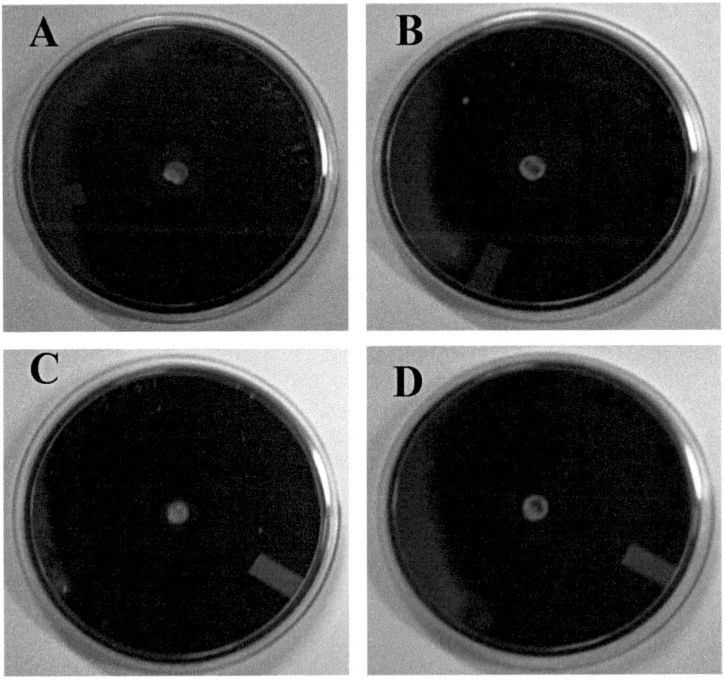

FIGURE 9.4 Hemolysis on blood agar medium by cell-free culture supernatant of bacterial strains. (A) *P. aeruginosa* OBP1; (B) *P. aeruginosa* OBP2; (C) *P. aeruginosa* OBP3; and (D) *P. aeruginosa* OBP4.

It is well known that microorganisms growing on hydrocarbons frequently produce BSs with emulsification or surfactant activity (Bora et al., 2011). Such microbial behavior has been considered as a biological strategy to facilitate the availability of hydrophobic substrates (Rojas-Avelizapa et al., 1999). In the present investigation, different assays based on biochemical and physical parameters were carried out with the objective to screen the potential hydrocarbon utilizing and BS producing bacterial strains from the petroleum-contaminated environments. Among the physical parameter-based screening techniques, rapid drop collapse test is considered as a high throughput screening technique (Muller et al., 2012) which assists in the isolation of potent bacterial isolates. The drop collapse test to be a simple and sensitive technique used to detect and quantify the BS production. For direct detection of BS production by the bacterial isolates, two different independent techniques were performed which include hemolytic assay and CTAB agar test. Hemolysis assay is widely used for the screening of BS producing bacteria and could also be used for the quantification of glycolipid-type BS produced by bacteria (Tuleva et al., 2002). The blood agar plate technique was used to screen a novel thermophilic hydrocarbon-degrading *P. aeruginosa* AP02-1 producing RL. The CTAB agar test is another technique used for the detection of glycolipid-type BS production by the bacterial colonies in the culture plate directly. Screening of RL producing thermophilic hydrocarbon-degrading *P. aeruginosa* APO2, AB4, and 2B by CTAB agar blue plate method from the soil contaminated with petroleum products.

9.5 BACTERIAL GROWTH CHARACTERISTICS AND BIOSURFACTANT (BS) PRODUCTION

The bacterial strains OBP1, OBP2, OBP3, and OBP4 could grow on MS medium supplemented with n-hexadecane causing reduction in surface tension of the culture medium from 68.5 mNm^{-1} to 31.1, 37.6, 35.5 and 33.2 mNm^{-1}, respectively between 84 and 96 h of incubation. Growth curve of the bacterial strains revealed the maximum biomass productions between 120 and 144 h of incubation. The BS concentration ($g.l^{-1}$) produced by *P. aeruginosa* strains in MSM supplemented with 2% n-hexadecane. The same are presented in Figures 9.5 and 9.6, correspondingly. The production of BS started after 36–48 h of incubation, the highest production was achieved towards the early stationary phase between 108 and 120 h. The biomass and BS production were in the range of 4.73–5.10 $g.l^{-1}$ and 2.83–4.57 $g.l^{-1}$, respectively. The Scanning Electron Micrographs of the bacterial strains grown on n-hexadecane supplemented medium are shown in Figure 9.7.

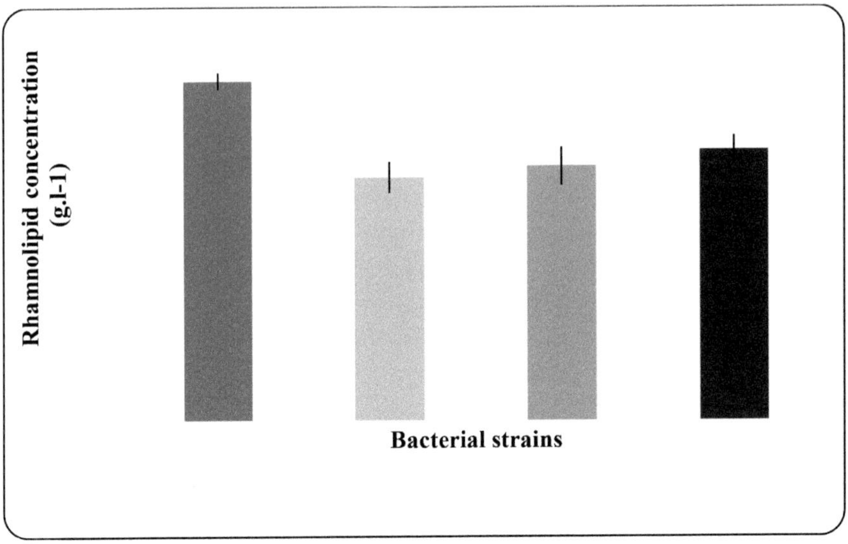

FIGURE 9.5 Biosurfactant concentration ($g.l^{-1}$) produced by *P. aeruginosa* strains in mineral salt medium supplemented with 2% n-hexadecane (mean of 3 experiments ± SD).

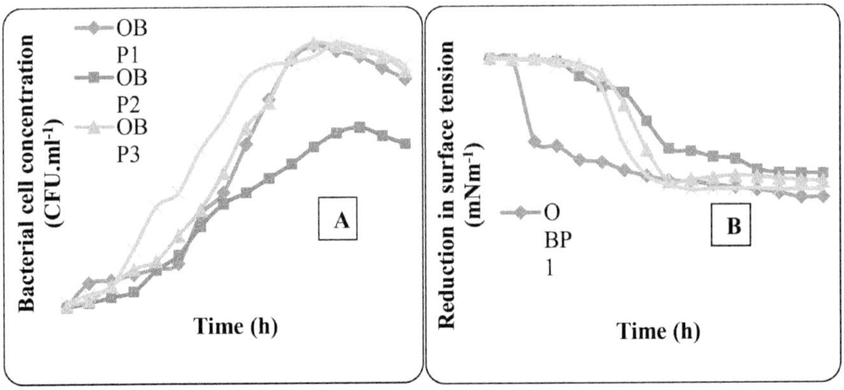

FIGURE 9.6 Time profile of (A) growth and (B) reduction in the surface tension of the culture broth by the *P. aeruginosa* strains in mineral salt medium supplemented with 2% n-hexadecane (mean of 3 experiments).

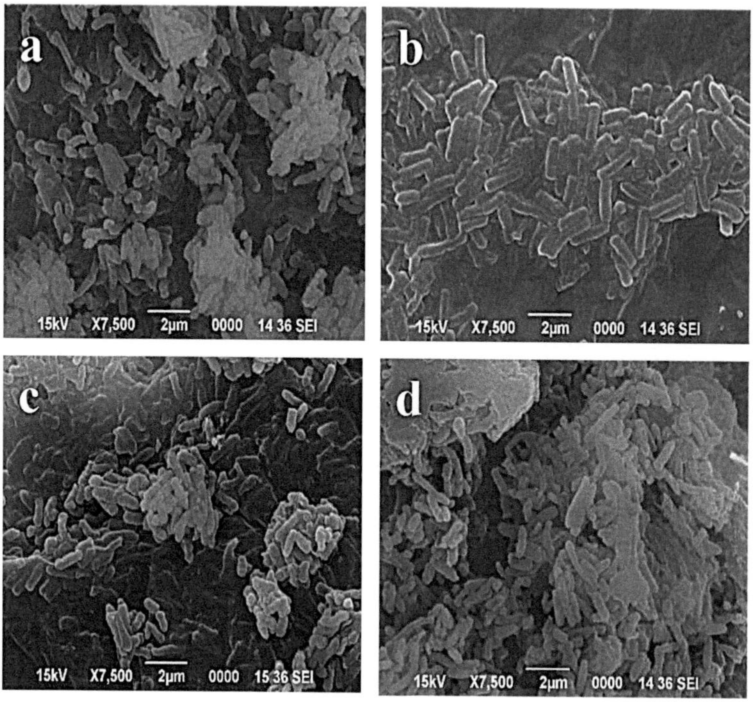

FIGURE 9.7 Scanning electron micrograph of *P. aeruginosa* strains showing growth on n-hexadecane. (a) *P. aeruginosa* OBP1; (b) *P. aeruginosa* OBP2; (c) *P. aeruginosa* OBP3; and (d) *P. aeruginosa* OBP4.

All four bacterial strains showed the requirement of similar nutrient and culture conditions which might be due to their isolation from the similar type of habitats that have the continuous exposure to the petroleum hydrocarbons. The growth curves of the bacterial strains in relation to duration of culture in MSM supplemented with n-hexadecane as sole source of carbon, $(NH_4)_2SO_4$ and urea together as the nitrogen source showed nearly a parallel relationship between cell growth, reduction in the surface tension and BS production. Results clearly suggest a growth-associated production of BSs by the bacterial strains. Moreover, the preference for carbon substrates was quite similar among the strains. Though n-hexadecane was found to be the best carbon source for BS production by the bacterial strains, but the pattern of growth behavior was relatively different from each other, which might be due to the intrinsic variability for utilizing hydrocarbons. The initial reduction in the surface tension was about 45.1–66.3 mNm^{-1} after 36–48 h of inoculation during the early exponential phase of bacterial growth. Results indicate the initiation of BS production by the bacterial cells in the culture broth. The possible cause for the initiation of BS production in the early exponential phase could be due to availability of hydrocarbons as the source of carbon through pseudo-solubilization. After 60–108 h of bacterial growth, the BS concentration increased rapidly, reaching its maximum after 120–132 h which was mainly towards the later stage of stationary phase of growth. Such behavior might be due to the release of cell-bound BS into the culture broth (Naik and Sakthivel, 2006). The maximum surface activity in 156–168 h culture indicated the optimum level of BS production by the bacterial strains towards the stationary phase. The production of BS by the bacterial strains in n-hexadecane supplemented medium took place during the late exponential phase to early stationary phase of growth, suggesting BS production as the secondary metabolite.

Various strains of *P. aeruginosa* were reported to show an overproduction of RL when cultures reached their stationary phase of growth (Muller et al., 2012). The depletion of nitrogen source in the culture medium mainly takes place during the stationary phase of cell growth (Clarke et al., 2010; Leon and Kumar, 2005; Adebusoye et al., 2007). N-limiting conditions do not favor RL production, but production starts with the exhaustion of nitrogen in the culture medium. Moreover, limitation of multivalent ions such as Mg^{2+}, Ca^{2+}, K^{+}, Na^{+} and trace elements were reported to cause an enhancement in the RL yield either in resting-cell cultures or during the stationary phase of growth (Plaza et al., 2006). The variation appeared in the pattern of RL production was due to the use of different *Pseudomonas* strains, cultivation conditions which include pH and temperature, and media composition. Hence there is a possibility that under the applied culture conditions, multiple limitations might occur having a positive influence on RL formation, thus leading to an increase in the specific productivity per bacterial cell (Perfumo et al., 2006).

9.6 CONCENTRATION EFFECT OF N-HEXADECANE ON BIOSURFACTANT (BS) PRODUCTION

BS production by the bacterial strains initially increased with the increasing concentration of n-hexadecane until it reached the maximum value and then leveled off. Moreover, the growth of the bacterial strains significantly reduced as the concentration of n-hexadecane exceeded more than 2.5% (v/v) resulting in insignificant production of BS in the culture medium. The same are presented in Tables 9.7 and 9.8.

TABLE 9.7 Influence of Different Concentrations of n-Hexadecane on Growth and Biosurfactant Production by *P. aeruginosa* Strains

Percentage of n-Hexadecane (v/v)	Properties	*P. aeruginosa* Strain			
		OBP1	OBP2	OBP3	OBP4
1.0	DB ($g.l^{-1}$)	4.47 ± 0.71	4.53 ± 0.67	4.58 ± 0.64	4.62 ± 0.93
	ST (mNm^{-1})	32.6 ± 0.85	38.2 ± 0.82	36.0 ± 0.42	34.3 ± 0.65
	BS ($g.l^{-1}$)	3.96 ± 1.00	2.48 ± 0.31	2.76 ± 0.97	2.92 ± 0.49
1.5	DB ($g.l^{-1}$)	4.65 ± 0.52	4.94 ± 0.36	4.73 ± 0.72	4.89 ± 0.56
	ST (mNm^{-1})	32.0 ± 0.95	37.9 ± 0.49	35.5 ± 0.38	33.9 ± 0.74
	BS ($g.l^{-1}$)	4.43 ± 0.74	2.70 ± 0.32	2.83 ± 0.43	3.03 ± 0.52
2.0	DB ($g.l^{-1}$)	4.87 ± 0.63	5.03 ± 0.37	4.66 ± 0.52	5.10 ± 0.21
	ST (mNm^{-1})	31.1 ± 0.88	37.6 ± 0.51	36.2 ± 0.86	33.2 ± 0.79
	BS ($g.l^{-1}$)	4.57 ± 0.53	2.86 ± 0.28	2.72 ± 0.73	3.17 ± 0.37
2.5	DB ($g.l^{-1}$)	4.70 ± 0.48	4.89 ± 0.62	4.28 ± 0.66	4.91 ± 0.52
	ST (mNm^{-1})	32.8 ± 0.62	38.0 ± 0.47	38.2 ± 0.29	33.8 ± 0.80
	BS ($g.l^{-1}$)	4.41 ± 0.71	2.77 ± 0.79	2.18 ± 0.57	3.04 ± 0.39
3.0	DB ($g.l^{-1}$)	3.20 ± 0.65	4.08 ± 0.36	3.73 ± 0.82	3.88 ± 0.75
	ST (mNm^{-1})	36.2 ± 0.48	39.9 ± 0.84	40.8 ± 0.37	38.6 ± 0.62
	BS ($g.l^{-1}$)	3.84 ± 0.92	2.02 ± 0.55	1.93 ± 0.77	2.53 ± 0.45

Results represent mean ± S.D of three experiments.

Abbreviations: DB: dry biomass; ST: surface tension; BS: yield of biosurfactant.

TABLE 9.8 Influence of Different Concentration of n-Hexadecane on the Colony Forming Unit (CFU) of *P. aeruginosa* Strains

Bacterial Strain	Concentration of n-Hexadecane				
	1.0%	1.5%	2.0%	2.5%	3.0%
P. aeruginosa OBP1	3.9×10^8	4.7×10^8	5.9×10^9	4.2×10^8	2.3×10^5
a OBP2	5.5×10^7	6.1×10^8	6.6×10^9	5.8×10^7	3.5×10^5
OBP3	4.8×10^8	5.4×10^9	3.9×10^8	2.6×10^5	1.1×10^3
OBP4	4.5×10^8	6.8×10^9	7.5×10^{11}	6.1×10^8	3.8×10^6

Results represent the mean of three experiments. CFU: colony-forming unit.

As shown in Table 9.8, the number of $cfu.ml^{-1}$ of the bacterial strains OBP1, OBP2, and OBP4 increased from 5.5×10^7 to 7.5×10^{11} with the increase in the concentration of n-hexadecane from 1.0 to 2.0%; however, a similar change from 4.8×10^8 to 5.4×10^9 $cfu.ml^{-1}$ in the case of OBP3 occurred from 1.0 to 1.5% n-hexadecane. The increase in the concentration of n-hexadecane above 2.0% showed a sharp reduction in the $cfu.ml^{-1}$ of the bacterial strains and the same in the case of OBP3 at above 1.5%.

The population of the CFU of OBP3 bacterial strain got reduced with increase in the concentration of n-hexadecane above 1.5%, whereas the same happened in the case of the other three bacterial strains at a higher concentration (2.0%). A maximum $cfu.ml^{-1}$ of OBP1, OBP2, and OBP4 was detected by the plate count technique in the culture medium supplemented with 2.0% n-hexadecane whereas OBP3 possessed a maximum $cfu.ml^{-1}$ in 1.5% n-hexadecane.

9.7 EFFECT OF NITROGEN SOURCES ON BIOSURFACTANT (BS) PRODUCTION

Nitrogen source plays a crucial role in the production of BSs by bacteria. Bacterial strains exhibited poor growth and surface activities in the nitrogen-deficient media. Both organic and inorganic nitrogen sources influenced the growth and BS production in the bacterial strains. Data represented in Table 9.9.

TABLE 9.9 Influence of Various Nitrogen Sources on Growth and Biosurfactant Production of *P. aeruginosa* Strain

Different Nitrogen Sources	Properties	*P. aeruginosa* Strain			
		OBP1	OBP2	OBP3	OBP4
Nitrogen Free	DB ($g.l^{-1}$)	0.51 ± 0.18	0.24 ± 0.10	0.27 ± 0.21	0.33 ± 0.14
	ST (mNm^{-1})	56.4 ± 0.23	58.8 ± 0.15	58.7 ± 0.27	59.7 ± 0.07
	BS ($g.l^{-1}$)	0.10 ± 0.27	0.05 ± 0.21	0.07 ± 0.31	0.07 ± 0.22
NH_4Cl	DB ($g.l^{-1}$)	3.17 ± 0.29	2.86 ± 0.78	3.27 ± 0.51	2.97 ± 0.80
	ST (mNm^{-1})	40.4 ± 0.71	42.5 ± 1.00	42.0 ± 0.24	42.5 ± 0.67
	BS ($g.l^{-1}$)	2.05 ± 0.45	1.04 ± 0.67	1.0 ± 0.22	1.28 ± 0.49
$(NH_4)_2SO_4$	DB ($g.l^{-1}$)	3.70 ± 0.51	1.10 ± 0.66	2.06 ± 0.74	4.23 ± 0.37
	ST (mNm^{-1})	36.7 ± 0.72	40.9 ± 0.24	39.8 ± 0.42	38.6 ± 0.65
	BS ($g.l^{-1}$)	3.14 ± 0.66	1.13 ± 0.30	1.07 ± 0.71	2.38 ± 0.43

TABLE 9.9 *(Continued)*

Different Nitrogen Sources	Properties	*P. aeruginosa* Strain			
		OBP1	OBP2	OBP3	OBP4
NH_4NO_3	DB ($g.l^{-1}$)	2.76 ± 0.56	3.17 ± 0.70	3.44 ± 0.65	3.28 ± 0.97
	ST (mNm^{-1})	39.2 ± 0.73	42.6 ± 0.93	41.0 ± 0.39	41.4 ± 0.71
	BS ($g.l^{-1}$)	2.03 ± 0.41	1.10 ± 0.72	1.06 ± 0.86	1.68 ± 0.64
$NH_4H_2PO_4$	DB ($g.l^{-1}$)	3.57 ± 0.42	3.38 ± 0.43	3.71 ± 0.51	3.62 ± 0.50
	ST (mNm^{-1})	38.2 ± 0.36	42.2 ± 0.72	40.8 ± 0.88	40.6 ± 0.62
	BS ($g.l^{-1}$)	2.34 ± 0.67	1.02 ± 0.56	1.10 ± 0.90	1.86 ± 0.47
KNO_3	DB ($g.l^{-1}$)	2.94 ± 0.56	2.13 ± 0.67	3.07 ± 0.79	2.26 ± 0.87
	ST (mNm^{-1})	42.4 ± 0.35	44.5 ± 0.56	43.6 ± 0.81	43.8 ± 0.56
	BS ($g.l^{-1}$)	1.06 ± 0.77	0.83 ± 0.38	0.78 ± 0.95	1.04 ± 0.33
H_2NCONH_4	DB ($g.l^{-1}$)	2.34 ± 0.72	2.48 ± 0.46	2.27 ± 0.67	2.61 ± 0.25
	ST (mNm^{-1})	37.3 ± 0.41	41.7 ± 0.29	40.5 ± 0.51	39.3 ± 0.65
	BS ($g.l^{-1}$)	2.32 ± 0.54	1.12 ± 0.52	1.15 ± 0.83	2.19 ± 0.43
Yeast extract	DB ($g.l^{-1}$)	2.63 ± 0.43	2.77 ± 0.38	2.58 ± 0.62	3.12 ± 0.93
	ST (mNm^{-1})	37.8 ± 0.20	40.9 ± 0.73	41.7 ± 1.02	38.8 ± 0.52
	BS ($g.l^{-1}$)	2.15 ± 0.76	1.10 ± 0.55	1.06 ± 0.83	2.07 ± 0.73
Beef extract	DB ($g.l^{-1}$)	2.42 ± 0.36	2.43 ± 0.81	2.27 ± 0.52	2.86 ± 0.29
	ST (mNm^{-1})	39.6 ± 0.72	42.3 ± 0.95	40.5 ± 0.63	41.5 ± 0.33
	BS ($g.l^{-1}$)	1.87 ± 0.29	1.11 ± 0.21	1.16 ± 0.47	1.14 ± 0.71
Peptone	DB ($g.l^{-1}$)	2.58 ± 0.81	2.82 ± 0.43	2.65 ± 0.59	3.04 ± 0.11
	ST (mNm^{-1})	38.4 ± 0.33	42.8 ± 0.62	42.4 ± 0.28	40.2 ± 0.52
	BS ($g.l^{-1}$)	2.08 ± 0.27	1.07 ± 0.83	1.10 ± 0.51	1.95 ± 0.76

Mean ± S.D of 3 experiments.

Abbreviations: DB: dry biomass; ST: surface tension; BS: yield of biosurfactant.

Different organic and inorganic nitrogen sources were tested for their role on BS production which included beef extract, yeast extract, peptone, ammonium dihydrogen orthophosphate ($NH_4H_2PO_4$), urea (H_2NCONH_4), ammonium chloride (NH_4Cl), potassium nitrate (KNO_3), ammonium nitrate (NH_4NO_3) and ammonium sulfate $(NH_4)_2SO_4$. The nitrogen sources were used in different concentrations and combinations to determine their effect on the production of BS. After the completion of fermentation, various parameters such as reduction in the surface tension, yield of BS and dry biomass were determined. Among the inorganic sources, $(NH_4)_2SO_4$ was found to be the best for growth and BS production as compared to the other sources. Among the organic nitrogen sources, urea was found to be efficient against yeast extract, beef extract and peptone promoting growth and BS production in all the bacterial strains.

9.8 EFFECT OF COMBINED NITROGEN SOURCES ON BIOSURFACTANT (BS) PRODUCTION

The type and concentration of inorganic nitrogen sources affected the production of BSs. The highest production was obtained with $(NH_4)_2SO_4$ at a concentration of 2.0 g.l^{-1} in all the bacterial strains except for OBP2 which exhibited optimal BS production at 1.0 g.l^{-1}. Similarly, urea at the concentration of 2.0 g.l^{-1} was proved to be the best organic nitrogen source. Further, the combination of both $(NH_4)_2SO_4$ and urea at a concentration of 2.0 g.l^{-1} each was found to be efficient for the growth and BS production. Data thus obtained are presented in Tables 9.10(a), (b), and (c), respectively.

TABLE 9.10(A) Influence of Different Concentration of $(NH_4)_2SO_4$ on Growth and Biosurfactant Production of *P. aeruginosa* Strains

Concentration of $(NH_4)_2SO_4$ (g.l^{-1})	Properties	*P. aeruginosa* Strain			
		OBP1	OBP2	OBP3	OBP4
1.0	DB (g.l^{-1})	3.53 ± 0.47	3.89 ± 0.65	3.78 ± 0.25	4.08 ± 0.63
	ST (mNm^{-1})	37.7 ± 0.36	37.7 ± 0.69	41.3 ± 0.52	39.8 ± 0.81
	BS (g.l^{-1})	2.97 ± 0.38	1.00 ± 0.30	0.87 ± 0.46	2.19 ± 0.45
2.0	DB (g.l^{-1})	3.78 ± 0.51	4.18 ± 0.56	3.96 ± 0.74	4.23 ± 0.37
	ST (mNm^{-1})	36.7 ± 0.72	40.9 ± 0.24	39.8 ± 0.42	38.6 ± 0.65
	BS (g.l^{-1})	3.14 ± 0.66	1.13 ± 0.30	1.07 ± 0.71	2.38 ± 0.43
3.0	DB (g.l^{-1})	3.64 ± 0.58	3.96 ± 0.52	3.83 ± 0.68	4.14 ± 0.35
	ST (mNm^{-1})	37.0 ± 0.36	38.5 ± 0.40	40.6 ± 0.50	39.4 ± 0.30
	BS (g.l^{-1})	3.03 ± 0.73	1.06 ± 0.28	0.96 ± 0.83	2.23 ± 0.44

Results represent mean ± S.D of three experiments. DB, dry biomass; ST: surface tension; BS: yield of biosurfactant

TABLE 9.10(B) Influence of Different Concentration of Urea (H_2NCONH_4) on the Growth and Biosurfactant Production of *P. aeruginosa* Strains

Percentage of H_2NCONH_4 (g.l^{-1})	Properties	*P. aeruginosa* Strain			
		OBP1	OBP2	OBP3	OBP4
1.0	DB (g.l^{-1})	2.22 ± 0.64	2.31 ± 0.38	2.13 ± 0.56	2.43 ± 0.54
	ST (mNm^{-1})	38.9 ± 0.28	42.5 ± 0.62	41.8 ± 0.72	41.6 ± 0.28
	BS (g.l^{-1})	2.18 ± 0.54	1.04 ± 0.58	1.04 ± 0.71	1.97 ± 0.83
2.0	DB (g.l^{-1})	2.34 ± 0.72	2.48 ± 0.46	2.27 ± 0.67	2.61 ± 0.25
	ST (mNm^{-1})	37.3 ± 0.41	41.7 ± 0.29	40.5 ± 0.51	39.3 ± 0.65
	BS (g.l^{-1})	2.32 ± 0.54	1.12 ± 0.52	1.15 ± 0.83	2.19 ± 0.43
3.0	DB (g.l^{-1})	2.28 ± 0.45	2.40 ± 0.84	2.20 ± 0.82	2.55 ± 0.80
	ST (mNm^{-1})	37.9 ± 0.38	42.2 ± 0.71	41.3 ± 0.39	40.4 ± 0.42
	BS (g.l^{-1})	2.27 ± 0.23	1.07 ± 0.56	1.09 ± 0.84	2.07 ± 0.74

Results represent mean ± S.D of three experiments. DB, dry biomass; ST: surface tension; BS: yield of biosurfactant.

TABLE 9.10(C) Combined Effect of Two Different Nitrogen Sources on the Growth and Biosurfactant Production of *P. aeruginosa* Strains*

Bacterial strain	MSM + H_2NCONH_4 (2.0 g.l^{-1})+ $NH_4(SO_4)_2$ (2.0 g.l^{-1})		
	DB (g.l^{-1})	ST (mNm^{-1})	BS (g.l^{-1})
P. aeruginosa OBP1	4.87 ± 0.63	31.1 ± 0.88	4.57 ± 0.53
OBP2	5.03 ± 0.37	37.6 ± 0.51	2.86 ± 0.28
OBP3	4.73 ± 0.72	35.5 ± 0.38	2.83 ± 0.43
OBP4	5.10 ± 0.21	33.2 ± 0.79	3.17 ± 0.37

Results represent mean ± S.D of three experiments. DB, dry biomass; ST: surface tension; BS: yield of biosurfactant.

*In case of OBP2, concentration of $NH_4(SO_4)_2$ was 1.0 g.l^{-1}.

9.9 BACTERIAL BIOSURFACTANT (BS) IN NITROGEN SOURCES

Nitrogen being a vital component of proteins is required for the microbial growth and production of enzymes in the fermentation process (Leon and Kumar, 2005). The nitrogen source in the culture medium plays a significant role in the RL production (Sim et al., 1997; Poremba et al., 1991) and contributes to pH control (Salihu et al., 2009). Several sources of nitrogen were reported to be used in the production of BSs, such as urea, ammonium sulfate (Bordoloi and Konwar, 2007; Aparna et al., 2012), ammonium nitrate (Radwan and Sorkhoh, 1993) sodium nitrate

(Lotfabad et al., 2009; Abdel-Mawgoud et al., 2010; Tahzibi et al., 2004), yeast extract (Abbasi et al., 2012; Tuleva et al., 2002), meat, and malt extract (Mata-Sandoval et al., 2001). The production of BSs by the strains of *P. aeruginosa* was examined in the presence of various organic and inorganic nitrogen sources. NH_4NO_3, $NaNO_3$, and $(NH_4)_2SO_4$ enhanced the growth and production of BSs; whereas KNO_3 lowered both growth and BS production (Table 9.9). The inorganic nitrogen sources such as NH_4NO_3 and $NaNO_3$ caused a similar pattern of growth of bacterial strains and BS production. $(NH_4)_2SO_4$ was found to be the best nitrogen source. Organic nitrogen source urea was found to be better than the yeast extract, leading to a high RL yield (Table 9.9). In the case of urea and $(NH_4)_2SO_4$ each with 2.0 $g.l^{-1}$, the BSs yield was 4.57, 2.86, 2.83 and 3.17 $g.l^{-1}$ in the case of OBP1, OBP2, OBP3, and OBP4, respectively which was much higher as compared to the individual nitrogen sources. The use of organic and inorganic nitrogen in combination was more effective in the production of BS. Bordoloi and Konwar (1998) obtained higher yield of RL when $(NH_4)_2SO_4$ and urea were used as nitrogen sources. Increase in RL production when $(NH_4)_2SO_4$ and trace metals were added throughout the fermentation process and the same supported our findings. Out of the defined media like Luria broth, M9, modified mineral salt and basal salt media, the modified MSM was found to be the best to produce BS.

9.10 EFFECT OF MACRO AND MICRO-NUTRIENTS ON BIOSURFACTANT (BS) PRODUCTION

Metal ions are known to play a crucial role in the growth and production of BSs as they participate in various metabolic pathways in the form of cofactors of many enzymes. Various concentrations of macronutrients like Na and K were added in the culture medium in the form of KH_2PO_4 and Na_2HPO_4 to determine their influence on the production of BS. Similarly, various concentrations of micronutrients were prepared and added to the culture media. The micronutrient solution included $FeSO_4{\cdot}7H_2O$ (1000 μg.l–1), $CuSO_4{\cdot}5H_2O$ (50 $\mu g.l^{-1}$), H_3BO_3 (10 $\mu g.l^{-1}$), $MnSO_4$ (10 $\mu g.l^{-1}$), $ZnSO_4{\cdot}7H_2O$ (70 $\mu g.l^{-1}$), and MoO_3 (10 $\mu g.l^{-1}$). A control experiment was run simultaneously without supplying any of the micronutrients. With the completion of the fermentation process, various parameters such as reduction in the surface tension, yield of BS and dry biomass were determined. The influence of macronutrients is presented in Tables 9.11(a)–(d).

TABLE 9.11(A) Influence of Magnesium Sulfate ($MgSO_4.7H_2O$) on Growth and Biosurfactant Production of *P. aeruginosa* Strains

Concentration of $MgSO_4.7H_2O$ (g.l^{-1})	Properties	*P. aeruginosa* Strains			
		OBP1	OBP2	OBP3	OBP4
0.0	DB (g.l^{-1})	4.23 ± 0.73	4.34 ± 0.38	4.21 ± 0.59	4.41 ± 0.83
	ST (mNm^{-1})	38.6 ± 0.35	43.4 ± 0.76	42.6 ± 0.38	45.0 ± 0.20
	BS (g.l^{-1})	3.95 ± 0.82	2.16 ± 0.13	2.30 ± 0.61	2.55 ± 0.47
0.1	DB (g.l^{-1})	4.59 ± 0.72	4.84 ± 0.37	4.50 ± 0.76	4.87 ± 0.43
	ST (mNm^{-1})	32.5 ± 0.41	38.4 ± 0.30	36.2 ± 0.73	34.1 ± 0.17
	BS (g.l^{-1})	4.39 ± 0.62	2.65 ± 0.76	2.71 ± 0.38	2.94 ± 0.52
0.2	DB (g.l^{-1})	4.87 ± 0.63	5.03 ± 0.37	4.73 ± 0.72	5.10 ± 0.21
	ST (mNm^{-1})	31.1 ± 0.88	37.6 ± 0.51	35.5 ± 0.38	33.2 ± 0.79
	BS (g.l^{-1})	4.57 ± 0.53	2.86 ± 0.28	2.83 ± 0.43	3.17 ± 0.37
0.3	DB (g.l^{-1})	4.79 ± 0.57	4.93 ± 0.25	4.67 ± 0.46	4.96 ± 0.60
	ST (mNm^{-1})	31.7 ± 0.50	38.1 ± 0.74	35.8 ± 0.90	33.7 ± 0.35
	BS (g.l^{-1})	4.48 ± 0.94	2.82 ± 0.67	2.78 ± 0.22	3.11 ± 0.71

Results represent mean ± S.D of three experiments. DB, dry biomass; ST: surface tension; BS: yield of biosurfactant.

TABLE 9.11(B) Influence of Calcium Chloride ($CaCl_2 \cdot 2H_2O$) on Growth and Biosurfactant Production of *P. aeruginosa* Strains

Concentration of $CaCl_2 \cdot 2H_2O$ (mg.l^{-1})	Properties	*P. aeruginosa* Strain			
		OBP1	OBP2	OBP3	OBP4
0	DB (g.l^{-1})	4.34 ± 0.71	4.56 ± 0.22	4.28 ± 0.53	4.60 ± 0.36
	ST (mNm^{-1})	37.2 ± 0.94	42.7 ± 0.12	41.5 ± 0.29	44.3 ± 0.50
	BS (g.l^{-1})	4.04 ± 0.42	2.20 ± 0.52	2.33 ± 0.73	2.39 ± 0.71
25	DB (g.l^{-1})	4.67 ± 0.49	4.93 ± 0.81	4.57 ± 0.72	4.95 ± 0.64
	ST (mNm^{-1})	31.8 ± 0.32	38.2 ± 0.73	35.9 ± 0.91	34.0 ± 0.37
	BS (g.l^{-1})	4.40 ± 0.63	2.68 ± 0.30	2.74 ± 0.54	3.01 ± 0.52
50	DB (g.l^{-1})	4.87 ± 0.63	5.03 ± 0.37	4.73 ± 0.72	5.10 ± 0.21
	ST (mNm^{-1})	31.1 ± 0.88	37.6 ± 0.51	35.5 ± 0.38	33.2 ± 0.79
	BS (g.l^{-1})	4.57 ± 0.53	2.86 ± 0.28	2.83 ± 0.43	3.17 ± 0.37
75	DB (g.l^{-1})	4.77 ± 0.33	4.97 ± 0.83	4.70 ± 0.54	5.04 ± 0.70
	ST (mNm^{-1})	32.0 ± 0.51	38.4 ± 0.65	36.0 ± 0.50	33.8 ± 0.22
	BS (g.l^{-1})	4.51 ± 0.39	2.80 ± 0.73	2.77 ± 0.61	3.09 ± 0.45

Results represent mean ± S.D of three experiments. DB, dry biomass; ST: surface tension; BS: yield of biosurfactant.

TABLE 9.11(C) Influence of Potassium Dihydrogen Phosphate (KH_2PO_4) on Growth and Biosurfactant Production of *P. aeruginosa* Strains

Concentration of KH_2PO_4 (g.l^{-1})	Properties	***P. aeruginosa*** **Strains**			
		OBP1	**OBP2**	**OBP3**	**OBP4**
0.0	DB (g.l^{-1})	3.87 ± 0.27	4.04 ± 0.55	3.73 ± 0.26	4.14 ± 0.42
	ST (mNm^{-1})	38.7 ± 0.41	43.6 ± 0.82	42.5 ± 1.03	41.0 ± 0.57
	BS (g.l^{-1})	2.78 ± 0.18	1.34 ± 0.54	1.41 ± 0.35	1.56 ± 0.26
0.875	DB (g.l^{-1})	4.52 ± 0.56	4.79 ± 0.98	4.36 ± 0.42	4.81 ± 0.70
	ST (mNm^{-1})	32.9 ± 0.74	38.8 ± 0.14	36.7 ± 0.30	34.2 ± 0.54
	BS (g.l^{-1})	4.32 ± 0.22	2.61 ± 0.63	2.70 ± 0.58	2.89 ± 0.31
1.75	DB (g.l^{-1})	4.87 ± 0.63	5.03 ± 0.37	4.73 ± 0.72	5.10 ± 0.21
	ST (mNm^{-1})	31.1 ± 0.88	37.6 ± 0.51	35.5 ± 0.38	33.2 ± 0.79
	BS (g.l^{-1})	4.57 ± 0.53	2.86 ± 0.28	2.83 ± 0.43	3.17 ± 0.37
3.5	DB (g.l^{-1})	4.76 ± 0.47	4.95 ± 0.73	4.65 ± 0.38	5.01 ± 0.52
	ST (mNm^{-1})	31.8 ± 0.75	38.2 ± 0.59	36.0 ± 0.27	33.9 ± 0.44
	BS (g.l^{-1})	4.47 ± 0.38	2.78 ± 0.96	2.75 ± 0.53	3.07 ± 0.90

Results represent mean ± S.D of three individual experiments. NB: DB, dry biomass; ST: surface tension; BS: yield of biosurfactant.

TABLE 9.11(D) Influence of Disodium Hydrogen Phosphate (Na_2HPO_4) on Growth and Biosurfactant Production of *P. aeruginosa* Strains

Concentration of Na_2HPO_4 (g.l^{-1})	Properties	***P. aeruginosa*** **Strains**			
		OBP1	**OBP2**	**OBP3**	**OBP4**
0.0	DB (g.l^{-1})	3.28 ± 0.63	3.54 ± 0.82	2.96 ± 0.37	3.66 ± 0.41
	ST (mNm^{-1})	39.5 ± 0.19	44.6 ± 0.41	43.3 ± 0.58	42.0 ± 0.62
	BS (g.l^{-1})	2.08 ± 0.37	1.05 ± 0.66	1.17 ± 0.29	1.26 ± 0.45
1.81	DB (g.l^{-1})	4.48 ± 0.71	4.72 ± 0.37	4.29 ± 0.96	4.75 ± 0.65
	ST (mNm^{-1})	33.7 ± 0.49	39.0 ± 0.80	37.1 ± 0.63	34.9 ± 0.23
	BS (g.l^{-1})	4.22 ± 0.52	2.49 ± 0.56	2.65 ± 0.48	2.83 ± 0.74
3.61	DB (g.l^{-1})	4.87 ± 0.63	5.03 ± 0.37	4.73 ± 0.72	5.10 ± 0.21
	ST (mNm^{-1})	31.1 ± 0.88	37.6 ± 0.51	35.5 ± 0.38	33.2 ± 0.79
	BS (g.l^{-1})	4.57 ± 0.53	2.86 ± 0.28	2.83 ± 0.43	3.17 ± 0.37
7.22	DB (g.l^{-1})	4.73 ± 0.84	4.90 ± 0.55	4.60 ± 0.39	4.87 ± 0.21
	ST (mNm^{-1})	32.2 ± 1.02	38.5 ± 0.94	36.2 ± 0.67	34.7 ± 0.83
	BS (g.l^{-1})	4.40 ± 0.76	2.76 ± 0.38	2.69 ± 0.25	3.00 ± 0.57

Results represent mean ± S.D of three experiments. DB, dry biomass; ST: surface tension; BS: yield of biosurfactant.

In addition of Na_2HPO_4 and KH_2PO_4 at concentrations of 3.61 and 1.75 $g.l^{-1}$, respectively, the bacterial strains showed higher cell growth and maximum production of BSs. The essential metal ions Mg^{2+}, Fe^{2+} and Ca^{2+} supplied in the form of $MgSO_4 \cdot 7H_2O$ ($g.l^{-1}$), $FeSO_4 \cdot 7H_2O$ ($g.l^{-1}$) and $CaCl_2 \cdot 2H_2O$ ($g.l^{-1}$), respectively attributed to significant enhancement of cell growth and BS production. The best BS production was obtained with $MgSO_4 \cdot 7H_2O$ and $CaCl_2 \cdot 2H_2O$ at a concentration of 0.2 $g.l^{-1}$ and 50 $mg.l^{-1}$, respectively. Addition of trace elements like Zn^{2+}, Mn^{3+} and BO_3^{3+} in the medium significantly enhanced the production of BSs. The concentration of metal ions plays a very important role in the production of BSs as they form important cofactors of many enzymes (Leon and Kumar, 2005). Limitation of Ca^{2+}, Fe^{2+}, K^+, Mg^{2+}, Na^+ and other trace minerals towards the late exponential phase of growth were also reported to enhance the production of RLs (Henkel et al., 2012; Perfumo et al., 2006). The highest RL production from *P. aeruginosa* DSM 7107 and DSM 7108 in Ca^{2+} free media.

The concentrations of Na_2HPO_4, KH_2PO_4, $MgSO_4 \cdot 7H_2O$, $FeSO_4 \cdot 7H_2O$, and $CaCl_2 \cdot 2H_2O$ at 3.6 $g.l^{-1}$, 1.75 $g.l^{-1}$, 0.2 $g.l^{-1}$, 1.0 $mg.l^{-1}$ and 50.0 $mg.l^{-1}$, respectively were found to be optimum for the production of BS by the bacterial strains. The use of negative control having no micronutrients in the culture media and application of 100 $\mu l.l^{-1}$ of each of the stock solution of $CuSO_4 \cdot 7H_2O$, $MnSO_4 \cdot 5H_2O$, H_3BO_3, $ZnSO_4 \cdot 7H_2O$, and MnO_3 in 1 L of MSM was found to be effective in growth and BS production of *P. aeruginosa* strains and the data are presented in Table 9.12.

TABLE 9.12 Influence of Trace Elements on Growth and Biosurfactant Production of *P. aeruginosa* Strains

Micronutrients* ($\mu l.l^{-1}$)	Properties	*P. aeruginosa* Strains			
		OBP1	OBP2	OBP3	OBP4
0	DB ($g.l^{-1}$)	4.54 ± 0.83	4.67 ± 0.52	4.38 ± 0.77	4.66 ± 0.19
	ST (mNm^{-1})	35.8 ± 0.20	41.0 ± 0.49	39.5 ± 0.41	38.7 ± 0.62
	BS ($g.l^{-1}$)	4.24 ± 0.62	2.38 ± 0.37	2.44 ± 0.83	2.81 ± 0.70
50	DB ($g.l^{-1}$)	4.79 ± 0.62	4.94 ± 0.27	4.67 ± 0.71	5.02 ± 0.58
	ST (mNm^{-1})	31.7 ± 0.80	38.2 ± 0.15	36.0 ± 0.98	33.8 ± 0.37
	BS ($g.l^{-1}$)	4.50 ± 0.31	2.80 ± 0.66	2.78 ± 0.43	3.11 ± 0.65
100	DB ($g.l^{-1}$)	4.87 ± 0.63	5.03 ± 0.37	4.73 ± 0.72	5.10 ± 0.21
	ST (mNm^{-1})	31.1 ± 0.88	37.6 ± 0.51	35.5 ± 0.38	33.2 ± 0.79
	BS ($g.l^{-1}$)	4.57 ± 0.53	2.86 ± 0.28	2.83 ± 0.43	3.17 ± 0.37

TABLE 9.12 *(Continued)*

Micronutrients* (µl.l^{-1})	Properties	*P. aeruginosa* Strains			
		OBP1	OBP2	OBP3	OBP4
200	DB (g.l^{-1})	4.83 ± 0.73	5.02 ± 0.50	4.70 ± 0.27	5.07 ± 0.83
	ST (mNm^{-1})	31.6 ± 0.40	38.0 ± 0.97	35.8 ± 0.63	33.7 ± 0.40
	BS (g.l^{-1})	4.54 ± 1.02	2.83 ± 0.58	2.78 ± 0.75	3.13 ± 0.81

**Stock* solution of micronutrients prepared.

Results represent mean ± S.D of three experiments. DB, dry biomass; ST: surface tension; BS: yield of biosurfactant.

9.11 EFFECT OF TEMPERATURE ON BIOSURFACTANT (BS) PRODUCTION PH AND AGITATION

To determine the effect of temperature on BS production, bacterial cultures were incubated at a temperature range from 30 to 42°C with shaking at 180 rpm. On completion of the fermentation process, various parameters such as reduction in the surface tension, yield of BS and dry biomass were determined. The bacterial strains could grow and produce BS in almost all temperatures applied and data thus obtained are presented in Table 9.13. The bacterial strains could grow and produce BS in almost all temperatures applied.

TABLE 9.13 Influence of Temperature (°C) on Growth and Biosurfactant Production of *P. aeruginosa* Strains

Temperature (°C)	Properties	*P. aeruginosa* Strains			
		OBP1	OBP2	OBP3	OBP4
35	DB (g.l^{-1})	4.45 ± 0.56	4.52 ± 0.62	4.33 ± 0.55	4.92 ± 0.18
	ST (mNm^{-1})	32.8 ± 0.24	38.8 ± 0.76	36.3 ± 0.39	35.0 ± 0.98
	BS (g.l^{-1})	4.11 ± 0.32	2.33 ± 0.51	2.51 ± 0.52	3.04 ± 0.36
37	DB (g.l^{-1})	4.87 ± 0.63	4.95 ± 0.38	4.73 ± 0.72	5.10 ± 0.21
	ST (mNm^{-1})	31.1 ± 0.88	38.1 ± 0.41	35.5 ± 0.38	33.2 ± 0.79
	BS (g.l^{-1})	4.57 ± 0.53	2.82 ± 0.82	2.83 ± 0.43	3.17 ± 0.37
40	DB (g.l^{-1})	4.85 ± 0.64	5.03 ± 0.37	4.68 ± 0.82	5.05 ± 0.28
	ST (mNm^{-1})	31.1 ± 0.28	37.6 ± 0.51	36.1 ± 0.45	34.3 ± 0.80
	BS (g.l^{-1})	4.50 ± 0.83	2.86 ± 0.28	2.78 ± 0.71	3.07 ± 0.31
42	DB (g.l^{-1})	1.08 ± 0.77	2.65 ± 0.80	1.02 ± 0.29	1.10 ± 0.40
	ST (mNm^{-1})	49.8 ± 0.41	40.4 ± 0.62	50.7 ± 1.04	49.2 ± 0.12
	BS (g.l^{-1})	0.95 ± 0.94	1.28 ± 0.27	0.78 ± 0.83	1.03 ± 0.56

Results represent mean ± S.D of three experiments. DB, dry biomass; ST: surface tension; BS: yield of biosurfactant.

BS production increased with an increase in temperature from 30–37°C, remained nearly constant at 37–40°C and then decreased when temperature was increased further to above 40°C. The optimal temperature for growth and BS production was found to be 37°C in the case of bacterial strains OBP1, OBP3, and OBP4, but the strain OBP2 exhibited maximum growth and production of BS at 40°C.

9.12 EFFECT OF PH ON BIOSURFACTANT (BS) PRODUCTION

To determine the effect of pH on the production of BSs by the bacterial strains, the pH of the media was adjusted between pH 6.5–7.2 and the production of BSs was observed with the assessment of parameters like reduction in the surface tension, yield of BS and dry biomass were determined. The influence of the initial pH of the culture medium on the production of BS by the bacterial strains was determined and data is presented in Table 9.14. The influence of the initial pH of the culture medium on the production of BS by the bacterial strains was determined.

TABLE 9.14 Influence of pH on Growth and Biosurfactant Production

pH	Properties	*P. aeruginosa* Strains			
		OBP1	OBP2	OBP3	OBP4
6.5	DB ($g.l^{-1}$)	4.66 ± 0.83	4.50 ± 0.28	4.57 ± 0.35	4.92 ± 0.38
	ST (mNm^{-1})	31.8 ± 0.39	39.8 ± 0.62	36.4 ± 0.82	34.7 ± 0.20
	BS ($g.l^{-1}$)	4.43 ± 0.27	2.24 ± 0.70	2.70 ± 0.80	3.03 ± 0.25
6.8	DB ($g.l^{-1}$)	4.87 ± 0.63	4.91 ± 0.56	4.73 ± 0.72	5.10 ± 0.21
	ST (mNm^{-1})	31.1 ± 0.88	38.5 ± 0.24	35.5 ± 0.38	33.2 ± 0.79
	BS ($g.l^{-1}$)	4.57 ± 0.53	2.60 ± 0.40	2.83 ± 0.43	3.17 ± 0.37
7.0	DB ($g.l^{-1}$)	4.80 ± 0.54	5.03 ± 0.37	4.68 ± 0.38	5.01 ± 0.35
	ST (mNm^{-1})	32.4 ± 0.71	37.6 ± 0.51	36.0 ± 0.75	33.8 ± 0.22
	BS ($g.l^{-1}$)	4.50 ± 0.19	2.86 ± 0.28	2.79 ± 0.49	3.06 ± 0.57
7.2	DB ($g.l^{-1}$)	4.48 ± 0.42	4.58 ± 0.72	4.25 ± 0.33	4.55 ± 0.81
	ST (mNm^{-1})	35.7 ± 0.86	38.4 ± 0.25	39.2 ± 0.71	36.2 ± 0.72
	BS ($g.l^{-1}$)	3.94 ± 0.55	2.77 ± 0.80	2.31 ± 0.29	2.85 ± 0.40

Results represent mean ± S.D of three individual experiments. NB: DB, dry biomass; ST: surface tension; BS: yield of biosurfactant.

The bacterial strains could grow in all the tested pH values; however, growth, and BS production got reduced at higher acidic and alkaline pH levels. Growth and production of BSs were better at slightly acidic to neutral pH values. The optimal pH was found to be 6.8 except for OBP2 which exhibited optimal growth and BS production at a pH of 7.0

9.13 EFFECT OF SHAKING ON BIOSURFACTANT (BS) PRODUCTION

Erlenmeyer flasks of 250 ml volume containing 100 ml of MSM inoculated with the bacterial strains were used to produce BS. The conical flasks were incubated in an orbital incubator shaker with the agitation rate set at 100, 120, 150, 180, 200, and 220 rpm to determine the impact of agitation on BS production. The influence of shaking on the production of BS under the different agitation rates is presented in Table 9.15. A flask without any agitation was maintained and considered to be at 0 rpm for mixing as the control. After the completion of the fermentation, parameters like reduction in the surface tension, yield of BS and dry biomass were determined.

TABLE 9.15 Influence of Agitation (rpm) on Growth and Biosurfactant Production of *P. aeruginosa* Strains

Agitation (rpm)	Properties	*P. aeruginosa* Strains			
		OBP1	OBP2	OBP3	OBP4
0	DB ($g.l^{-1}$)	2.63 ± 0.73	2.94 ± 0.83	2.56 ± 0.15	2.19 ± 0.59
	ST (mNm^{-1})	40.7 ± 0.63	44.7 ± 0.49	43.6 ± 0.62	44.3 ± 0.35
	BS ($g.l^{-1}$)	0.94 ± 0.37	0.73 ± 0.56	0.65 ± 0.39	0.76 ± 0.27
100	DB ($g.l^{-1}$)	4.16 ± 0.52	4.25 ± 0.35	4.27 ± 0.58	4.05 ± 0.43
	ST (mNm^{-1})	39.8 ± 0.66	42.7 ± 0.48	40.8 ± 0.31	41.6 ± 0.52
	BS ($g.l^{-1}$)	1.21 ± 0.42	0.83 ± 0.30	0.98 ± 0.92	0.90 ± 0.61
120	DB ($g.l^{-1}$)	4.22 ± 0.73	4.48 ± 0.41	4.34 ± 0.89	4.37 ± 0.26
	ST (mNm^{-1})	34.9 ± 0.52	39.2 ± 0.20	38.5 ± 0.42	39.2 ± 0.47
	BS ($g.l^{-1}$)	4.23 ± 0.35	2.47 ± 0.28	2.43 ± 0.56	2.47 ± 0.39
150	DB ($g.l^{-1}$)	4.65 ± 0.39	4.94 ± 0.72	4.51 ± 0.53	4.76 ± 0.80
	ST (mNm^{-1})	32.7 ± 0.92	38.8 ± 0.38	36.8 ± 0.71	37.5 ± 0.45
	BS ($g.l^{-1}$)	4.39 ± 0.70	2.73 ± 0.47	2.69 ± 0.95	2.88 ± 0.37
180	DB ($g.l^{-1}$)	4.87 ± 0.34	5.03 ± 0.56	4.73 ± 0.74	4.90 ± 0.35
	ST (mNm^{-1})	31.1 ± 0.85	37.6 ± 0.24	35.5 ± 0.42	34.8 ± 0.82
	BS ($g.l^{-1}$)	4.57 ± 0.85	2.86 ± 0.30	2.83 ± 0.71	3.04 ± 0.39
200	DB ($g.l^{-1}$)	4.71 ± 0.61	4.87 ± 0.99	4.58 ± 0.37	5.10 ± 0.37
	ST (mNm^{-1})	31.8 ± 0.38	38.4 ± 0.52	36.7 ± 0.61	33.2 ± 0.65
	BS ($g.l^{-1}$)	4.45 ± 0.22	2.70 ± 0.49	2.65 ± 0.30	3.17 ± 0.43

TABLE 9.15 *(Continued)*

Agitation (rpm)	Properties	*P. aeruginosa* Strains			
		OBP1	OBP2	OBP3	OBP4
220	DB ($g.l^{-1}$)	3.93 ± 0.47	4.04 ± 0.29	3.86 ± 0.39	4.82 ± 0.48
	ST (mNm^{-1})	33.0 ± 0.31	39.3 ± 0.42	38.5 ± 0.15	35.7 ± 1.22
	BS ($g.l^{-1}$)	3.68 ± 0.55	2.32 ± 0.71	2.24 ± 0.44	2.90 ± 0.81

Results represent mean ± S.D of three experiments. DB, dry biomass; ST: surface tension; BS: yield of biosurfactant.

With the increase in shaking from 100 to 180 rpm, BS production increased sharply with higher cell growth. However, increasing of shaking above 180 rpm caused heavy foaming and reduced the level of BS production. The bacterial strains showed higher growth and BS production at the optimum shaking speed of 180 rpm except for OBP4 which exhibited optimal growth and BS production at 200 rpm.

Temperature was reported to influence the production of BSs (Salihu et al., 2009; Leon and Kumar, 2005). Temperature exhibited a noticeable influence on the production of RLs, possibly due to its effect on the physiology of the bacterial strains. The optimum temperature (Table 9.13) obtained in the present investigation was in the range of 37–40°C. Increase of temperature beyond the range caused a drastic reduction in the production of BS. The optimum activity of RL production by *P. aeruginosa* J4 between 30 and 37°C which decreased with further increase in the temperature. Temperatures between 28 and 40°C were reported in producing RL by various strains of *P. aeruginosa* (Henkel et al., 2012; Perfumo et al., 2006; Leon and Kumar, 2005). Such variations in temperature clearly indicate the physiological variations among *P. aeruginosa* strains. The temperature range between 32 and 34°C resulted in higher RL production by *P. aeruginosa* DSM 2659 strain. The maximum RL production by *P. aeruginosa* cultured in mannitol (20 $g.l^{-1}$) supplemented medium at 34.5°C and further increase in the temperature above 36C caused a significant reduction in the production of BS. The effect of two different temperatures, 28 and 37°C on the production of BS and observed no variations between both temperatures regarding the surface tension values and BS concentration.

The pH of culture medium exhibited a clear influence on RL production (Leahy and Colwell, 1990) which might be due to its effect on the cellular metabolism. The inoculated culture medium was found to be less turbid below pH 6.0, indicating less growth in the MSM and was further confirmed by the reduction in the bacterial biomass. However, the culture medium became turbid because of dense bacterial growth with the increase in pH of the medium from 6.5–7.2. An increase in the pH of the medium above 7.2 enhanced the bacterial growth but caused considerable reduction in the surface activity and the level of BS production. Similarly, a decrease in the pH below 6.8 reduced the bacterial growth and affected BS production (Table 9.14). The optimal pH (6.8) obtained in the present investigation agrees with the pH values for RL production by *P. aeruginosa* strains (Wei et al., 2005; Chayabutra and Ju, 2000). The maximum production of BS at a neutral pH of 6.8 and further decreased as the pH of the medium moved towards alkalinity and the same has supported the present findings. Moreover, pH is known to have a profound impact on the behavior of surface-active molecules (Silva et al., 2010). At lower pH (< 6.0), the RL moiety, which at pH5.6, remained at least 50% unchanged, but when the pH increased above 6.8, the RL moiety became negatively charged and surface activity reached its maximum.

Oxygen is necessary for microbial metabolism. Among the facultative microorganisms, *P. aeruginosa* are reported to be growing in the environments having low oxygen concentration. However, the production of surface-active compounds by *P. aeruginosa* involves the stages of oxidation of the substrate (Maier and Soberon-Chavez, 2000). Thus, agitation plays an important role on the production of BS by the bacterial strains promoting phase mixture and/at adequate oxygen transfer rate. The agitation velocity of the culture medium was a determining factor in mixing both aqueous and hydrophobic phases as well as the mass transfer of oxygen into the bacterial cultures. The secondary function of agitation is to keep the microorganisms in suspension. Moreover, agitation is a vital factor for the bacteria, especially when the carbon sources are complex hydrocarbons.

The utilization of n-alkanes in the shake-flask cultures are related with the availability of dissolved oxygen. The utilization or degradation of hydrocarbons was an oxygen-intensive metabolic process. When the fermentation process was carried out at 180 rpm for 15 days, the bacterial strains produced the maximum BS. The production of RL by the bacterial strains increased when the agitation rate was increased from 100 to 200 rpm (Table 9.15). However, the maximum production was achieved between 180 and 200 rpm. Agitation speed above 200 rpm was reported to be unfavorable for the bacterial growth due to the sheer damage even though it provided enough dissolved oxygen. On the other hand, agitation at lower speed of 100–150 rpm caused stagnant regions in the fermentor due to improper mixing (Subasioglu and Cansunar, 2008). Moreover, the required agitation rate that favors the optimum production of BS varies from strain to strain, suggesting differences in the oxygen consumption capacity for the metabolic processes. The increased agitation velocity caused a negative effect on the reduction in the surface tension by the produced BSs during the cultivation of *Serratia sp.* SVGG16 on ethanol-blended gasoline and the optimum result was obtained with the lowest surface tension value of 34 mNm^{-1} at 100 pm. The agitation affecting the mass transfer efficiency of oxygen, components of the culture medium, and was crucial for the cell growth and BS formation by the aerobic bacterium *P. aeruginosa.*

9.14 EFFECT OF LOW-COST RENEWABLE CARBON SUBSTRATES IN BIOSURFACTANT (BS) PRODUCTION

Different carbon substrates such as agro-industrial wastes, non-edible vegetable oils, kitchen wastes, and petroleum refinery wastes were screened for their role in the production of BS. Agro-industrial wastes containing de-oiled mustard seed cakes, waste raw glycerol of biodiesel, waste residual molasses and sugarcane bagasse. Jatropha seed oil (*Jatropha curcas*), Nahor seed oil (*Mesua ferrea*), Castor seed oil (*Ricinus communis*), Pongamia seed oil (*Pongamia glabra*) and Sesame seed oil (*Sesamum indicum*) were selected as non-edible vegetable oils for the experiment. Waste residual kitchen oil and oily sludge produced by the petroleum refinery were also used for the purpose. With the completion of the fermentation process, parameters like reduction in surface tension, yield of BS and dry biomass were determined for each carbon substrate. Data obtained from the experiment are presented in Table 9.16.

The bacterial strains were able to utilize vegetable oils. The inedible vegetable oils such as sesame seed oil followed by Nahor seed oil proved to be promising to produce BSs. Among the other tested substrates waste glycerol, followed by petroleum refinery sludge and waste residual kitchen oil were found to be suitable for the production of BS. However, in terms of bacterial biomass and BS production, the strain OBP1 was found to be the best. Hence, for further studies, the strain OBP1 was selected for screening the different inedible vegetable oils as carbon substrate to produce BS.

TABLE 9.16 Influence of Various in-Edible Vegetable Oil as Carbon Source on Biosurfactant Production of *P. aeruginosa* OBP1

Vegetable Oil	DB ($g.l^{-1}$)	ST (mNm^{-1})	BS ($g.l^{-1}$)
Jatropha curcas	1.88 ± 0.6	40.6 ± 1.2	1.02 ± 0.9
Mesua ferrea	3.84 ± 0.3	36.4 ± 0.4	2.34 ± 0.3
Ricinus communis	3.47 ± 0.8	38.3 ± 0.8	2.01 ± 0.6
Sesamum indicum	4.42 ± 0.5	37.1 ± 0.2	2.57 ± 0.7
Pongamia glabra	2.56 ± 0.3	39.8 ± 0.8	1.76 ± 0.5

Results represent mean ± S.D of three experiments. DB, dry biomass; ST: surface tension; BS: yield of biosurfactant.

In the present investigation, agro-industrial wastes like residual glycerol and residual kitchen oil were found to be promising to produce RL as compared to other carbon substrates. It was estimated that 10 kg of glycerol wastes are produced for every 100 kg of biodiesel (Yateem et al., 2002). Moreover, due to the presence of impurities, further purification of the biodiesel-derived glycerol for its industrial applications was reported to be unprofitable. Non-edible vegetable oils such as sesame seed oil and Nahor seed oil were found to be efficient. The variation in RL production in the different vegetable oils might be associated with the composition of saturated and unsaturated fatty acids as well as the number of carbon atoms of the oils. Vegetable oils have been reported to be more efficient in RL production by *P. aeruginosa* as compared to glucose, glycerol, and hydrocarbons (Maier and Soberon-Chavez, 2000; Mata-Sandoval et al., 2001). They include palm seed oil, olive oil, sunflower oil, safflower oil, canola oil, soybean oil and corn oil (Benincasa, 2009). However, based on cost and global food supply, the use of food-grade oils for producing RL is not economically significant. Besides, the recent food storage crisis, limited land availability for crop cultivation and food industry with increasing food demand have persuaded the price of edible plant-based oils to increase (Ishigami et al., 1987). In this perspective, non-edible vegetable oils may be an alternate substrate for RL production. Therefore, industrial wastes such as waste glycerol, petroleum refinery sludge, kitchen waste oil and non-edible oils such as Nahor (Indian Iron Wood) and sesame seed oil could be good carbon sources to produce RLs.

The vegetable oils such as sesame seed oil followed by Nahor seed oil proved to be promising in the production of BSs, and Table 9.16 shows the same. Among the other tested substrates, waste glycerol followed by petroleum refinery sludge and waste residual kitchen oil were found to be suitable to produce BS and the data thus obtained are presented in Table 9.17.

TABLE 9.17 Influence of Various Low-Cost Carbon Substrates on Biosurfactant Production of *P. aeruginosa* Strains

Carbon Sources	Properties	*P. aeruginosa* Strains			
		OBP1	OBP2	OBP3	OBP4
Waste glycerol of biodiesel	DB ($g.l^{-1}$)	3.28 ± 0.76	2.75 ± 0.51	2.63 ± 0.28	3.40 ± 0.62
	ST (mNm^{-1})	37.6 ± 0.40	33.7 ± 0.80	39.5 ± 0.63	37.0 ± 0.38
	BS ($g.l^{-1}$)	1.85 ± 0.56	3.90 ± 0.42	1.52 ± 0.51	2.24 ± 0.18
De-oiled mustard seed cakes	DB ($g.l^{-1}$)	2.54 ± 0.80	1.16 ± 0.37	1.03 ± 0.90	0.95 ± 0.74
	ST (mNm^{-1})	39.8 ± 0.32	44.5 ± 0.59	49.5 ± 0.72	49.3 ± 0.58
	BS ($g.l^{-1}$)	1.48 ± 0.64	0.47 ± 0.83	0.27 ± 0.25	0.28 ± 0.95
Waste residual molasses	DB ($g.l^{-1}$)	5.07 ± 0.55	5.02 ± 0.83	4.97 ± 0.56	5.04 ± 0.51
	ST (mNm^{-1})	50.4 ± 0.37	50.6 ± 0.59	51.2 ± 0.75	51.7 ± 0.43
	BS ($g.l^{-1}$)	0.14 ± 0.45	0.10 ± 0.35	0.18 ± 0.62	0.16 ± 0.59
Sugarcane bagasse	DB ($g.l^{-1}$)	2.46 ± 0.57	2.52 ± 0.70	2.48 ± 0.45	2.41 ± 0.52
	ST (mNm^{-1})	51.8 ± 0.38	52.4 ± 0.47	52.7 ± 0.80	51.5 ± 0.35
	BS ($g.l^{-1}$)	0.13 ± 0.40	0.12 ± 0.59	0.07 ± 0.19	0.18 ± 0.22
Waste residual kitchen oil	DB ($g.l^{-1}$)	2.98 ± 0.55	1.48 ± 0.38	1.28 ± 0.50	1.07 ± 0.41
	ST (mNm^{-1})	37.3 ± 0.36	40.3 ± 0.20	45.6 ± 0.71	42.7 ± 0.35
	BS ($g.l^{-1}$)	2.26 ± 0.70	0.91 ± 0.56	0.44 ± 0.43	0.68 ± 0.70
Petroleum refinery sludge	DB ($g.l^{-1}$)	3.10 ± 0.25	3.27 ± 0.81	4.02 ± 0.60	3.22 ± 0.95
	ST (mNm^{-1})	37.0 ± 0.32	39.8 ± 0.70	37.6 ± 0.83	36.5 ± 0.52
	BS ($g.l^{-1}$)	1.96 ± 0.60	1.03 ± 0.46	1.85 ± 0.79	2.33 ± 0.23

Results represent mean ± S.D of three experiments. DB, dry biomass; ST: surface tension; BS: yield of biosurfactant.

KEYWORDS

- **ammonium chloride**
- **colony-forming unit**
- **disodium hydrogen phosphate**
- **mineral salt medium**
- **polyaromatic hydrocarbons**
- **potassium dihydrogen phosphate**
- **potassium nitrate**

REFERENCES

Adebusoye, S. A., et al., (2007). Microbial degradation of petroleum in a polluted tropical stream. *World J. Microbiol. Biotechnol., 23*, 1149–1159.

Alexander, M., (2000). Aging, bioavailability, and over stimulation of risk from environmental pollutants. *Environ. Sci. Technol., 34*, 4259–4265.

Batista, S. B., et al., (2006). Isolation and characterization of biosurfactant/bioemulsifier producing bacteria from petroleum contaminated sites. *Bioresour. Technol., 97*, 868–875.

Benincasa, M., & Accorsini, F. R., (2008). *Pseudomonas aeruginosa* LBI production as an integrated process using the wastes from sunflower-oil refining as a substrate. *Bioresour. Technol., 99*, 3843–3849.

Bora, T. C., et al., (2011). Bioprospecting microbial diversity from northeast gene pool. *Sci. Cult., 77*, 446–450.

Bordoloi, N. K., & Konwar, B. K., (2009). Bacterial biosurfactant in enhancing solubility and metabolism of petroleum hydrocarbons. *J. Hazardous Materials, 170*(1), 495–505.

Chaerun, S. K., et al., (2004). Interaction between clay minerals and hydrocarbon utilizing indigenous microorganisms in high concentrations of heavy oil: Implications for bioremediation. *Environ. Inter., 30*, 911–922.

Chaillan, F., et al., (2004). Identification and biodegradation potential of tropical aerobic hydrocarbon-degrading microorganisms. *Res. Microbiol., 155*(7), 587–595.

Chayabutra, C., & Ju, L. K., (2000). Degradation of n-hexadecane and its metabolites by *Pseudomonas aeruginosa* under microaerobic and anaerobic conditions. *Appl. Environ. Microbiol., 66*, 493–498.

Clarke, K., et al., (2010). Enhanced rhamnolipid production by *Pseudomonas aeruginosa* under phosphate limitation. *World J. Microbiol. Biotechnol., 26*, 2179–2184.

Das, K., & Mukherjee, A. K., (2007). Crude petroleum-oil biodegradation efficiency of *Bacillus subtilis* and *Pseudomonas aeruginosa* strains isolated from a petroleum oil-contaminated soil from north-east India. *Bioresour. Technol., 98*, 1339–1345.

Desai, J. D., & Banat, I. M., (1997). Microbial production of surfactants and their commercial potential. *Microbiol. Mol. Bio. Rev., 61*, 47–64.

Desai, J. D., & Desai, A. J., (1993). Production of biosurfactants. In: Kosaric, N., (eds.), *Biosurfactants, Production, Properties, Applications* (pp. 65–97). Marcel Dekker, New York.

Deziel, E., et al., (1999). Liquid chromatography/mass spectrometry analysis of mixtures of rhamnolipids produced by *Pseudomonas aeruginosa* strain 57RP grown on mannitol or naphthalene. *Biochim. Biophys. Acta, 1440*, 244–252.

Haba, E., et al., (2000). Screening and production of rhamnolipids by *Pseudomonas aeruginosa* 47T2 NCIB 40044 from waste frying oils. *J. Appl. Microbiol., 88*, 379–387.

Henkel, M., et al., (2012). Rhamnolipids as biosurfactants from renewable resources: Concepts for next-generation rhamnolipid production. *Process Biochem., 47*, 1207–1219.

Ishigami, Y., et al., (1987). Surface-active properties of succinoyl trehalose lipids as microbial biosurfactants. *J. JPN Oil Chem. Soc., 36*, 847–85.

Iwabuchi, N., et al., (2002). Extracellular polysaccharides of *Rhodococcus rhodochrous* S-2 stimulate the degradation of aromatic components in crude oil by indigenous marine bacteria. *Appl. Environ. Microbiol., 68*, 2337–2343.

Janbandhu, A., & Fulekar, M. H. J., (2011). Biodegradation of phenanthrene using adapted microbial consortium isolated from petrochemical contaminated environment. *Hazard. Mater., 187*, 333–340.

Karanth, N. G. K., et al., (1999). Microbial production of biosurfactants and their importance. *Curr. Sci., 77*, 116–126.

Kullen, M. J., et al., (2001). Use of the DNA sequence of variable regions of the 16S rRNA gene for rapid and accurate identification of bacteria in the *Lactobacillus acidophilus* complex. *J. Appl. Microbiol., 89*, 511–516.

Leahy, J. G., & Colwell, R. R., (1990). Microbial degradation of hydrocarbons in the environment. *Microbiol. Rev., 54*, 305–315.

Leon, V., & Kumar, M., (2005). Biological upgrading of heavy crude oil. *Biotechnol. Bioprocess Eng., 10*, 471–481.

Maier, R. M., & Soberon-Chavez, G., (2000). *Pseudomonas aeruginosa* rhamnolipids: biosynthesis and potential applications. *Appl. Microbiol. Biotechnol., 54*, 625–633.

Manresa, A., et al., (1991). Kinetic studies on surfactant production by *Pseudomonas aeruginosa* 44Tl. *J. Ind. Microbiol., 8*, 133–136.

Mata-Sandoval, J. C., et al., (1999). High-performance liquid chromatography method for the characterization of rhamnolipid mixtures produced by *Pseudomonas aeruginosa* UG2 on com oil. *Chromatogr., 864*, 211–220.

Mata-Sandoval, J. C., et al., (2001). Influence of rhamnolipids and Triton X-100 on the biodegradation of three pesticides in aqueous and soil slurries. *J. Agric. Food Chem., 49*, 3296–303.

Mehdi, H., & Giti, E., (2008). Investigation of alkane biodegradation using the microtiter plate method and correlation between biofilm formation, biosurfactant production and crude oil degradation. *Inter. Biodeter. Biodegr., 62* 170–178.

Muller, M. M., et al., (2012). Rhamnolipids-next generation surfactants. *J. Biotechnol., 162*, 366–380.

Naik, R. P., & Sakthivel, N., (2006). Functional characterization of a novel hydrocarbon clastic *Pseudomonas* sp. strain PUP6 with plant-growth-promoting traits and antifungal potential. *Res. Microbiol., 157*, 538–546.
Patel, R. M., & Desai, A. J., (1997). Biosurfactant production by *Pseudomonas aeruginosa* GS3 from molasses. *Lett. Appl. Microbiol., 25*, 91–94.
Perfumo, A., et al., (2006). Rhamnolipid production by a novel thermotolerant hydrocarbon-degrading *Pseudomonas aeruginosa* AP02-1. *J. Appl. Microbiol., 75*, 132–138.
Plaza, G. A., et al., (2006). Use of different methods for detection of thermophilic biosurfactant producing bacteria from hydrocarbon-contaminated and bioremediated soils. *J. Petro. Sci. Eng., 50*, 71–77.
Rojas-Avelizapa, N. G., et al., (1999). Isolation and characterization of bacteria degrading polychlorinated biphenyls from trans-former oil. *Folia Microbiol., 44*, 317–322.
Saikia, J. P., et al., (2013). Possible protection of silver nanoparticles against salt by using rhamnolipid. *Colloids Surf. B. Biointerfaces, 104*, 330–332.
Saikia, R. R., et al., (2012). Isolation of biosurfactant-producing *Pseudomonas aeruginosa* RS29 from oil-contaminated soil and evaluation of different nitrogen sources in biosurfactant production. *Ann. Microbiol., 62*, 753–763.
Salihu, A., et al., (2009). An investigation for potential development on biosurfactants. *Biotechnol. Mol. Biol. Rev., 3*, 111–117.
Silva, S. N. R. L., et al., (2010). Glycerol as substrate for the production of biosurfactant by *Pseudomonas aeruginosa* UCP 0992. *Colloids Surf. B. Biointerfaces, 79*, 174–183.
Subasioglu, T., & Cansunar, E., (2008). Nutritional factors effecting rhamnolipid production by a nosocomial *Pseudomonas aeruginosa*. *J. Biol. Chern., 36*, 77–81.
Syldatk, C., & Wagner, F., (1987). Production of biosurfactants. In: Kosaric, N., et al., (eds.), *Biosurfactants and Biotechnology* (pp. 89–120). Marcel Dekker, New York.
Tuleva, B. K., et al., (2002). Biosurfactant production by a new *Pseudomonas putida* strain. *Z. Naturforsch, 57C*, 356–360.
Vasileva-Tonkova, E., & Galabova, D., (2003). Hydrolytic enzymes and surfactants of bacterial isolates from lubricant-contaminated wastewater. *Z. Naturforsch. C., 58*, 87–92.
Vasileva-Tonkova, E., et al., (2006). Production and properties of biosurfactants from a newly isolated *Pseudomonas* fluorescens HW-6 growing on hexadecane. *Z. Naturforsch. C., 61*, 553–559.
Wei, Y. R., et al., (2005). Rhamnolipid production by indigenous *Pseudomonas aeruginosa* J4 originating from petrochemical wastewater. *Biochem. Eng., 127*, 146–154.
Whiteley, A. S., & Bailey, M. J., (2000). Bacterial community structure and physiological state within an industrial phenol bioremediation system. *Appl. Environ. Microbial., 66*, 2400–2407.
Wonga, P., et al., (2004). Isolation and characterization of novel strains of *Pseudomonas aeruginosa* and Serratia marcescens possessing high efficiency to degrade gasoline, kerosene, diesel oil, and lubricating oil. *Curro. Microbiol., 49*, 415–422.
Wu, J. Y., et al., (2008). Rhamnolipid, production with indigenous *Pseudomonas aeruginosa* EMI isolated from oil-contaminated site. *Bioresour. Technol., 99*, 1157–1164.
Yalcin, E., & Ergene, A., (2010). Preliminary characterization of biosurfactants produced by microorganisms isolated from refinery wastewaters. *Environ. Technol., 31*, 225–232.
Yateem, A., et al., (2002). Isolation and characterization of biosurfactant-producing bacteria from oil-contaminated soil. *Soil Sediment Contamin., 11*, 41–55.

FURTHER READING

Bharali, P., & Konwar, B. K., (2011). Production and physicochemical characterization of a biosurfactant produced by *Pseudomonas aeruginosa* OBP1 isolated from petroleum sludge. *Appl. Biochem. Biotechnol., 164*, 1444–1460.

Bharali, P., Das, S., Konwar, B. K., & Thakur, A. J., (2011). Crude biosurfactant from thermophilic *Alcaligenes faecalis*: Feasibility in Petro-spill bioremediation. *Inter. Biodeter. Biodegr., 65*, 682–690.

Das, S., Kalita, S. J., Bharali, P., Konwar, B. K., Das, B., & Thakur, A. J., (2013). Organic reactions in "green surfactant": An avenue to bisuracil derivative. *ACS Sustainable Chem. Eng.,* (p. 301). doi: 10.1021/sc4002774, Publication Date (Web).

Pranjal, B., (2015). *Bioremediation of Crude Oil Contaminated Soil.* (Thesis: Supervisor B K Konwar), Dept. of Mol. Biol. and Biotechnology, Tezpur University (Central), Napaam – 784028, Assam, India.

CHAPTER 10

Physical Characterization of Biosurfactant

10.1 REDUCTION IN SURFACE, INTERFACIAL TENSION (IFT), AND CMC

An aliquot of 5.0 ml culture broth was collected at a regular interval of 12 h for a period of 168 h to determine the reduction in surface tension. The culture broth of each bacterial strain was centrifuged at 8,000 rpm for 15 min at 4°C. The surface tension of the cell-free culture supernatant was measured by using a digitalized tensiometer (Krűss Tensiometer K9 ET/25) at 25 ± 1°C. Before using the platinum ring, it was thoroughly washed three times with distilled water followed by acetone and then allowed to dry at room temperature (RT) (Bodour and Miller-Maier, 1998). The interfacial tension (IFT) was measured against diesel. Each experiment was repeated thrice for taking the average value. The biosurfactants (BSs) produced by *P. aeruginosa* strains were able to reduce the surface tension of the culture medium. Surface tension of the culture medium as acted by the inoculated bacterial strains was drastically reduced from 68.5 to about 31.1 mNm^{-1}. Reductions in the IFT of diesel containing culture supernatant of four bacterial strains as compared to the control culture medium without any bacteria are in Table 10.1.

TABLE 10.1 Properties of Biosurfactant Produced by *P. aeruginosa* Strains

Bacterial Strains	BS ($g.l^{-1}$)	ST (mNm^{-1})	IFT[a] (mNm^{-1})	CMC ($mg.l^{-1}$)
P. aeruginosa OBP1	4.57 ± 0.65	31.1 ± 0.88	1.5 ± 0.65	45 ± 0.86
OBP2	2.86 ± 0.79	37.6 ± 0.51	3.1 ± 0.48	105 ± 0.34
OBP3	2.83 ± 0.63	35.5 ± 0.38	2.8 ± 0.93	90 ± 0.58
OBP4	3.17 ± 0.37	33.2 ± 0.79	2.2 ± 0.28	65 ± 0.94

Results represent mean ± S.D of three experiments.

Abbreviations: BS: Yield of biosurfactant; ST: surface tension; IFT[a]: interfacial tension in diesel; CMC: critical micelle concentration.

Reductions in the IFT of diesel containing culture supernatant of four bacterial strains as compared to the control culture medium without any bacteria are presented in Table 10.1. The minimum IFT of the culture supernatant containing the BS of OBP1, OBP2, OBP3, and OBP4 was 1.5, 3.4, 2.8 and 2.2 mNm^{-1}, respectively.

The BS produced by the bacterial strains was able to reduce the surface tension of the culture medium significantly from 68.5 to 31.1 mNm^{-1}. However, the behavior of BS production by the bacterial strains was quite different. The effectiveness is measured by the minimum value to which the surface tension could be reduced, whereas efficiency is measured by the BS concentration required to produce a considerable reduction in the surface tension of water. The CMC values of the BSs produced by the bacterial strains were in the range of 45–105 $mg.l^{-1}$. Based on CMC, the BS of bacterial strain OBP1 could be considered as both efficient and effective, while BS produced by OBP3 and OBP4 could be adjudged effective only. Efficient surfactants are known to have very low CMC values, i.e., less surfactant is required to decrease surface tension (George and Jayachandran, 2008) and exhibit some of the physical functions such as emulsification, solubilization, and foaming even at a relatively low concentration (Lee et al., 2004). Previously, a range of CMC values between 10 and 230 $mg.l^{-1}$ were reported for the rhamnolipids (RL) isolated from the different microbial sources (Van Hamme et al., 2006). The CMC values obtained in the present investigation differed within the strains as well as the other reported strains of *P. aeruginosa*. Such variation in the CMC values might be due to the differences in purity and composition of RLs (Monterio et al., 2007). Further, due to the intrinsic variability of the RLs accumulated and the complexity of its composition, number, and proportions of homologs, presence of unsaturated bonds, branching, and length of the aliphatic chain of the RL collectively affect the CMC and surface tension values between the RLs produced (Silva et al., 2010). The RL homologs could also differ with the bacterium, medium, and cultivation conditions (Wu et al., 2008). The CMC values obtained in the present study were much lower than the chemical surfactants such as sodium dodecyl sulfate (SDS) having a CMC value of 2,100 $mg.l^{-1}$ (Monterio et al., 2000).

10.2 CMC AND CMD

Critical micelle concentration (CMC) of the isolated BS samples was determined by measuring the surface tension of the aqueous BS solutions at different dilution concentrations up to the constant value of surface tension. Further, the surface tension of aqueous BS samples was determined at their critical micelle dilutions (CMDs) of 10 and 100 times, i.e., CMD^{-1} and CMD^{-2}, respectively. For the calibration of the instrument, the surface tension of the pure water was measured before each set of experiment. The platinum ring was washed thrice with distilled water followed by acetone and then allowed to dry at RT. Each measurement was repeated thrice, and the average value was taken (Bodour and Miller-Maier, 1998; Vasileva-Tonkova and Gesheva, 2007).

The CMC values were determined by diluting the isolated BS solution of the bacterial strains to several times in sterile distilled water. The values were found to be in the range of 45–105 $mg.l^{-1}$ and are presented in Table 10.2 and Figure 10.1. The surface tension values at the CMC for the BS of OBP1, OBP2, OBP3, and OBP4 were in the order of 31.1, 37.6, 35.5 and 33.2 mNm^{-1}, respectively. After the attainment of the CMC, no further reduction in the surface tension was observed.

It was reported that concentration of the surfactant below their CMC level reduces the surface and IFT between air/water, oil/water and soil/water systems. Such reduction in the IFT between crude oil and the soil particles, the capillary force that holds them together gets reduced, resulting in their separation. IFT is considered as an important factor in oil recovery because capillary number increases with decrease in the IFT. Capillary number is determined by the ratio of viscous force to the capillary force. The IFT against diesel decreased from 29 to 2.3 mNm^{-1} by the bacterial strains. Similar IFT values of 1.0 mNm^{-1} against kerosene was found in the case of *P. aeruginosa* 47T2 and 44T1 strain (Bailey and Ollis, 1986); 1.3 mNm^{-1} in the case of *P. aeruginosa* LBI strain (Choi et al., 2009); 2.0 mNm^{-1} against n-hexadecane in the case of *P. aeruginosa* UCP0992 strain (Silva et al., 2010) 1.85 mNm^{-1} against petroleum crude oil in the case of *P. aeruginosa* strain (Xia et al., 2011).

CMD is an indirect means of measuring the surfactant production related to the range of the CMC (Reddy et al., 1958). CMD is a very crucial factor for the oil recovery process. During the application of BSs in MEOR technology, there is a higher chance of dilution of the introduced surfactant into the oil well due to the leakage of water from the water tables present nearby the oil reservoir. Therefore, it is very much important to determine the surface activity of the surfactant at different dilutions levels such as CMD^{-1} (1:10) and CMD^{-2} (1:100). Reduction in the surface tension at CMD^{-1} was almost like that of normal, whereas the CMD^{-2} caused a slight increase in the surface tension of the system due to the higher dilution.

10.3 EFFECT OF TEMPERATURE, PH, SALINITY, AND METAL ION ON CMC AND CMD OF BIOSURFACTANT (BS)

The CMC values were determined by diluting the isolated BS solution of the bacterial strains to several times in sterile distilled water. The values were found to be in the range of 45–105 $mg.l^{-1}$. The surface tension values at the CMC for the BS of OBP1, OBP2, OBP3, and OBP4 were in the order of 31.1, 37.6, 35.5 and 33.2 mNm^{-1}, respectively. After the attainment of the CMC, no further reduction in the surface tension was observed. The cell-free culture supernatant containing the BS showed almost stable surface activity over a wide range of pH values. The effects of pH on BSs were determined at normal concentration, CMD^{-1} (10 times dilution) and CMD^{-2} (100 times dilution) (Tables 10.3–10.5).

BSs at normal and CMD^{-1} concentrations showed no significant difference in their surface activity at all the tested pH levels. However, the concentration at CMD^{-2} exhibited a reduction in the surface activity due to the lower concentration, which leads to an increase in surface tension. The activity of the BSs produced by the bacterial strains was found to be optimum between the pH of 5–8. Extreme pH below 5 and above 8 caused increased surface tension. The cell-free culture supernatant of the bacterial strains retained the surface activity even after 60 min of incubation at temperatures ranging from 4–100°C. The cell-free culture supernatants remained effective even after autoclaving at 121°C for 30 min. Concentrations of the supernatant at CMD^{-1} and CMD^{-2} on exposure to temperatures 4–100°C for 60 min exhibited almost stable surface tension.

At CMD^{-2}, the cell-free culture supernatant exhibited comparatively lesser surface activity than CMD^{-1} due to the lowering of surfactant concentration. The cell-free culture supernatants at CMD^{-1} and CMD^{-2} remained effective similar to that of normal concentration even after autoclaving at 121°C for 30 min.

The BS retained its surface activity by reducing the surface tension up to a concentration of 4% NaCl and the effect of salinity on BSs was determined at normal, CMD^{-1} and CMD^{-2} concentrations. The reduction in surface tension of the cell-free culture supernatant was almost like that of CMD^{-1}, referring to the intact efficiency of BS at CMD^{-1}. At higher dilution of CMD^{-2}, the cell-free culture supernatant exhibited significant activity indicating their efficiency even at lower concentrations. However, the level of surface activity was comparatively lower than that of normal and CMD^{-1}. The cell-free culture supernatant of the bacterial strains was treated with different metal ions like K^+, Ca^{2+}, Mg^{2+}, Fe^{2+} and Al^{3+} (2% w/v). The culture supernatant at CMD^{-1} and CMD^{-2} on exposure to the different metal ions exhibited almost stable surface tensions.

The cell-free culture supernatant containing the BS showed almost stable surface activity over a wide range of pH values. The effects of pH on BSs were determined at normal concentration, CMD^{-1} and CMD^{-2}. Extreme pH below 5 and above 8 caused increased surface tension.

The cell-free culture supernatant of the bacterial strains retained the surface activity even after 60 min of incubation at temperatures ranging from 4–100°C. The cell-free culture supernatants remained effective even after autoclaving at 121°C for 30 min. Concentrations of the supernatant at CMD^{-1} and CMD^{-2} on exposure to temperatures 4–100°C for 60 min exhibited almost stable surface tension.

TABLE 10.2 Influence of pH on the Surface Activity of Biosurfactant Produced by *P. aeruginosa* Strains in 2% n-Hexadecane Supplemented Medium at Normal and Critical Micelle Dilutions (CMD^{-1} and CMD^{-2})

Bacterial Strains	pH	Surface Tension (mN/m)		
		Cell-Free Culture Supernatant	CMD^{-1}	CMD^{-2}
P. aeruginosa OBP1	2	34.8 ± 1.00	35.6 ± 0.32	48.2 ± 0.46
	5	31.8 ± 0.42	32.0 ± 0.56	42.5 ± 0.49
	7	31.1 ± 0.37	31.8 ± 0.47	41.9 ± 0.55
	8	31.2 ± 0.22	31.9 ± 0.28	42.8 ± 0.42
	11	32.6 ± 0.28	33.8 ± 0.14	45.6 ± 0.65
P. aeruginosa OBP2	2	41.5 ± 0.58	48.9 ± 0.78	63.7 ± 0.31
	5	37.8 ± 0.35	41.0 ± 0.23	58.3 ± 0.46
	7	37.6 ± 0.11	40.7 ± 0.83	57.6 ± 0.29
	8	38.1 ± 0.34	41.5 ± 0.69	58.8 ± 0.36
	11	39.7 ± 0.53	43.2 ± 0.45	61.0 ± 0.53
P. aeruginosa OBP3	2	40.4 ± 0.36	46.3 ± 0.93	63.5 ± 0.29
	5	35.7 ± 0.31	38.7 ± 0.67	56.8 ± 0.49
	7	35.4 ± 0.19	38.2 ± 0.30	55.6 ± 1.00
	8	36.5 ± 0.83	39.5 ± 0.47	57.4 ± 0.19
	11	38.7 ± 0.62	42.5 ± 0.61	60.8 ± 0.38
P. aeruginosa OBP4	2	38.5 ± 0.43	43.3 ± 0.60	57.1 ± 0.51
	5	33.7 ± 0.76	39.8 ± 0.71	50.3 ± 0.94
	7	33.1 ± 0.93	39.5 ± 0.35	48.7 ± 0.54
	8	33.6 ± 0.51	39.7 ± 0.29	51.4 ± 0.66
	11	35.8 ± 0.23	41.4 ± 0.72	54.9 ± 0.39

The data shown in the table are mean values of triplicates.

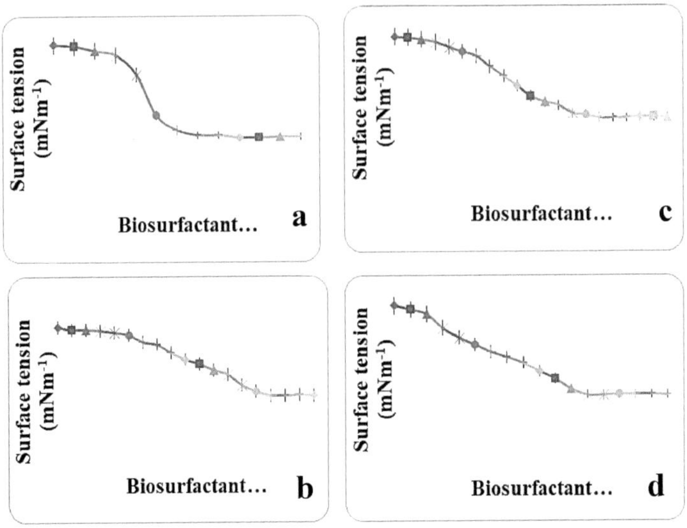

FIGURE 10.1 Determination of critical micelle concentration (CMC) of the biosurfactants produced by *P. aeruginosa* strains. Results represent the mean of three independent experiments ± standard deviation. (a) *P. aeruginosa* OBP1; (b) *P. aeruginosa* OBP2; (c) *P. aeruginosa* OBP3; and (d) *P. aeruginosa* OBP4.

TABLE 10.3 Influence of Temperature on the Surface Activity of Biosurfactant Produced by *P. aeruginosa* Strains in 2% n-Hexadecane Supplemented Medium at Normal and Critical Micelle Dilutions (CMD^{-1} and CMD^{-2})

Bacterial Strains	Exposure to 60 min at the Temperature (°C)	Surface Tension (mN/m)		
		Cell-Free Culture Supernatant	CMD–1	CMD–2
P. aeruginosa OBP1	4	32.3 ± 0.23	33.5 ± 0.39	42.7 ± 0.92
	25	31.2 ± 0.62	31.3 ± 0.19	41.9 ± 0.36
	37	31.1 ± 0.56	31.5 ± 0.17	41.6 ± 1.20
	50	31.5 ± 0.72	31.8 ± 0.39	41.9 ± 0.39
	75	31.8 ± 0.18	32.3 ± 0.40	42.4 ± 0.41
	100	32.0 ± 1.23	36.3 ± 0.94	47.2 ± 0.55
	121 (for 30 min)	32.3 ± 0.81	35.9 ± 0.40	47.7 ± 0.56
P. aeruginosa OBP2	4	38.3 ± 0.41	43.0 ± 0.25	58.1 ± 0.29
	25	37.5 ± 0.38	40.3 ± 0.48	57.6 ± 0.38
	37	37.6 ± 0.28	40.8 ± 0.39	57.5 ± 0.65
	50	37.8 ± 0.37	41.1 ± 0.52	57.8 ± 0.75
	75	38.0 ± 0.51	41.6 ± 0.41	58.2 ± 0.28
	100	39.4 ± 0.20	43.9 ± 0.40	59.3 ± 0.47
	121 (for 30 min)	39.2 ± 0.51	44.2 ± 0.35	59.1 ± 0.76

TABLE 10.3 *(Continued)*

Bacterial Strains	Exposure to 60 min at the Temperature (°C)	Surface Tension (mN/m)		
		Cell-Free Culture Supernatant	CMD–1	CMD–2
P. aeruginosa OBP3	4	36.8 ± 0.39	39.6 ± 0.77	56.8 ± 0.87
	25	35.5 ± 0.38	38.4 ± 0.93	55.3 ± 0.54
	37	35.3 ± 0.40	38.3 ± 0.38	55.5 ± 0.39
	50	35.3 ± 0.19	38.7 ± 0.84	55.9 ± 0.82
	75	35.9 ± 0.36	39.4 ± 0.62	56.4 ± 0.39
	100	38.3 ± 0.72	41.5 ± 0.65	59.2 ± 0.35
	121 (for 30 min)	38.2 ± 0.10	41.8 ± 0.94	59.6 ± 0.62
P. aeruginosa OBP4	4	34.7 ± 0.39	40.6 ± 0.54	49.5 ± 0.82
	25	33.2 ± 0.91	39.8 ± 0.77	48.7 ± 0.22
	37	33.2 ± 0.32	39.7 ± 0.83	48.8 ± 0.91
	50	33.5 ± 0.17	39.7 ± 0.39	49.2 ± 0.30
	75	33.7 ± 0.63	40.2 ± 0.98	49.7 ± 0.45
	100	34.0 ± 0.72	43.8 ± 0.74	51.4 ± 0.76
	121 (for 30 min)	34.6 ± 0.83	44.1 ± 0.59	51.2 ± 0.98

The data shown here are mean values of triplicates.

Environmental factors such as pH, temperature, salinity, and metal ions are known to influence BS's activity and stability (Aparna et al., 2012; Sarubbo et al., 2007; Xia et al., 2011). BSs produced by the bacterial strains at normal CMD^{-1} and CMD^{-2} concentrations were found to stable and showed optimum surface activity at pH 5–8. The emulsification activity (E_{24}) of the cell-free culture supernatant of the bacterial strains against diesel was quite stable at pH 5–8, though the optimum emulsifying activity was observed between pH 7–8. Considerable reduction in the stability as well as in the surface activity of BSs was observed beyond pH 8. Such a decrease in the surface activity might be due to the alteration of surfactant structures at the extreme pH conditions. An increase in pH from 5–8 caused an increase in the negative charge on the polar head of the RL molecule (pKa 5.6) enhancing its solubility in water.

However, below pH 5, surface activity decreased due to the protonation of the RL molecules affecting their precipitation. RL molecules contain a single free carboxylic acid group at the β-hydroxyl fatty acid moiety and are responsible for the anionic nature of RL (Silva et al., 2010). An increase in pH from 5.5–8.0 leads to an increase in the negative charge on the polar head of RL. A charge repulsion between the adjacent polar heads leads to the formation of larger head diameter and sequentially change the morphology of RL molecules from bilayer sheets to vesicles and then to micelles. BSs produced by the different strains of *P. aeruginosa* showed optimum surface activity with higher emulsification between the pH ranges of 4–12 (Aparna et al., 2012; Xia et al., 2011; Silva et al., 2010). BSs maintained their surface activity after exposure to temperatures from 4–100°C and showed appreciable thermostability at both normal and diluted conditions (CMD^{-1} and CMD^{-2}). The cell-free culture supernatants also exhibited stable surface activity even after autoclaving at 12°C for 30 min. Such thermal stability of BSs indicates its utility in those industries where wet sterilization is of principal importance. Extreme stability of BSs was reported by various authors for *P. aeruginosa* isolate Bs 20; also for the marine bacterium *Brevibacterium aureum* MSA13 and the bacterial strain of *Alcaligenes faecalis*. The present study clearly indicates the thermostability properties of the BSs produced by four bacterial strains. The emulsification activity (E_{24}) of the culture supernatant of the bacterial strains against diesel was quite stable at all the tested temperatures. Emulsions were found to be stable up to one month at RT. It is interesting to note that BSs retained about 53% of their original emulsifying properties even after autoclaving, indicating their thermal stability, which broadens the scope of their application in MEOR.

The BSs retained their surface activity up to the addition of 5% NaCl at both normal and CMD^{-1} conditions. The culture supernatants retained significant surface activity at CMD^{-2} indicating their efficiency even at lower concentration. Emulsification activity (E_{24}) of culture supernatants against diesel remained almost unchanged up to the addition of 3% NaCl. Bognolo reported that chemical surfactants usually get deactivated at around 2–3% of salt concentrations. An increase in the salt concentration beyond 4%, caused a significant reduction in the emulsification activity (E_{24}) of culture supernatants, indicating their ineffectiveness at such elevated salinity. When NaCl is added, the negative charge of carboxylic acid groups of RL molecules is shielded by the Na^+ ions in the electrical double layer that leads to the formation of a close-packed monolayer. Thus, the formation of Na^+-RL complex reduces surface tension values (Silva et al., 2010). The stability of culture supernatant against higher pH and salinity suggests its applicability in bioremediation of marine environments and in industries where high salinities and pH prevail.

BSs showed appreciable surface activity in the presence of the different metal ions, but not in the case of Al^{3+} ions. This could be due to the complex formation between the available RL with Al^{3+} ions present in the cell-free culture supernatant, as a result, free RL molecules in the solution gets reduced. Consequently, the RL molecules available at the interface are lesser, which in turn reduces its surface activity. RLs tends to form complexes with heavy metal cations more preferably than that of other non-toxic metal ions such as Ca^{2+}, Mg^{2+}, K^{+}, etc., for which they possess much lesser affinity. The addition of trivalent ions (Al^{3+}) cause an increase in the surface activity of the BSs produced by *P. aeruginosa*. Cations of lowest to highest affinity for RL BSs were $K^{+} < Mg^{2+} < Mn^{2+} < Ni^{2+} < CO^{2+} < Ca^{2+} < Hg^{2+} < Fe^{2+} < Zn^{2+} < Cd^{2+} < Pb^{2+} < Cu^{2+} < Al^{3+}$. The anionic nature of the RLs enables them to absorb metals ions from the soil such as arsenic, cadmium, copper, lanthanum, lead, and zinc due to their complexation ability (Christofi and Ivshina, 2002) and help in the remediation of toxic metal ions.

At CMD^{-2}, the cell-free culture supernatant exhibited comparatively lesser surface activity than CMD^{-1} due to lowering of surfactant concentration. The cell-free culture supernatants at CMD^{-1} and CMD^{-2} remained effective similar to that of normal concentration even after autoclaving at 121°C for 30 min. The BS retained its surface activity by reducing the surface tension up to a concentration of 4% NaCl and the effect of salinity on BSs was determined at normal, CMD^{-1} and CMD^{-2} concentrations (Table 10.4).

TABLE 10.4 Influence of Salinity on the Surface Activity of Biosurfactant Produced by *P. aeruginosa* Strains in 2% n-Hexadecane Supplemented Medium at Normal and Critical Micelle Dilutions (CMD^{-1} and CMD^{-2})

Bacterial Strains	NaCl Concentration (g%)	Surface Tension (mN/m)		
		Cell-Free Culture Supernatant	CMD^{-1}	CMD^{-2}
P. aeruginosa OBP1	0	31.2 ± 0.87	31.4 ± 0.56	41.7 ± 0.92
	2	31.4 ± 0.36	31.3 ± 0.20	41.3 ± 0.39
	3	31.7 ± 0.90	31.8 ± 0.54	42.2 ± 0.49
	4	34.5 ± 0.13	39.3 ± 0.73	48.2 ± 0.28
	5	40.4 ± 0.64	42.6 ± 0.29	53.7 ± 0.36
OBP2	0	37.4 ± 0.43	40.6 ± 0.48	57.7 ± 0.26
	2	37.6 ± 0.26	40.9 ± 0.71	57.5 ± 0.40
	3	37.9 ± 0.39	41.0 ± 0.39	57.8 ± 0.40
	4	40.5 ± 0.48	43.7 ± 0.51	59.6 ± 1.00
	5	46.6 ± 0.65	50.3 ± 0.30	61.9 ± 0.49

TABLE 10.4 *(Continued)*

Bacterial Strains	NaCl Concentration (g%)	Surface Tension (mN/m)		
		Cell-Free Culture Supernatant	CMD^{-1}	CMD^{-2}
OBP3	0	35.4 ± 0.28	38.3 ± 0.83	55.5 ± 0.72
	2	35.4 ± 0.76	38.5 ± 0.62	55.6 ± 0.49
	3	35.7 ± 0.39	39.9 ± 0.96	55.6 ± 0.49
	4	37.3 ± 0.49	41.0 ± 0.70	58.3 ± 0.02
	5	43.1 ± 0.65	48.8 ± 0.52	59.8 ± 0.39
OBP4	0	33.1 ± 0.38	39.7 ± 0.30	48.6 ± 0.56
	2	33.3 ± 0.29	39.6 ± 0.73	48.6 ± 0.75
	3	33.6 ± 0.48	39.7 ± 0.40	48.9 ± 0.95
	4	36.0 ± 0.20	42.5 ± 0.67	50.4 ± 0.34
	5	39.6 ± 0.29	48.2 ± 0.93	53.7 ± 0.94

The data shown here are mean values of triplicates.

The culture supernatant at CMD^{-1} and CMD^{-2} on exposure to the different metal ions exhibited almost stable surface tensions. Data obtained are presented in Table 10.5.

TABLE 10.5 Influence of Metal Ions on the Surface Activity of Biosurfactant Produced by *P. aeruginosa* Strains in 2% n-Hexadecane Supplemented Medium at Normal and Critical Micelle Dilutions (CMD^{-1} and CMD^{-2})

Bacterial Strains	Exposure to Metal Ions (2% w/v)	Surface Tension (mN/m)		
		Cell-Free Culture Supernatant	CMD^{-1}	CMD^{-2}
P. aeruginosa OBP1	K^+	31.2 ± 0.40	37.4 ± 0.75	45.7 ± 0.60
	Ca^{2+}	31.1 ± 0.58	36.3 ± 0.37	45.6 ± 0.45
	Mg^{2+}	31.3 ± 0.67	36.5 ± 0.53	45.7 ± 0.60
	Fe^{2+}	31.2 ± 0.22	36.3 ± 0.63	45.5 ± 0.45
	Al^{3+}	34.7 ± 0.36	38.9 ± 0.52	48.8 ± 0.69
OBP2	K^+	37.6 ± 0.54	43.2 ± 0.56	58.2 ± 0.45
	Ca^{2+}	37.5 ± 0.65	42.4 ± 0.23	57.5 ± 0.23
	Mg^{2+}	37.4 ± 0.83	42.7 ± 0.75	57.4 ± 0.34
	Fe^{2+}	37.7 ± 0.31	43.4 ± 0.32	57.8 ± 0.62
	Al^{3+}	39.3 ± 0.53	46.5 ± 0.69	60.7 ± 0.42

TABLE 10.5 *(Continued)*

Bacterial Strains	Exposure to Metal Ions (2% w/v)	Surface Tension (mN/m)		
		Cell-Free Culture Supernatant	CMD^{-1}	CMD^{-2}
OBP3	K^{+}	35.7 ± 0.67	41.5 ± 0.98	52.5 ± 0.51
	Ca^{2+}	35.5 ± 0.34	40.5 ± 0.54	51.6 ± 0.11
	Mg^{2+}	35.3 ± 0.37	40.5 ± 0.29	51.5 ± 0.27
	Fe^{2+}	35.2 ± 0.83	40.6 ± 0.78	51.7 ± 0.39
	Al^{3+}	37.7 ± 0.56	43.2 ± 0.43	53.8 ± 0.83
OBP4	K^{+}	33.5 ± 0.29	39.7 ± 0.93	47.0 ± 0.36
	Ca^{2+}	33.2 ± 0.65	39.4 ± 0.53	46.9 ± 0.73
	Mg^{2+}	33.3 ± 0.56	39.6 ± 0.77	46.7 ± 0.29
	Fe^{2+}	33.4 ± 0.93	39.5 ± 0.49	46.9 ± 0.60
	Al^{3+}	35.8 ± 0.37	41.6 ± 0.61	49.2 ± 0.35

The data shown here are mean values of triplicates.

In the case of trivalent metal (Al^{3+}) there was a significant reduction in the surface activities of BSs at all concentrations.

10.4 EMULSIFICATION ACTIVITY (E_{24})

The emulsification index was measured using the standard method as described by Cooper and Goldenberg. Different hydrocarbons were used for testing the emulsification efficiency. Three ml of the test hydrocarbon was added to 2 ml of cell-free culture supernatant in a glass test tube and homogenized in a vortex at high speed for 2 min. The resulting mixture was kept at RT for 24 h and emulsification index (E_{24}) was calculated by the following formula:

$$\text{Emulsification Index (E24\%)} = \frac{\text{Height of the emulsion layer}}{\text{Total height of the mixture}} \times 100$$

The bacterial strains showed a wide difference in the emulsification activity against the test hydrocarbons. However, the cell-free culture supernatant of the bacterial strains exhibited appreciable emulsification indices against diesel, n-hexadecane, n-octadecane, crude oil, n-dodecane, lubricating oil, n-paraffin, and kerosene were in the range of 64–82%, 67–77%, 64–74%, 57–73%, 63–70%, 52–67%, 49–68%, and 50–64%, respectively. Retention of emulsions even after 30 days indicates the formation of a relatively stable emulsion. It was observed that the cell-free culture supernatants of the bacterial strains couldn't emulsify the iso-octane up to a significant level. The emulsification activity of the BSs present in the culture supernatant of the bacterial strains was assessed against the different hydrocarbons and is presented in Figure 10.2.

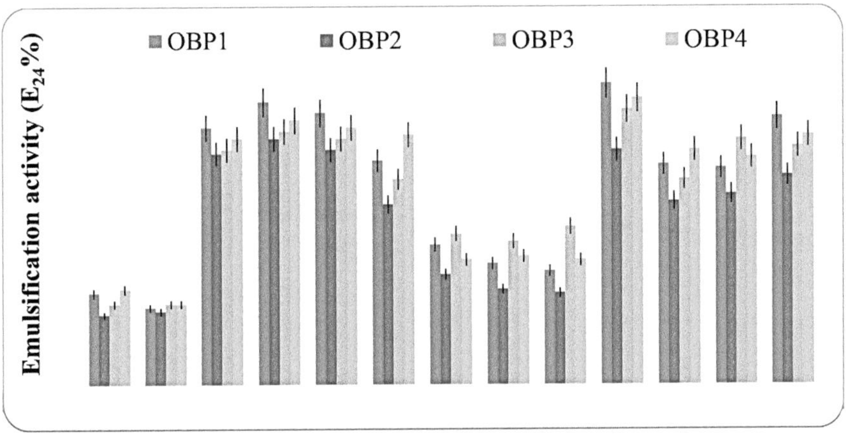

FIGURE 10.2 Emulsification indices (E_{24}%) exhibited by the culture supernatants of *P. aeruginosa* strains with various hydrophobic substrates. Results represent the mean of three experiments ± SD.

It was observed that the cell-free culture supernatants of the bacterial strains couldn't emulsify the iso-octane up to a significant level.

The emulsifying power is an important character of surfactants. Hence, E_{24} was determined for all four BSs using cell-free culture supernatant against the different hydrophobic substrates. The cell-free culture supernatants showed appreciable emulsification indices (E_{24}) with diesel, n-hexadecane, n-octadecane, crude oil, n-dodecane, lubricating oil, n-paraffin, and kerosene. In addition to the surface and IFTs, stabilization of an oil-water emulsion is usually used as an indicator of surface activity (Kim et al., 1999). The preference for hydrophobic substrates and the behavior of E_2 were quite different among the four strains of *P. aeruginosa*. Most of the microbial surfactants are substrate-specific causing solubilization or emulsification of hydrocarbons at different rates (Cunha et al., 2004). E_{24} was calculated using oil/water ratio 3:2, which indicates the constitution of the oil phase to be 60% of the total mixture volume. This signifies that the value of E_{24} greater than or equal to 60 is responsible for the complete emulsification of the oil phase (Abdel-Mawgoud et al., 2009).

Such conditions were observed in the case of cell-free culture supernatant of the bacterial strains against diesel and n-hexadecane. The water-oil emulsions were found to be compact and remained stable at RT for more than one month, suggesting possible application of the BSs in the bioremediation process for enhancing the availability of the recalcitrant hydrocarbons. Moreover, the ability of the BSs to emulsify specifically the crude oil products such as n-hexadecane, octadecane, kerosene, diesel, and lubricating oil might facilitate their microbial assimilation which could be useful for the bioremediation of petroleum contaminated environments (Abdel-Mawgoud et al., 2009). Further, the ability of the BSs to emulsify the vegetable oil also suggests their potential application in the pharmaceutical and cosmetic industries (Silva et al., 2010).

10.5 INFLUENCE OF TEMPERATURE, PH, AND SALINITY ON THE EMULSIFICATION ACTIVITY (E_{24})

The E_{24} of the cell-free culture supernatant of the bacterial strains against diesel was quite stable at all pH levels, but the maximum activity was shown in the pH range of 5–8 (Figure 10.3(A)). E_{24} of the diesel supplemented cell-free culture supernatant of the bacterial strains was quite stable at all temperatures from 4–100°C (Figure 10.3(B)). The E_{24} was also stable up to 30 days even when stored at 4°C. It is interesting to note that the BS retained its emulsifying activity even after heating at the autoclaving temperature of 121°C for 30 min, indicating the thermal stability of the cell-free culture supernatant of the bacterial strains. Further, The E_{24} against diesel also remained unchanged over the tested NaCl concentrations from 2–4% (Figure 10.3(C)).

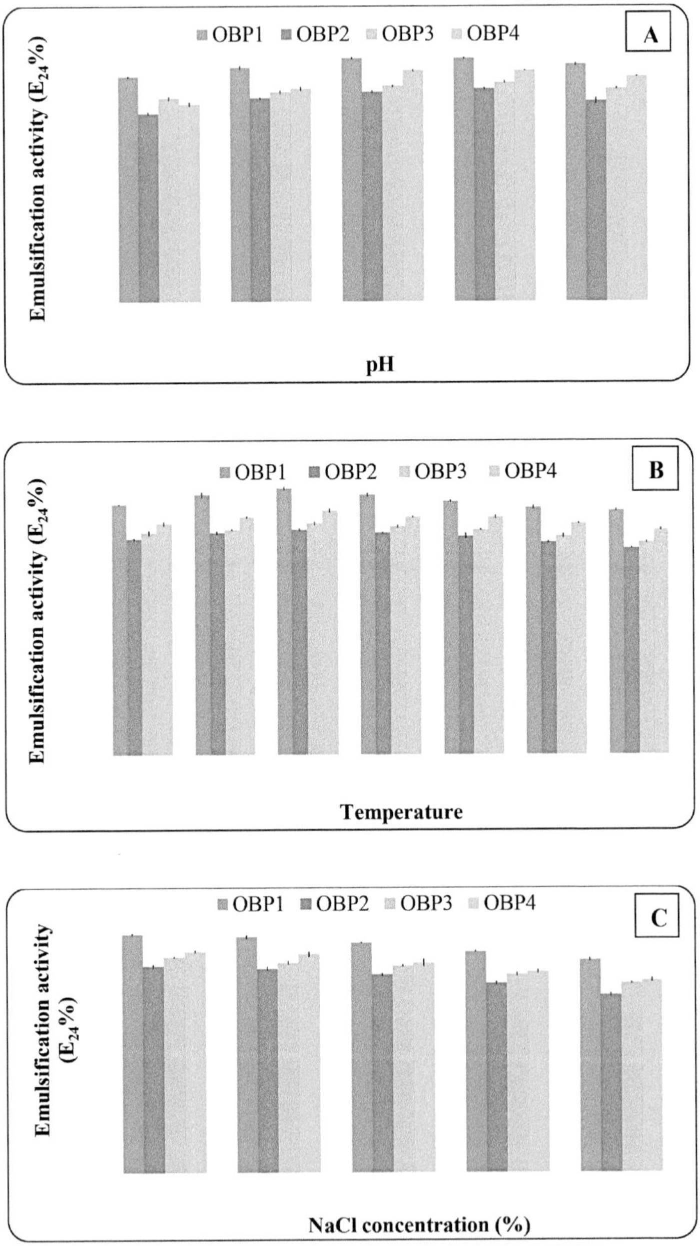

FIGURE 10.3 Effect of (A) pH; (B) temperature; and (C) salinity (NaCl) on the emulsifying properties (E_{24}%) of culture supernatants of *P. aeruginosa* strains against diesel. Results represent the mean of 3 experiments ± SD.

10.6 FOAMING INDEX (F_{24}%) AND ITS STABILITY

An aliquot of 20 ml of the BS solution (1 g.l^{-1}) was transferred to a glass measuring cylinder of 50 ml volume and compressed N_2 gas was passed through the solution at a flow rate of 0.5 l.min^{-1} for 2 min (Abbasi et al., 2012). The foaming index of the BS was calculated after 24 h by using the following equation:

$$\text{Emulsification Index } (E_{24}\%) = \frac{\text{Height of the foam}}{\text{Total height of the liquid and foam}} \times 100$$

The cell-free culture supernatants of the bacterial strains produced stable foam with the foaming index (F_{24}%) in the range of 50.4–65.5%. F_{24}% is presented in Figure 10.4.

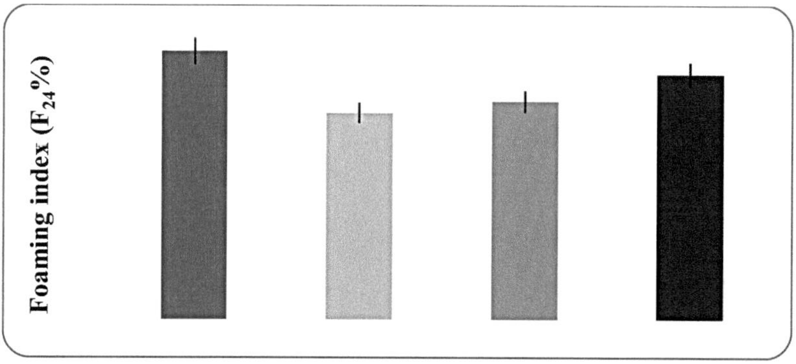

FIGURE 10.4 Foaming indices (F_{24}%) exhibited by the culture supernatant of *P. aeruginosa* strains. Results represent the mean of 3 experiments ± SD.

Foams produced by the cell-free culture supernatants of all four bacterial strains remained relatively stable up to 24 h. The cell-free culture supernatant containing BS produced by each of the four bacterial strains caused stable foam with F_{24}% in the range of 50.4–65.5%. The foam produced during the experiment was relatively stable up to 24 h. Such characteristics of BSs indicate their possible application in coal and mineral froth-flotation as a frothing and co-frothing agent (Abbasi et al., 2012).

10.7 BIOSURFACTANT (BS) STABILITY

Stability studies were carried out using cell-free culture supernatant obtained after centrifuging the bacterial cultures at 8,000 rpm for 15 min at

4°C. Aliquots of 10 ml cell-free culture supernatant were subjected to 4°, 25°, 37°, 50°, 75°, 100°C for 60 and 121°C for 30 min and cooled to RT after which the surface tension of the culture supernatant was measured at normal concentration, at CMD^{-1} and CMD^{-2} and emulsification index against diesel was measured. To study the pH stability of the BS, the pH of the culture supernatant was adjusted to different pH values 2–11, and the surface tension of the treated culture supernatant was measured at normal concentration, at CMD^{-1} and CMD–2 and emulsification index against diesel was measured. Similarly, the effect of NaCl concentration (1–5%) (w/v) on the surface tension of the culture supernatant at normal concentration, CMD^{-1} and CMD^{-2} and emulsification activity was determined. To investigate the effect of metal ions on the surface activity, the culture supernatants of the bacterial strains were mixed with different metal ions of K^{+}, Ca^{2+}, Mg^{2+}, Fe^{2+}, and Al^{3+} (20,000 $mg.l^{-1}$) separately. Changes in the surface tension of the culture supernatant at normal concentration, CMD^{-1} and CMD^{-2} for each case were measured. In this study, the synthetic surfactant SDS was used as a standard. The assays were carried out in triplicates.

10.8 CELL SURFACE HYDROPHOBICITY

The cell surface hydrophobicity or BATH assay of the bacterial cells was measured by using the standard procedure. Briefly, bacterial strains were grown on n-hexadecane (2% v/v) and glucose (2% w/v) separately in mineral salt medium (MSM). Bacterial cells were harvested from culture medium by centrifugation at 8,000 rpm for 15 min at 4°C and washed twice in PUM buffer and suspended in the same buffer to give an optical density of approximately 0.5 – 0.6 at 600 nm. The cell suspension (2.0 ml) with 0.5 ml of test hydrocarbon was vortexed in a test tube vigorously for 3 min and allowed to settle down for 15 min at RT. The bottom aqueous phase was carefully removed, and the optical density was measured at 600 nm in a spectrophotometer. The cell surface hydrophobicity was expressed as the percentage of adherence to hydrocarbon and calculated by the following formula:

$$\frac{\text{Optical density of the aqueous phase}}{\text{Optical density of the initial cell suspension}} \times 100$$

Hydrophobicity of the bacterial cell surfaces growing on two different carbon sources such as glucose and n-hexadecane was determined and is presented in Figure 10.5. All the bacterial strains possessed a wide extent of variability in their surface hydrophobicity against the tested hydrocarbons. The surface hydrophobicity of four strains of *P. aeruginosa* cells growing on n-hexadecane was observed to be much higher as compared to the bacterial strains growing on glucose containing medium. Further, it was observed that the cell surface hydrophobicity of the bacterial cells at the exponential growth phase was much lower than that of the bacterial cells at the stationary phase of growth.

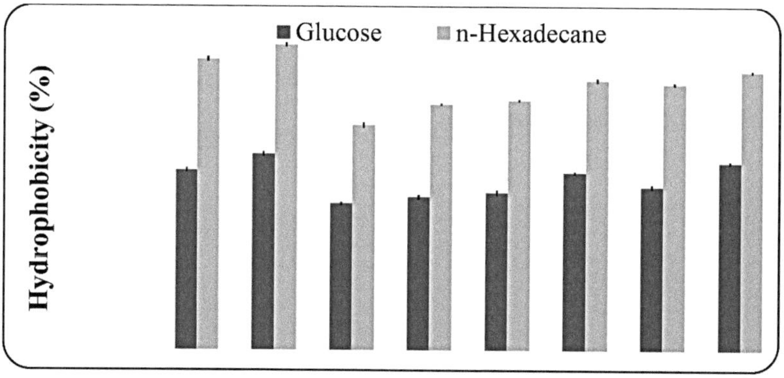

FIGURE 10.5 Hydrophobicity of bacterial strains at exponential and stationary phase of growth when cultivated in mineral salt medium supplemented with 2% n-hexadecane or glucose. Values are the mean of 3 experiments ± SD.

The bacterial strains possessed wide variability in their surface hydrophobicity, and the same was found to increase with the complexity of carbon sources. In the case of n-hexadecane grown cells, higher level of hydrophobicity and BS production were observed, which indicated the bacterial strain's ability to have BS-mediated uptake of alkane. The cell surface hydrophobicity of bacteria in the exponential phase was much lower than that of the stationary phase, which also proved higher levels of BS production in the late exponential phase of growth, indicating the accumulation as secondary metabolite. Cells in the early or mid-exponential growth phases were less hydrophobic than those in stationary phase, which suggested that the production of BSs contributed to cell surface hydrophobicity. The RL at a concentration above CMC causes a reduction of the total cellular lipopolysaccharide (LPS) content of the bacterial cells of *P. aeruginosa* and concentration below CMC causes alteration in the composition of the outer membrane protein which contributes to the enhancement of surface hydrophobicity of the bacterial cells. BSs produced by the bacterial strains during growth on the immiscible carbon sources modify their cell surface physiology and make them more hydrophobic, which improves the adhesion of complex hydrocarbon to their surface or makes it available to them (Ron and Rosenberg, 2002, 2010).

KEYWORDS

- **critical micelle concentration**
- **critical micelle dilutions**
- **interfacial tension**
- **lipopolysaccharide**
- **mineral salt medium**
- **sodium dodecyl sulfate**

REFERENCES

Abbasi, H., et al., (2012). Biosurfactant-producing bacterium, *Pseudomonas aeruginosa* MA01 isolated from spoiled apples: Physicochemical and structural characteristics of isolated biosurfactant. *J. Biosci. Bioengg., 113*, 211–219.

Abdel-Mawgoud, A. M., et al., (2010). Rhamnolipids: Diversity of structures, microbial origins and roles. *Appl. Microbiol. Biotechnol., 86*, 1323–1336.

Aparna, A., et al., (2012). Production and characterization of biosurfactant produced by a novel *Pseudomonas* sp. 2B. *Colloids Surf. B. Biointerfcaes, 95*, 2–2 9.

Bailey, J. E., & Ollis, D. F., (1986). *Biochemical Engineering Fundamentals* (2nd edn.). New McGraw-Hill, York.

Bodour, A. A., & Miller-Maier, R. M., (1998). Application of a modified drop collapse technique for surfactant quantitation and screening of biosurfactant-producing microorganisms. *J. Microbiol. Method., 32*, 273–280.

Choi, Y. J., et al., (2009). Enhancement of aerobic biodegradation in an oxygen-limiting environment using a saponin-based microbubble suspension. *Environ. Pollut., 157*, 2197–2202.

Christofi, N., & Ivshina, I. B., (2002). Microbial surfactants and their use in field studies of soil remediation. *J. Appl. Microbiol., 93*, 915–929.

Cooper, D. G., & Paddock, D. A., (1984). Production of biosurfactants from *Torulopsis bombicola*. *Appl. Environ. Microbiol, 47*, 173–176.

Costa, S. G. V. A. O., et al., (2010). Structure, properties, and applications of rhamnolipids produced by *Pseudomonas aeruginosa* L2-1 from cassava wastewater. *Process Biochem., 45*, 1511–1516.

Cunha, C. D., et al., (2004). Serratia sp. SVGG 16: A promising biosurfactant producer isolated from tropical soil during growth with ethanol-blended gasoline. *Process Biochem., 39*, 2277–2282.

George, S., & Jayachandran, K., (2008). Analysis of rhamnolipid· biosurfactants produced through submerged fermentation using orange fruit peelings as sole carbon source. *Appl. Biochem. Biotechnol., 58*, 428–434.

Herman, D. C., et al., (1995). Removal of cadmium lead and zinc from soil by a rhamnolipid biosurfactant. *Environ. Sci. Technol., 29*, 2280–2285.

Kim, H. S., et al., (1999). Characterization of a biosurfactant, mannosylerythritol lipid produced from *Candida* sp. SY 16. *Appl. Microbiol. Biotechnol., 52*, 713–721.

Lee, K. M., et al., (2004). Rhamnolipid production in batch and fed-batch fermentation using *Pseudomonas aeruginosa* BYK-2 KCTC 18012P. *Biotech. Bioprocess Engg., 9*, 267–273.

Maier, R. M., & Soberon-Chavez, G., (2000). *Pseudomonas aeruginosa* rhamnolipids: biosynthesis and potential applications. *Appl. Microbiol. Biotechnol., 54*, 625–633.

Monterio, S. A., et al., (2007). Molecular and structural characterization of the biosurfactant produced by *Pseudomonas aeruginosa* DAUPE 614. *Chem. Phys. Lipids, 147*, 1–13.

Palleroni, N. J., (1984). Family 1. *Pseudomonadaceae*. In: Krieg, N. R., (ed.), *Bergey's Manual of Systematic Bacteriology* (pp. 141–219). Williams & Wilkin, Baltimore.

Reddy, P. G., et al., (1983). Isolation and functional characterization of hydrocarbon emulsifying and solubilizing factors produced by a *Pseudomonas species*. *Biotechnol. Bioeng., 25*, 387–401.

Ron, E., & Rosenberg, E., (2002). Biosurfactants and oil bioremediation. *Curr. Opinion Biotechnol., 13*, 249–252.

Silva, S. N. R. L., et al., (2010). Glycerol as substrate for the production of biosurfactant by *Pseudomonas aeruginosa* UCP 0992. *Colloids Surf. B. Biointerfaces, 79*, 174–183.

Van, H. J. D., et al., (2006). Physiological aspects: Part 1 in a series of papers devoted to surfactants in microbiology and biotechnology. *Biotechnol. Adv., 24*, 604–620.

Vasileva-Tonkova, E., & Gesheva, V., (2007). Biosurfactant production by Antarctic facultative anaerobe *Pantoea* sp. during growth on hydrocarbons. *Curr. Microbiol., 54*, 136–141.

Wu, J. Y., et al., (2008). Rhamnolipid, production with indigenous *Pseudomonas aeruginosa* EMI isolated from oil-contaminated site. *Bioresour. Technol., 99*, 1157–1164.

Xia, W., et al., (2011). Comparative study of biosurfactant produced by microorganisms isolated from formation water of petroleum reservoir. *Colloids Surf. A Physicochem. Eng. Aspects, 392*, 124–130.

Zhang, Y., & Miller, R. M., (1994). Effect of a *Pseudomonas* rhamnolipid biosurfactant on cell hydrophobicity and biodegradation of octadecane. *Appl. Environ. Microbiol., 60*, 2101–2106.

FURTHER READING

Bharali, P., & Konwar, B. K., (2011). Production and physicochemical characterization of a biosurfactant produced by *Pseudomonas aeruginosa* OBP1 isolated from petroleum sludge. *Appl. Biochem. Biotechnol., 164*, 1444–1460.

Bharali, P., Das, S., Konwar, B. K., & Thakur, A. J., (2011). Crude biosurfactant from thermophilic *Alcaligenes faecalis*: Feasibility in Petro-spill bioremediation. *Inter. Biodeter. Biodegr., 65*, 682–690.

Das, S., Kalita, S. J., Bharali, P., Konwar, B. K., Das, B., & Thakur, A. J., (2013). Organic reactions in "green surfactant": An avenue to bisuracil derivative. *ACS Sustainable Chem. Eng.* (p. 301). doi: 10.1021/sc4002774, Publication Date (Web).

Pranjal, B., (2015). *Bioremediation of Crude Oil Contaminated Soil.* (Thesis: Supervisor B K Konwar), Dept. of Mol. Biol. and Biotechnology, Tezpur University (Central), Napaam – 784028, Assam, India.

CHAPTER 11

Characterization of Potential Biosurfactant-Producing Bacteria

11.1 GROWTH CHARACTERS

The pure culture of each bacterial isolate was prepared by using Luria Bertani (LB) broth and incubated overnight in an orbital incubator shaker at 37°C with 180 rpm. An aliquot of 100 µl of the above fresh culture broth containing 1×108 ml^{-1} microbes (McFarland turbidity method) was inoculated to the 250 ml Erlenmeyer flask containing 100 ml of MSM. All the cultures were supplemented with 1% (v/v) of n-hexadecane and incubated at 37°C with 180 rpm on an orbital incubator shaker. The growth of the bacterial isolates was monitored by determining the cell forming unit (cfu. ml^{-1}) at a time interval of 12 h for 180 hours.

11.2 BIOMASS DETERMINATION

Biomass of the bacterial isolates was determined by centrifuging the culture broths at 8,000 rpm for 15 min at 4°C followed by washing twice with phosphate-buffered saline. The biomass was dried overnight at 45°C and weighed. In the case of aliphatic hydrocarbons (pentane, n-hexane, heptane, iso-octane, dodecane, tridecane, n-hexadecane, octadecane, eicosane, triacontane, and liquid paraffin) and petroleum products (phenol, benzene, toluene, xylene, kerosene, diesel, lubricating oil and crude oil), 1% (v/v) of each carbon source was added to the culture separately and in the case of polycyclic aromatic hydrocarbons (PAH) (pyrene, anthracene, naphthalene, fluorene, and phenanthrene), the culture medium was supplemented with 50 µg of each type of carbon source (Bordoloi and Konwar, 2009).

11.3 BIOCHEMICAL TESTS

11.3.1 ESTIMATION OF CARBOHYDRATE CONTENT

The total carbohydrate content in the isolated biosurfactant (BS) samples was quantified by standard phenol-sulfuric acid method (Dubois et al., 1956) using D-glucose as standard. The carbohydrate content of the unknown samples was calculated from the standard curve obtained by plotting optical density (OD490) versus concentration of D-glucose (0.1 $mg.ml^{-1}$).

11.3.2 ESTIMATION OF LIPID CONTENT

The lipid content in the isolated BS samples was estimated gravimetrically using a standard protocol of Folch et al. (423). Briefly, 50 μg of isolated BS was homogenized with a chloroform-methanol mixture (2:1, v/v). The crude extract was then mixed thoroughly with 1.0 ml of water and could separate into two distinct phases. The upper aqueous phase was removed with the help of a micropipette. Finally, the lower organic phase was collected, dried, and the total lipid content was determined gravimetrically.

11.4 MORPHOLOGICAL TESTS

The taxonomic identification of the selected bacterial isolates was carried out using the standard morphological, physiological, and biochemical tests. The bacterial isolates were taxonomically identified up to genus level with the help of standard procedures described in Bergey's Manual of Systematic Bacteriology (Buchanan and Gibbons, 1974).

11.5 DETECTION OF BIOSURFACTANT (BS)

11.5.1 CETYL TRIMETHYL AMMONIUM BROMIDE (CTAB) AGAR TEST

The bacterial strains were spread over the mineral salt medium (MSM) containing CTAB (0.2 $g.l^{-1}$), methylene blue (5 $mg.l^{-1}$), agar (16 $g.l^{-1}$) as solidifying agent and 0.1% (v/v) n-hexadecane as sole source of carbon. BS production could be detected by the formation of dark blue halos around the bacterial colonies (Siegmund and Wagner, 1991).

11.5.2 BLOOD HEMOLYSIS AGAR TEST

Hemolytic activity was determined on nutrient blood agar media plates containing mammalian blood (goat). With the help of a sterile cork-borer a single well of 6.0 mm diameter was made at the center of the blood agar plate. An aliquot of 100 µl of cell-free culture supernatant was inoculated into the well and incubated at 37°C for 48 h. Hemolytic activity indicates complete lysis of red blood cells surrounding the well containing bacterial culture supernatant. The diameter of the clear zone depends on the concentration of the BS produced by the bacteria (Siegmund and Wagner, 1991' Mulligan et al., 1984).

11.5.3 ORCINOL ASSAY

An aliquot of 0.5 ml of cell-free culture supernatant was extracted twice with 1 ml of diethyl ether. The ether fraction was collected, allowed to dry under fume hood and dissolved in 0.5 ml H_2O. Sample was further diluted to 10^{-1} dilution in 0.19% orcinol solution prepared in 53% (v/v) of concentrated H_2SO_4. The sample was then placed in boiling water for 30 min, cooled at room temperature (RT) for 15 min, and the absorbance at 421 nm was measured. Glycolipid concentration was calculated from a standard curve prepared with L-rhamnose and expressed as rhamnolipid (RL) values by multiplying rhamnose values with a coefficient of 3.4 obtained from the correlation of pure RL/rhamnose (Chandrasekaran and BeMiller, 1980).

11.6 QUANTIFICATION OF BIOSURFACTANT (BS)

11.6.1 ISOLATION OF BIOSURFACTANT (BS)

The culture supernatant was first centrifuged at 8,000 rpm for 20 min at 4 °C to remove the bacterial cells. The cell-free culture supernatant was then acidified to pH 2 with 6 N HCl and allowed to stand overnight at 4°C to precipitate the BS. The precipitate was harvested by centrifugation at 12,000 rpm for 15 min at 4°C. The recovered precipitate was extracted thrice with ethyl acetate at RT. The organic phase was collected in a round-bottom flask and connected to a rotary evaporator (Eyela, CCAS-1110, Rikakikai Co. Ltd., Tokyo) to remove the solvent. The process yielded a viscous honey-colored residue. The residue was then washed twice with n-hexane to remove any residual n-hexadecane. Finally, the yellowish product was dissolved in ethyl acetate, filtered, and concentrated using a rotary evaporator (Abdel-Mawgoud et al., 2009; Wu et al., 2008). It was weighted and expressed as $g.l^{-1}$.

11.6.2 PURIFICATION OF BIOSURFACTANT (BS)

Purification of the isolated crude BSs was carried out using thin-layer chromatography (TLC) with slight modifications. Briefly, aqueous slurry of silica gel 60 was prepared and was used for making preparative TLC glass plates (20×20 cm). Crude BS 2 g was dissolved in 4 ml chloroform and loaded onto the activated TLC plates with the help of glass capillary. The purification was carried out using chloroform: methanol: H_2O (65:15:2, v/v/v) as mobile phase. After the competition of the separation process, the plates were exposed to the iodine vapor to make visible the separated fraction. The separated fractions were then collected by scrapping out the separated fractions from the preparative plates. Finally, the separated compound was recovered from the silica gel by washing the collected TLC fractions with chloroform and methanol (2:1, v/v) through a glass column. During the purification steps, analytical TLC (using chloroform: methanol: H_2O, 65:15:2 v/v/v as mobile phase) and the reduction in the surface tension of water by the separated TLC fraction was carried out to check the purity and surface activity of the isolated fractions.

11.7 BIOCHEMICAL CHARACTERIZATION OF BIOSURFACTANT (BS)

Biochemical characterization of isolated crude BSs from four bacterial strains is presented in Table 11.1. The isolated BSs were glycolipid in nature.

TABLE 11.1 Biochemical Characterization of Biosurfactant Produced by *P. aeruginosa* Strains in MSM Supplemented with 2% n-Hexadecane

Biosurfactant Sample	Carbohydrate Content (%)	Lipid Content (%)	Protein Content (%)
OBP1	48.0 ± 0.3	28.7 ± 0.2	23.3 ± 0.2
OBP2	51.7 ± 0.1	28.3 ± 0.6	20.0 ± 0.1
OBP3	47.8 ± 0.6	30.8 ± 0.6	21.4 ± 0.1
OBP4	49.3 ± 0.2	29.0 ± 0.5	21.7 ± 0.4

Results represented mean ± S.D of three individual experiments.

11.7.1 THIN LAYER CHROMATOGRAPHY (TLC)

The isolated BS samples were dissolved in chloroform to result in a concentration of 0.3 $g.l^{-1}$. An aliquot of 100 μl of the BS solution was applied on 20×20 cm TLC plates (TLC Silica gel 60 F254, Merck, India). The plates were developed in a solvent system consisting of chloroform: methanol: acetic acid (65:10:2, v/v/v). For detecting the carbohydrate components in the separated fraction, the TLC plates were sprayed with anthrone reagent prepared in concentrated H_2SO_4. For the detection of lipid components, the TLC plates were exposed to the iodine fume. The presence of protein or amino acid components in the separated fractions was determined by spraying the TLC plates with ninhydrin solution (George and Jayachandran, 2008; Tahzibi et al., 2004). Qualitative analysis of the partially purified BS samples isolated from OBP1, OBP2, OBP3, and OBP4 was done by TLC. On spraying with orcinol reagents, brown-colored spots indicative of carbohydrate units was detected in silica plates. While exposing to the similar plates with iodine vapor, yellow spots indicative of lipids giving the same R_f value as that of glycosyl units were observed on the same region, and the same is presented in Figure 11.1.

The BS of OBP1 showed the presence of four spots having R_f values: 0.26, 0.52, and 0.67. In the case of OBP2, only three spots with R_f values 0.33, 0.40, and 0.71 appeared. Three spots with almost similar R_f values 0.34, 0.55, and 0.73 were observed in the BS of OBP3 and OBP4. For further purification of the isolated BSs, preparative TLC plates were used to collect those fractions which exhibited the highest values.

For further purification of the isolated BSs, preparative TLC plates were used to collect those fractions which exhibited the high surface activity in water and are shown in Table 11.2.

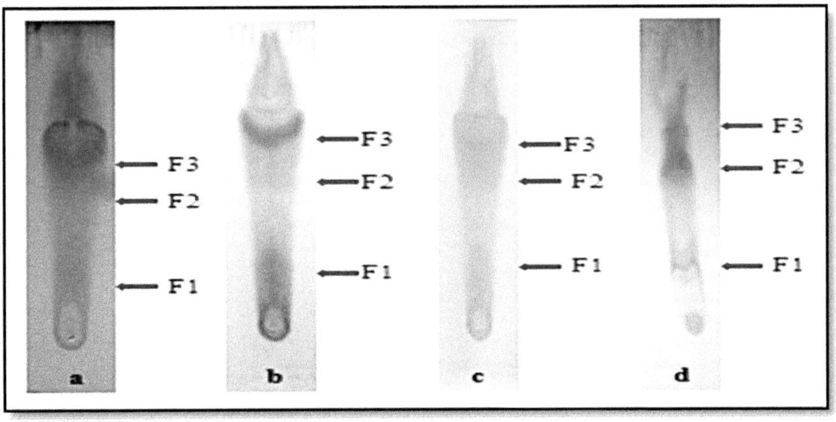

FIGURE 11.1 Thin-layer chromatogram of biosurfactants produced by *P. aeruginosa* strains when cultivated in mineral salt medium supplemented with n-hexadecane. (a) *P. aeruginosa* OBP1; (b) *P. aeruginosa* OBP2; (c) *P. aeruginosa* OBP3; and (d) *P. aeruginosa* OBP4. F1–F3 represents fractions that exhibit positive results of surface activity.

TABLE 11.2 TLC Separation of Partially Purified Biosurfactant Produced by *P. aeruginosa* Strains

Bacterial Strain	TLC Fractions	R_f Values	Surface Tension Reduction (mNm^{-1})
P. aeruginosa OBP1	S1	0.26	52.7 ± 0.83
	S2	0.52	30.8 ± 0.39
	S3	0.67	32.3 ± 0.52
OBP2	S1	0.33	56.3 ± 0.45
	S2	0.40	37.4 ± 0.72
	S3	0.71	43.2 ± 0.55
OBP3	S1	0.34	54.3 ± 0.43
	S2	0.55	39.3 ± 0.53
	S3	0.73	35.1 ± 0.12
OBP4	S1	0.33	51.5 ± 0.55
	S2	0.52	33.0 ± 0.34
	S3	0.74	38.7 ± 0.41

Results represented mean ± S.D of three individual experiments.

11.7.2 FOURIER TRANSFORM INFRARED SPECTROSCOPY (FTIR)

For elucidating the chemical bonds or the functional groups present, the isolated BS samples were subjected to FTIR analysis. The lyophilized BS sample of 1.0 mg was ground with 1.0 g of KBr and pressed with 7,500 kg for 30 S to result in a translucent KBr pellet. The IR spectra of the samples were recorded on a Nicolos Impact I 410, FTIR system, USA, with a spectral resolution and wavenumber accuracy of 4 and 0.01 cm^{-1}, respectively, and 32 scans with correlation for atmospheric CO_2 (Tuleva et al., 2002). The molecular composition of the freeze-dried BSs of the bacterial strains on n-hexadecane supplemented medium was analyzed by FTIR and is presented in Figure 11.2(a–d). The FTIR spectra of all four BSs showed different characteristic peaks with the presence of amino, carboxyl, hydroxyl, and carbonyl groups. All four spectra showed the same essential adsorption bands; only the relative areas under the various absorption bands are slightly different.

The FTIR spectrum of OBP1 band at 3430.19 cm^{-1} were caused by -CH stretching bands of 1725.32 cm^{-1} showed the presence of carbonyl groups. Similarly, the FTIR spectrum of freeze-dried BSs of the bacterial BSs showed the presence of amino, carboxyl, hydroxyl, and carbonyl bands; only the relative BSs produced by *P. aeruginosa* OBP2 in MSM n-hexadecane. In the case of indicated -OH group cm^{-1} represented -CH spectrum of OBP4 surfactant had an intense peak at a frequency of 3416.40 cm^{-1} referring to the presence of cm^{-1} represent stretching bands of at the frequency of 1732.60 In most of the FTIR spectra possessed absorption at 1534.92–1543.75 and 1638.10 amide II and PII band: protein and polysaccharide like substance.

The FTIR spectra of the BSs produced by OBP3 supplemented with n-hexadecane (Figure 11.2) exhibited the intense characteristic peak at -OH groups. The intense stretching bands at cm^{-1} corresponded to cm^{-1} -CH_2, -CH_3 OBP3 BS, the peaks at groups. Other characteristic peaks at 2924.99, 2858.66 and 1725.97 CH_2, -CH_3 and -C=O stretching bands, respectively. The BS possessed a characteristic representing -OH groups. Bands at 2927.03 and 2860.3 represented -CH_2 and -CH_3 group, respectively. Another characteristic showed the presence of carbonyl stretching. 1638.10–1646.38 cm^{-1} representing -C=O amide I, polysaccharide respectively which indicated the presence of in the isolated BSs.

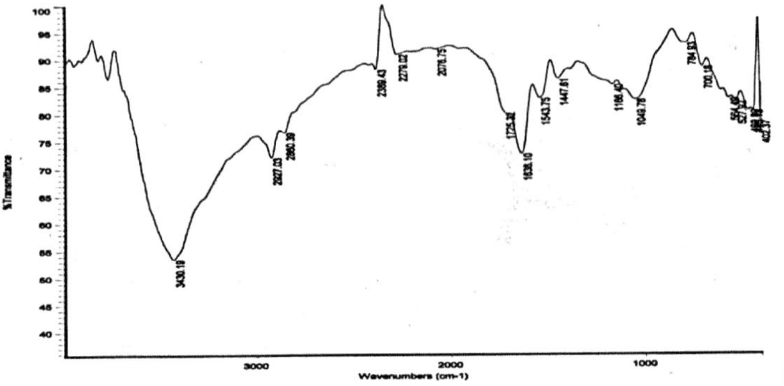

(a)

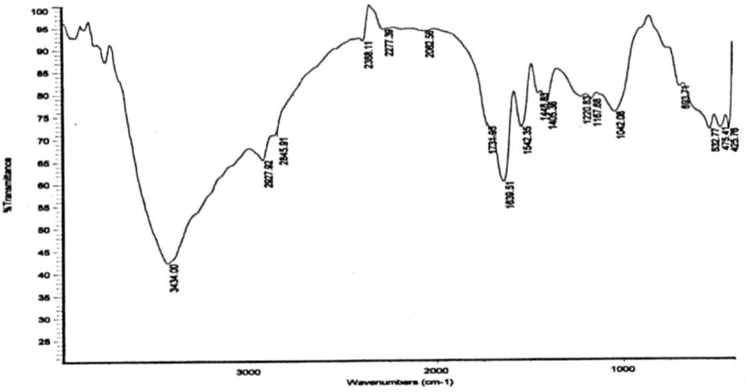

(b)

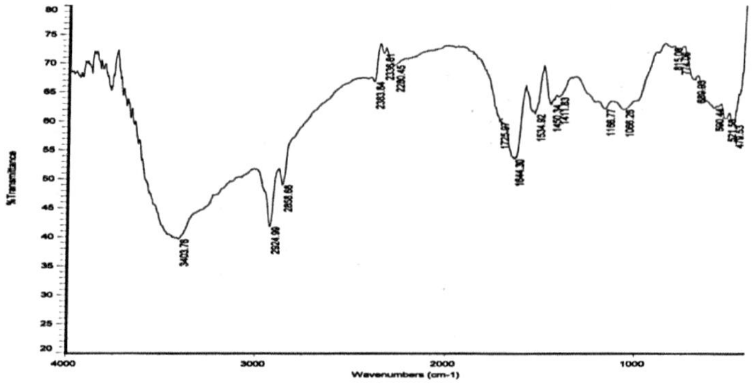

(c)

(Continued)

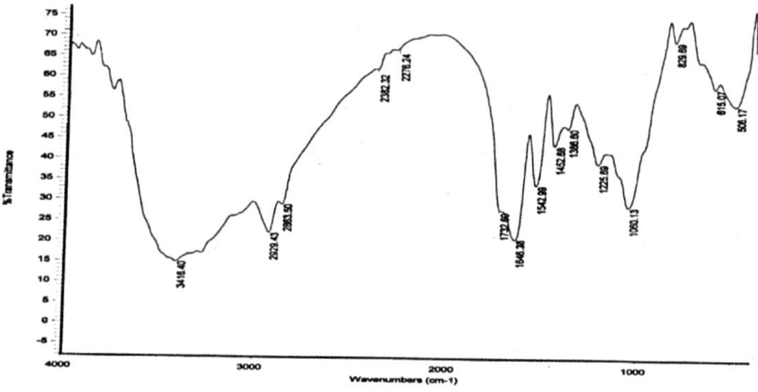

(d)

FIGURE 11.2 The FTIR spectra of the biosurfactants produced by (a) *P. aeruginosa* OBP1; (b) *P. aeruginosa* OBP2; (c) *P. aeruginosa* OBP3; and (d) *P. aeruginosa* OBP4 in mineral salt medium supplemented with n-hexadecane.

FTIR spectrum of OBP1 BS (Figure 11.2(a)) possessed a characteristic band at 3430.19 cm^{-1} representing -OH groups. Bands at 2927.03 and 2860.39 cm^{-1} were caused by-CH stretching bands of $-CH_2$ and $-CH_3$ groups. Stretching bands at 1725.32 cm^{-1} showed the presence of carbonyl groups. Similarly, the FTIR spectrum of OBP2 BS exhibited the intense characteristic peak at 3434.00 cm^{-1} representing -OH groups. The intense stretching bands at 2927.92, 2845.91, and 1731.95 cm^{-1} corresponded to cm^{-1} $-CH_2$, $-CH_3$ and -C=O groups, respectively.

In the case of OBP3 BS (Figure 11.2(b)), the peaks at 3403.76 cm^{-1} indicated -OH groups. Other characteristic peaks at 2924.99, 2858.66, and 1725.97 cm^{-1} represented $-CH_2$, $-CH_3$ and -C=O stretching bands, respectively. The FTIR spectrum of OBP4 surfactant had an intense peak at a frequency of 3416.40 cm^{-1} referring to the presence of -OH groups. Peaks at 2929.43 and 2863.50 cm^{-1} represent stretching bands of $-CH_2$ and $-CH_3$ groups. Another characteristic peak at the frequency of 1732.60 cm^{-1} showed the presence of carbonyl stretching.

In most of the FTIR spectra possessed absorption at 1042.08–1066.26, 1534.92–1543.75 and 1638.10–1646.38 cm^{-1} representing -C=O amide I, –N/–C=O amide II and PII band: polysaccharide, respectively, which indicated the presence of protein and polysaccharide like substance in the isolated BSs.

11.7.3 LIQUID CHROMATOGRAPHY AND MASS SPECTROSCOPY (LC-MS)

RL mixtures were separated from the isolated BS samples and identified by LC-MS, and the same are presented in Figure 11.3(a) and (b). BSs isolated from the bacterial strains OBP1 and OBP2 were characterized using a UPLC-ESI-MS (Waters) while the BSs from the bacterial strains OBP3 and OBP4 were characterized using LC-MS-MS (Agilent 6520). Samples were prepared by diluting with methanol at a concentration of 10 mg l^{-1}; 100 μl of the same was injected into a C8 WP-300 (5 μm) 150×4.6 mm column. The LC flow rate was 1.0 $ml.min^{-1}$. For the mobile phase, an acetonitrile-water gradient was used; starting with 30% of acetonitrile for 4 min, followed by 30–100% acetonitrile for 40 min then standby for 5 min and return to the initial condition. MS was performed with a single quadrupole mass spectrometer, equipped with a pneumatically assisted electrospray (ES) source and negative ion mode was used. The capillary was held at a potential of –3.5 KV and extraction voltage at –75 V. Full scan data were obtained by scanning from m/z 100 to 750 in the centroid mode using scan duration of 2.0 S and an inter-scan time of 0.2 S (Haba et al., 2003).

The purified TLC fractions of OBP1 and OBP2 BS were further purified on ultra-pure liquid chromatography (UPLC) system to separate the RL mixture and are presented in Figure 11.4(a) and (b). In the case of OBP3 and OBP4, the TLC fraction showing the highest surface activity was subjected to the UPLC system.

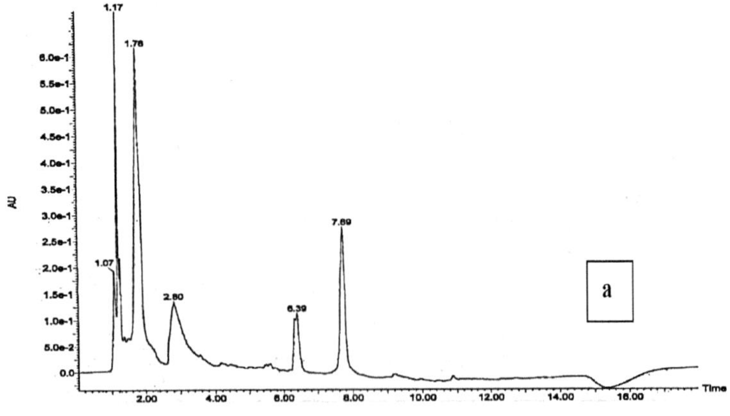

FIGURE 11.3 LC-MS profile of purified fraction of (a) *P. aeruginosa* OBP1; and (b) *P. aeruginosa* OBP2.

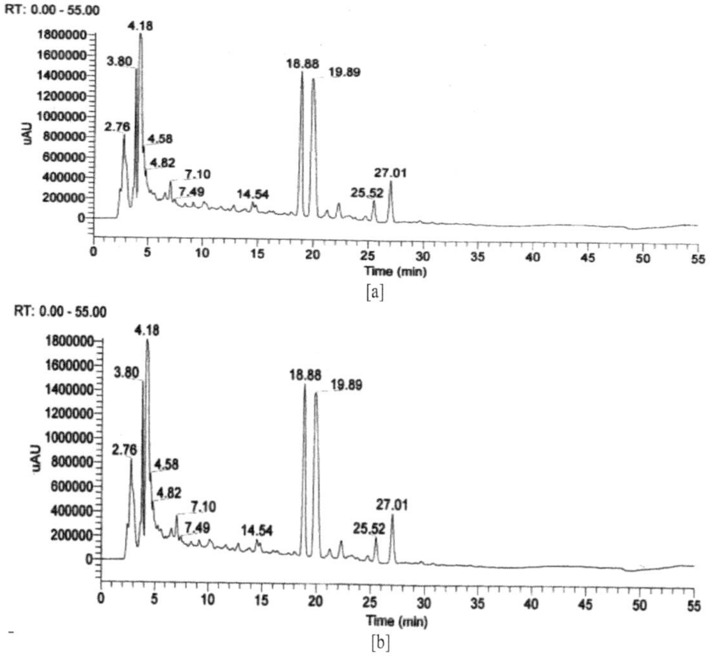

FIGURE 11.4 UPLC profile of purified fraction of (a) *P. aeruginosa* OBP3; and (b) *P. aeruginosa* OBP4.

The mass spectra of the purified RL from the bacterial strain OBP3 and OBP4 showed the presence of RL congers with multiple molecular ions, and the same are presented in Table 11.3.

TABLE 11.3 Chemical Composition of Rhamnolipid Mixture Produced by *P. aeruginosa* Strains as Determined by Mass Spectroscopic Analysis

Bacterial Strains	Rhamnolipid Congeners	Pseudomolecular Ion (m/z)
P. aeruginosa OBP1	Rha-C_{10}	331
	Rha-$C_{12:2}$	359
	Rha-C_8–C_{10}	479
	Rha-C_{10}–C_{10}	505
	Rha-C_{10}–$C_{12:1}$	528
	Rha-C_{12}–C_{10}	531
	Rha-C_{10}–C_{12}	531
	Rha-Rha-C_8–C_{10}	621
	Rha-Rha-C_{10}–C_{10}	648
	Rha-Rha-C_{10}–$C_{12:1}$	674

TABLE 11.3 *(Continued)*

Bacterial Strains	Rhamnolipid Congeners	Pseudomolecular Ion (m/z)
OBP2	Rha-$C_{8:2}$	302
	Rha-$C_{12:2}$	357
	Rha-C_{10}–C_{10}	501
	Rha-C_{10}–$C_{12:1}$	527
	Rha-Rha-C_{10}–C_{12}	531
	Rha-Rha-C_{12}–C_{10}	531
	Rha-Rha-C_{10}–C_{10}	648
	Rha-Rha-C_{10}–$C_{12:1}$	675
OBP3	Rha-$C_{8:2}$	302
	Rha-C_{10}	333
	Rha-C_{8}–C_{10}	477
	Rha-Rha-C_{10}	480
	Rha-C_{10}–C_{10}	502
	Rha-C_{10}–$C_{12:1}$	529
	Rha-Rha-C_{10}–C_{12}	531
	Rha-Rha-C_{12}–C_{10}	531
	Rha-Rha-C_{8}–C_{10}	622
OBP4	Rha-$C_{8:2}$	302
	Rha-C_{10}	332
	Rha-$C_{12:2}$	358
	Rha-C_{8}–C_{10}	477
	Rha-C_{10}–C_{10}	502
	Rha-C_{10}–C_{12}	532
	Rha-Rha-C_{8}–C_{10}	622
	Rha-Rha-C_{10}–C_{10}	648
	Rha-Rha-C_{10}–$C_{12:1}$	673
	Rha-Rha-C_{10}–C_{12}	679

11.7.4 *THERMOGRAVIMETRIC ANALYSIS (TGA)*

The thermal stability and decomposition profile of the BS samples were determined by using a thermogravimetric analyzer (Shimadzu TGA-50, Japan) operated at a heating rate of 10°C.min^{-1} under nitrogen flow rate of 30 ml.min^{-1}. The BS samples were heated from 30 to 600°C. TGA analyzes of the four isolated four BSs were carried out (weight loss versus temperature) and are presented in Figure 11.5. Thermogravimetric analysis (TGA) of dried BSs produced by *P. aeruginosa* strains during growth in MSM supplemented with n-hexadecane. Degradation study was done by heating the samples from 25 to 600°C.

Degradation study was done by heating the samples from 25 to 600°C. The thermogram of all the BS samples exhibited a two-step degradation pattern. The initial degradation of the BSs of OBP1, OBP2, OBP3, and OBP4 occurred at around 156.6, 135.6, 135.6 and 145.3°C, respectively, and its corresponding weight loss were in the order of 7.95, 6.26, 7.5 and 6.9%, respectively. The second step degradation of the BS samples were observed at 284.4, 276.3, 257.0 and 269.9C, respectively, and determined as degradation temperature (T_d). In the second step degradation, there were 24.1, 28.8, 26.1 and 26.9% weight loss in the BS samples of OBP1, OBP2, OBP3, and OBP4, respectively. The retention of 43–44% weight by the BS samples even after heating at 600°C reveals their thermo-stability.

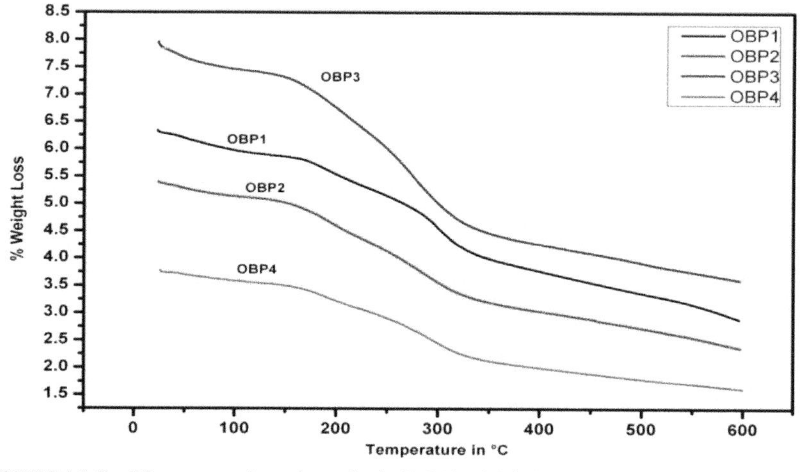

FIGURE 11.5 Thermogravimetric analysis (TGA) of dried biosurfactants produced by *P. aeruginosa* strains during growth in mineral salt medium supplemented with n-hexadecane.

11.7.5 DIFFERENTIAL SCANNING CALORIMETRY (DSC)

Differential scanning calorimetry (DSC) was used to assess the thermal properties of BS samples. 10 mg of dried BS was weighed into the aluminum pans and sealed hermetically prior to analysis. Sample was then heated at 10°C.min^{-1} under a dry nitrogen purge (50 ml.min^{-1}) from

30°C–400°C. Analysis of data for obtaining the onset peak and melting temperature as well as enthalpy was carried out using the universal V3.5 BTA instrument software. The melting temperature (Tm) was determined from the DSC endotherms. DSC thermograms of the four BSs are presented in Figure 11.6.

The thermogram of the BS samples revealed two different endothermic peaks. BS of the bacterial strain OBP1 exhibited two distinct endothermic peaks at around 16°C and 131°C. Also, a third smaller endothermic peak was observed nearby 140°C. The first peaks represent the enthalpy of dehydration and the second peak indicates the enthalpy of decomposition. DSC of dried BS produced by *P. aeruginosa* strains during growth in MSM supplemented with n-hexadecane.

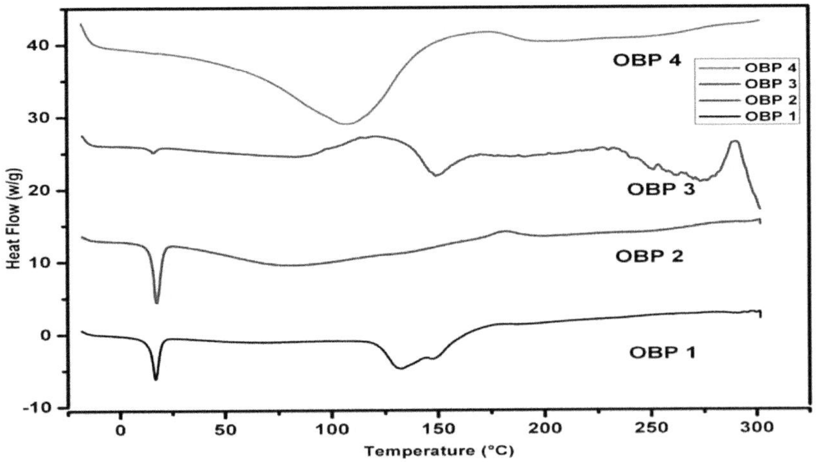

FIGURE 11.6 Differential scanning calorimetry (DSC) of dried biosurfactant produced by *P. aeruginosa* strains during growth in mineral salt medium supplemented with n-hexadecane.

11.7.6 TLC, FTIR, LC-MS, TGA, AND DSC-BASED BIOCHEMISTRY OF BACTERIAL BIOSURFACTANTS (BSS)

The TLC-derived fractions showed a positive reaction to sugars with orcinol reagents and lipids with iodine vapors, indicating the presence of both glycosyl units and lipid moieties on the same spots, but possessed negative reaction to amino groups with ninhydrin. The lower fraction with the average R*f* value of 0.32 indicated the presence of di-rhamnolipids while other with an average R*f* value of 0.71 for mono-rhamnolipid molecules. These observations were found to be consistent with the reports of Monteiro et al. (2007); Silva et al. (2010); Abbasi et al. (2012); and Lotfabad et al. (2010). The results of TLC clearly confirmed the glycolipid nature of the BS produced by four bacterial strains. Moreover, two independent assays comprising of CTAB agar test (Siegmund and Wagner, 1991) and orcinol assay (Chandrasekaran and BeMiller, 1980) along with TLC confirmed the production of RL by the bacterial strains in the n-hexadecane supplemented medium.

The infrared spectroscopic analysis revealed the presence of glycolipid-type compounds in the isolated BS samples. The appearance of additional bands in the spectra might be the result of contamination of polypeptides and polysaccharides from cell debris during the extraction of BS from the culture supernatant (Lotfabada et al., 2010). Most of the known BSs are glycolipids in nature (Desai and Banat, 1997), carbohydrates in combination with long-chain aliphatic acids or hydroxyaliphatic acids. RLs, trehalolipids, and sorpholipids are among the best-known glycolipids. *P. aeruginosa* (Robert et al., 1989), *R. erythropolis* (Rapp, 1979), *Mycobacterium sp.* (Cooper and Zajic, 1980) and *Torulopsis bombicola* (Gobbert et al., 1984) were reported to produce BSs that are glycolipid in nature. Based on biochemical, TLC, and FTIR results, it was assumed that the BSs produced by the bacterial strains had a common glycolipid type structure.

The LC-MS analysis of the BS samples produced by *P. aeruginosa* strains confirmed the results of the TLC with peak values appearing at 171, 193, 205, 339 m/z indicated the presence of lipids and peak value at 163 for carbohydrate moiety (Aparna et al., 2012; Naik and Sakthivel, 2006). The m/z values obtained were consistent with the molecular structure of Rha-$C_{8:2}$, Rha-C_{10}, Rha-$C_{12:2}$, Rha-C_8-C_{10}, Rha-C_{10}-C_{10}, Rha-C_{10}-C_{12}, Rha-C_{10}-$C_{12:1}$, Rha-Rha-C_8-C_{10}, Rha-Rha-C_{10}-C_{10}, Rha-Rha-C_{10}-$C_{12:1}$ and Rha-Rha-C_{12}-C_{10}. A total of twelve RL homologs were identified from the BSs produced by the bacterial strains, which included both mono and di-rhamnolipids. In the literature, the number of RL homologs reported varies from 4 to 28 (Lang and Wullbrandt, 1999; Benincasa and Accorsini, 2008). The predominant RL components present in the BSs of the bacterial strains were Rha-C10, Rha-C12:2, Rha-C8-C10, Rha-C10-C10, Rha-C10-C12, Rha-C10-C12:1,

Rha-Rha-C8-C10, Rha-Rha-C10-C10 and Rha-Rha-C10-C12:1. Results clearly showed the predominance of di-rhamnolipids over the mono-rhamnolipids and typically the main RLs were found to be Rha-C10-C10, Rha-Rha-C8-C12 and Rha-Rha-C10-C12. The results of MS as described by Aparna et al. (2012); Haba et al. (2003); Deziel et al. (1999); Lotfabad et al. (2010); and Yin et al. (2009) were found to be quite close to our findings with the higher abundance of di-rhamnolipids over mono-rhamnolipids. The difference between the RL composition and predominance of a particular type of congener in the present investigation was probably due to the factors like type of carbon substrate (Naik and Sakthivel, 2006; Choi et al., 2009), culture conditions (Leahy and Colwell, 1990), age of the culture (Zukerberg et al., 1979) and the strains of *P. aeruginosa* used (Naik and Sakthivel, 2006). Moreover, the RLs congeners identified were also found to depend on the nature of the analytical methodology used (Monterio et al., 2007). Benincasa et al. (2008); Haba et al. (2003); Costa et al. (2010); Déziel et al. (2003); and Lotfabad et al. (2010) used HPLC/MS, ESI-MS, and LC-MS; and reported the complete elucidation of RL mixtures produced by the different strains of *P. aeruginosa*. However, no one used n-hexadecane as the sole source of carbon. Further, the variation in the culture composition and the differences in the strains could explain the differences between their results and the present work.

For all four BSs, the weight loss corresponding to the temperature of 65–110°C was due to the loss of adsorbed moisture, consequently suggesting the samples being not completely anhydrous. Due to the presence of different types of moieties and linkages in the surfactant structures, they display a two-step degradation pattern as confirmed from the thermogram. The first step of degradation started at 159°C and ended at 353C resulting in a weight loss of about 19%, which might be due to the dehydration of the ester or the acid groups of the lipids. The same resulted in the generation of water or carbon dioxide as volatiles. The oxidative reactions are mainly responsible for the accelerated degradation. The second stage degradation might be due to the breaking of C-O linkages of the main structure.

Finally, at high temperature, there could be formations of heat-stable structures as shown by the TGA thermogram. A higher amount of 37–43% residues were observed at 600°C that might be due to the formation of fused complex C-C linkages or char, which is highly thermostable in nature. Deprotonation of the glycosyl head at higher temperature consequently resulted in the depolymerization of RL and subsequently lead to the generation of a substantial amount of char. TGA clearly indicated the presence of di-RL as the predominant form in the BS samples. Various factors such as interactions between the congeners of RL molecules, appearance of a steric hindrance because of branched aliphatic chains, and acyclic structures collectively might attribute to the thermo-stability of the BSs. Moreover, interaction and extensive H-bonding between the BS molecules were also evident from the FTIR studies.

The DSC study was performed on the BSs produced by the bacterial strains. All the samples except OBP4 exhibited two endothermic peaks at around 16°C and 131°C showing a lower enthalpy with lower temperature pre-transition and higher enthalpy with higher temperature main transitions. The first transition might be due to the formation of a crystalline phase by the aliphatic branching groups, which could be inter or intramolecular. At low temperatures, there was a cessation in the molecular motion of the molecules. The lower enthalpy transitions were mainly due to the chain unfolding of the lipids. The presence of flexible symmetric moieties and polar linkages in BS structures might form partially crystalline phases at lower temperatures which could be verified from the DSC thermogram. The second transition could be referred to as the melting of those crystalline regions. Melting of crystalline phase at high temperature was due to the stability of the crystalline structure, which might be because of the polar nature of the moieties present in the BS structures. However, it was observed that the lower endothermic peak Disappears in OBP4-BS, suggesting a higher concentration of di-rhamnolipids.

However, there are instances at which the heat capacity profile displays only one maximum. This is because the DSC could describe the bulk behaviour of one single event, which might not be detectable since they show only a small contribution to the overall melting. In all the cases, the increase in the di-rhamnolipid concentrations resulted in the broadening of the higher enthalpy transitions and shifting of the peak to lower values.

KEYWORDS

- **cetyl trimethyl ammonium bromide**
- **differential scanning calorimetry**
- **electrospray**
- **Fourier transform infrared spectroscopy**
- **liquid chromatography and mass spectroscopy**
- **thermogravimetric analysis**
- **thin-layer chromatography**

REFERENCES

Abbasi, H., et al., (2012). Biosurfactant-producing bacterium, *Pseudomonas aeruginosa* MA01 isolated from spoiled apples: Physicochemical and structural characteristics of isolated biosurfactant. *J. Biosci. Bioengg., 113*, 211–219.

Abdel-Mawgoud, A. M., et al., (2009). Characterization of rhamnolipid produced by *Pseudomonas aeruginosa* isolate BS20. *Appl. Biochem. Biotechnol., 2*, 329–345.

Abdel-Mawgoud, A. M., et al., (2010). Rhamnolipids: Diversity of structures, microbial origins and roles, *Appl. Microbiol. Biotechnol., 86*, 1323–1336.

Aparna, A., et al., (2012). Production and characterization of biosurfactant produced by a novel *Pseudomonas* sp. 2B. *Colloids Surf. B. Biointerfcaes, 95*, 2–2 9.

Benincasa, M., & Accorsini, F. R., (2008). *Pseudomonas aeruginosa* LBI production as an integrated process using the wastes from sunflower-oil refining as a substrate. *Bioresour. Technol., 99*, 3843–3849.

Benincasa, M., et al., (2004). Chemical structure, surface properties, and biological activities of the biosurfactant produced by *Pseudomonas aeruginosa* LBI from soap-stock. *Antonie Van Leeuwenhoek, 85*, 1–8.

Bordoloi, N. K., & Konwar, B. K., (2009). Bacterial biosurfactant in enhancing solubility and metabolism of petroleum hydrocarbons. *Journal of Hazardous Materials, 170*(1), 495–505.

Buchanan, R. E., & Gibbons, N. E., (1974). *Bergey's Manual of Determinative Bacteriology* (8th edn.). Williams and Wilkins, Baltimore.

Cappuccino, J. G., & Shermam, N., (1999). *Microbiology: A Laboratory Manual* (4th edn.). The Benjamin/Cummins Publishing Company Inc., California, USA.

Chandrasekaran, E. V., & BeMiller, J. N., (1980). Constituents' analysis of glycosaminoglycans. In: Whistler, R. L., (ed.), *Methods in Carbohydrate Chemistry* (p. 89). Academic, New York.

Choi, Y. J., et al., (2009). Enhancement of aerobic biodegradation in an oxygen-limiting environment using a saponin-based microbubble suspension. *Environ. Pollut., 157*, 2197–2202.

Cooper, D. G., & Zajic, J. E., (1980). Surface-active compounds from microorganisms. *Adv. Appl. Microbiol., 26*, 229–253.

Costa, S. G. V. A. O., et al., (2006). Production of *Pseudomonas aeruginosa* LBI rhamnolipids following growth on Brazilian native oils. *Process Biochem., 41*, 483–488.

Costa, S. G. V. A. O., et al., (2009). Cassava wastewater as· a substrate for the simultaneous production of rhamnolipids and, polyhydroxyalkanoates by·*Pseudomonas aeruginosa. J. Ind. Microbiol. Biotechnol., 36*, 1063–1072.

Costa, S. G. V. A. O., et al., (2010). Structure, properties, and applications of rhamnolipids produced by *Pseudomonas aeruginosa* L2-1 from cassava wastewater. *Process Biochem., 45*, 1511–1516.

Desai, J. D., & Banat, I. M., (1997). Microbial production of surfactants and their commercial potential. *Microbiol. Mol. Bio. Rev., 61*, 47–64.

Deziel, E., et al., (1996). Biosurfactant production by a soil *Pseudomonas* strains growing on polycyclic aromatic hydrocarbons. *Appl. Environ. Microbiol., 62*, 1908–1912.

Deziel, E., et al., (1999). Liquid chromatography/mass spectrometry analysis of mixtures of rhamnolipids produced by *Pseudomonas aeruginosa* strain 57RP grown on mannitol or naphthalene. *Biochim. Biophys. Acta, 1440*, 244–252.

Deziel, E., et al., (2003). rhlA is required for the production of a novel biosurfactant promoting swarming motility in *Pseudomonas aeruginosa:* 3-(3-hydroxyalkanoyloxy) alkanoic acids (HAAs), the precursors of rhamnolipids. *Microbiol., 149*, 2005–2013.

Dubois, M., et al., (1956). Colorimetric method for determination of sugars and related substances. *Anal. Chem., 28*, 350–356.

Folch, J. M., et al., (1956). A simple method for the isolation and purification of total lipids from animal tissues. *J. Biol. Chern., 226*,497–509.

George, S., & Jayachandran, K., (2008). Analysis of rhamnolipid· biosurfactants produced through submerged fermentation using orange fruit peelings as sole carbon source. *Appl. Biochem. Biotechnol., 58*, 428–434.

Gobbert, U., et al., (1984). Sophorose lipids formation by resting cells of *Torulopsis bombicola. Biotechnol. Lett., 6*, 225–230.

Haba, E., et al., (2003). Physicochemical characterization and antimicrobial properties of rhamnolipids produced by *Pseudomonas aeruginosa* 47T2 NCBIM 40044. *Biotechnol. Bioengg., 81*, 316–322.

Haba, E., et al., (2003). Use of liquid chromatography-mass spectrometry for studying the composition and properties of rhamnolipids produced by different strains of *Pseudomonas aeruginosa. Surf. Deterg., 6*, 155–161.

Lang, S., & Wullbrandt, D., (1999). Rhamnose lipids-biosynthesis, microbial production and application potential. *Appl. Microbiol. Biotechnol., 51*, 22–32.

Leahy, J. G., & Colwell, R. R., (1990). Microbial degradation of hydrocarbons in the environment. *Microbiol. Rev., 54*, 305–315.

Lotfabad, T. B., et al., (2009). An efficient biosurfactant-producing bacterium *Pseudomonas aeruginosa* MR01, isolated from oil excavation areas in south of Iran. *Colloids Surf. B. Biointerfaces, 69*, 183–193.

Lotfabada, T. B., et al., (2010). Structural characterization of a rhamnolipid-type biosurfactant produced by *Pseudomonas aeruginosa* MR01: Enhancement of di-rhamnolipid proportion using gamma irradiation. *Colloids Surf. B. Biointerfaces, 81*, 397–405.

Monterio, S. A., et al., (2007). Molecular and structural characterization of the biosurfactant produced by *Pseudomonas aeruginosa* DAUPE 614. *Chem Phys Lipids, 147*, 1–13.

Mulligan, C., et al., (1984). Selection of microbes producing biosurfactants in media without hydrocarbons. *J. Ferment. Technol., 62*, 311–314.

Naik, R. P., & Sakthivel, N., (2006). Functional characterization of a novel hydrocarbon clastic *Pseudomonas* sp. strain PUP6 with plant-growth-promoting traits and antifungal potential. *Res. Microbiol., 157*, 538–546.

Rapp, P., et al., (1997). Formation, isolation and characterization of trehalose dimycolates from *Rhodococcus erythropolis* grown on n-alkanes. *J. Gen. Microbiol., 115*, 491–503.

Robert, M., et al., (1989). Effect of the carbon source on biosurfactant production by *Pseudomonas aeruginosa* 44T1. *Biotechnol. Lett., 11*, 871–874.

Siegmund, I., & Wagner, F., (1991). New method for detecting rharnnolipids excreted by *Pseudomonas* species during growth on mineral agar. *Biotechnol. Techn., 5*, 265–268.

Silva, S. N. R. L., et al., (2010). Glycerol as substrate for the production of biosurfactant by *Pseudomonas aeruginosa* UCP 0992. *Colloids Surf. B. Biointerfaces, 79*, 174–183.

Tahzibi, A., et al., (2004). Improved production of rhamnolipids by a *Pseudomonas aeruginosa* mutant. *Iran. Biomed. J.,* 8, 25–31.

Tuleva, B. K., et al., (2002). Biosurfactant production by a new *Pseudomonas putida* strain. *Z. Naturforsch., 57*C, 356–360.

Wu, J. Y., et al., (2008). Rhamnolipid, production with indigenous *Pseudomonas aeruginosa* EMI isolated from oil-contaminated site. *Bioresour. Technol., 99*, 1157–1164.

Yin, H., et al., (2009). Characteristics of biosurfactant produced by *Pseudomonas aeruginosa* S6 isolated from oil-contaminating wastewater. *Process Biochem., 44*, 302–308.

Zukerberg, A., et al., (1979). Emulsifier of *Arthrobacter* RAG-1: Chemical and physical properties. *Appl. Environ. Microbiol., 37*, 414–420.

CHAPTER 12

Application of Bacterial Strains and Biosurfactants in Bioremediation

12.1 SOIL WASHING EXPERIMENT

Acid washed and dried sand was mixed with crude oil (10%, w/w) and left at room temperature (RT) for 12 days. The crude oil-contaminated sand samples weighing 20 g was transferred to each of 250 ml Erlenmeyer flasks containing 100 ml of aqueous biosurfactant (BS) solution at various concentrations (0.001, 0.005, 0.007, 0.01, and 0.1% w/v) and kept at 150 rpm for 24 h at RT. Similarly, the cell-free culture supernatants of the bacterial strains were used in place of BS solution. The contaminated sand samples were separated, dried, and washed twice with dichloromethane. The solvent part was removed, and the residual oil was determined gravimetrically. The percentage of oil removed was calculated using the equation:

$$\text{Crude oil removed in \%} = \frac{(\text{Oi} - \text{Or})}{\text{Oi}} \times 100$$

where; Oi is the initial oil in the sand sample (g) before washing and Or is the oil remaining in the sand sample (g) after washing (Wentzel et al., 2007). The mean and standard deviation of triplicates for each treatment were calculated.

12.2 REDUCTION OF VISCOSITY BY THE BIOSURFACTANT (BS)

The culture broths of the bacterial strains after 4 weeks of incubation with crude oil were extracted thrice with the equal volume of dichloromethane, dried over anhydrous sodium sulfate, filtered, and concentrated in vacuum. Viscosity measurement was performed on Ostwald viscometers, which allows the determination of viscosity of the control/treated crude oil. All determinations were carried out at 25°C using a concentration of 1 $mg.ml^{-1}$ of the extracted crude oil dissolved in hexane.

12.3 DEGRADATION OF PETROLEUM PRODUCTS AND RESIDUAL PETROLEUM

Verification experiment of hydrocarbon degradation was carried out on the residual petroleum products by using combined solvent extraction and column chromatographic techniques as described by Queiroga et al. (431). The residual crude oil was separated from the culture broth with the help of glass separating funnel and were recovered by washing twice with 30 ml of hexane, dried over anhydrous sodium sulfate, filtered, and concentrated in vacuum. Residual petroleum product 100 mg was fractionated on a chromatographic column (1.2 mm×30 cm) packed with 6.0 g silica gel (60–80 mesh size) and saturated with hexane for an overnight period. The aliphatic hydrocarbons were eluted with 30 ml of hexane, aromatic hydrocarbons with 30 ml of hexane-diethyl ether mixture solution (9:1, v/v) and resins and asphaltenes with 30 ml of chloroform-methanol-water mixture solution (21:8.4:0.6, v/v/v). All three extracts were dried at RT over anhydrous sodium sulfate and concentrated in a vacuum. The residual hydrocarbon components were determined by measuring the weight of the dry extract. The pattern of biodegradation and abiotic loss were evaluated by comparing their weights against the control samples as well as the fresh weight of petroleum products.

12.4 GC ANALYSIS OF DEGRADED PETROLEUM PRODUCTS

The saturate hydrocarbons were analyzed using a PerkinElmer gas chromatograph-mass spectrometer (GC-MS; model CP-3800 gas chromatography and Saturn 2200 mass spectrometer, Varian Technologies Japan, Inc.) equipped with a capillary column (TC-1, length 30 m, ID 0.25 mm, film thickness 0.1 lm) obtained from GL Science. For crude oil analysis, column temperature was first held at 50°C for 5 min, and then raised to 280°C. All analyzes were carried out with the split ratio of 20:1. Helium was used as the carrier gas with a flow rate of 0.8 $ml.min^{-1}$. Injector temperature was set at 250°C. The treatment of crude petroleum with the bacterial strains showed a noticeable reduction in the viscosity values as compared to the untreated control. The results are presented in Figures 11.3 and 11.4.

However, bacterial strain OBP4 and OBP2 had a very clear impact on the viscosity of crude oil, which significantly decreased from 48.7 Pa.sec before treatment to 34.6 Pa.sec after 30 days of the treatment. The reduction in the viscosity of crude oil following the treatment with *P. aeruginosa* strains (OBP1, OBP2, OBP3, and OBP4) has been presented in Figure 12.1.

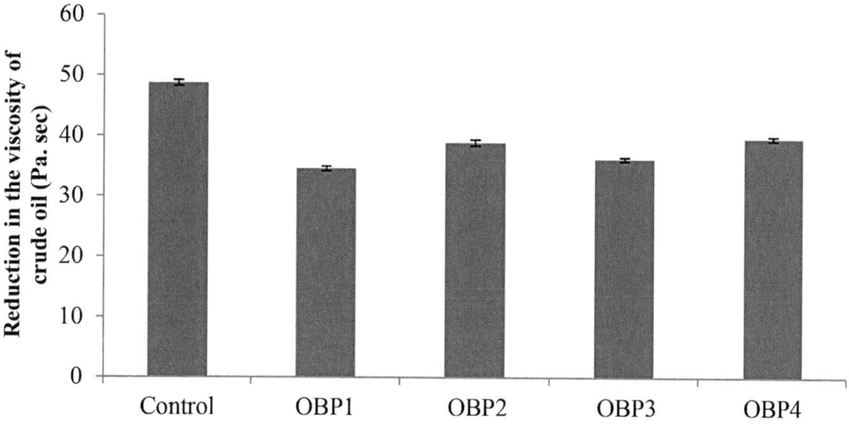

FIGURE 12.1 Reduction in the viscosity of crude oil after treatment with *P. aeruginosa* strains (mean of 3 experiments ± SD).

However, the bacterial strain OBP4 and OBP2 had a very clear impact on the viscosity of crude oil which significantly decreased from 48.7 Pa.sec before the treatment to 34.6 Pa.sec after 30 days of the treatment.

12.5 SOLUBILIZATION OF POLYAROMATIC HYDROCARBON (PAH) BY BIOSURFACTANT (BS)

The solubilization assay was carried out as described by Barkay et al. (287). Three polyaromatic hydrocarbons (PHA) anthracene, phenanthrene, and naphthalene, were selected. Stock solution of all the three PAHs (6 $mg.ml^{-1}$) were prepared in hexane. From the stock, 1 μl was distributed in glass test tubes to achieve 0.6 μg PHA in each. The test tubes were kept open in a fume hood to remove the solvent. This was followed by addition of 3.0 ml of assay buffer (20 mM Tris-HCl, pH 7.0) and 1.0 ml of BS solution in the increasing concentrations (100–800 $\mu g.l^{-1}$) was added to the above test tubes. Test tubes were capped and incubated overnight at 30°C in an orbital incubator shaker at 150 rpm in the dark. The solutions in the test tubes were filtered through a membrane filter (pore size 1.2 μm), and 2.0 ml of the filtrate was extracted with an equal volume of hexane. The resulting mixture of filtrate and hexadecane was centrifuged at 10,000 rpm for 10 min to separate the aqueous and hexane phase. PAHs in the hexane extracts were measured spectrophotometrically at 250, 253, and 273 nm (Barkay et al., 1999). Test tubes containing the assay buffer with PAHs but without BS served as the positive control while test tubes with assay buffer and BS, but without PAHs were used as blank. The data thus obtained are presented in Figure 12.2(a–d).

The effect of BSs on the solubility of PAHs such as phenanthrene, anthracene, and naphthalene were determined in the presence of BS. As shown in Figures 11.5 and 11.6, BSs possessed highly noticeable effect on the solubilization of the three tested PAHs. The solubility of PAHs in water was found to be higher in the presence of BSs as compared to the control having no BS. BSs either below or above its CMC were effective in solubilization of PAHs. Nevertheless, solubilization was much more pronounced when the concentration of BSs produced by the bacterial strains were increased above their respective CMC values. The BSs from all four bacterial strain exhibited appreciable solubilization of phenanthrene. The highest solubilization of phenanthrene was observed in OBP3 BS, followed by OBP4. In the case of anthracene, the BS from the bacterial strainOBP4 showed a significant level of solubilization as compared to the BS from the OBP3 and OBP1. Naphthalene was found to be solubilized significantly only by the BS from OBP4.

12.6 BIODEGRADATION OF CRUDE OIL BY THE BACTERIAL STRAINS

The bacterial strains were assessed for their ability to degrade crude oil components in the culture medium. The medium was supplemented with 2.0 ml (1.9 g) of crude oil and inoculated with the individual bacterial strains separately. After 30 days of incubation, the residual fractions of aliphatic, aromatic, and NSO compounds were determined and are presented in Table 12.1.

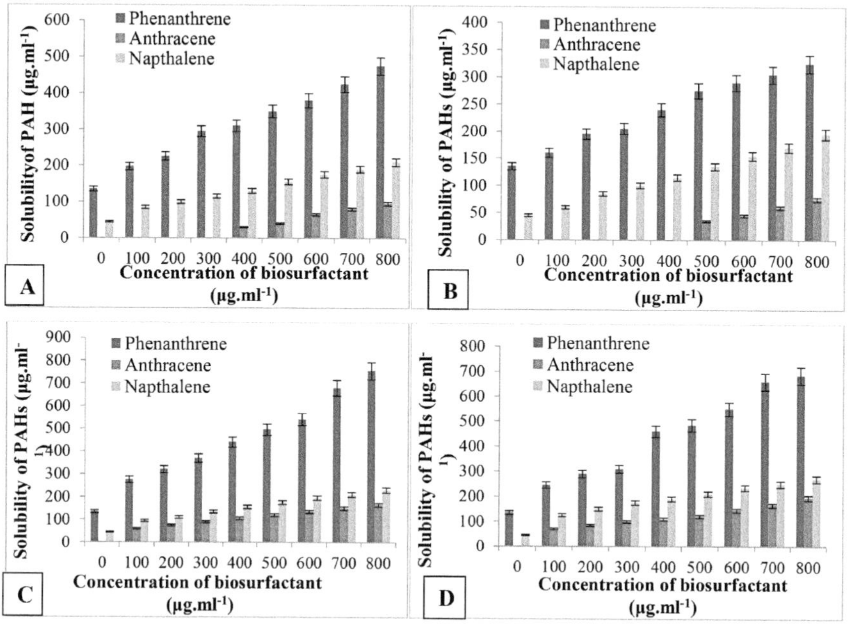

FIGURE 12.2 PAH solubilization assay showing the decrease in the available phenanthrene, anthracene, and naphthalene concentration with increasing concentration of biosurfactants produced by (A) *P. aeruginosa* OBP1; (B) *P. aeruginosa* OBP2; (C) *P. aeruginosa* OBP3; and (D) *P. aeruginosa* OBP4. Mean of 3 experiments ± SD.

TABLE 12.1 Degradation of Aliphatic, Aromatic, and NSO Fractions of Crude Oil by *P. aeruginosa* Strains 30 Days of Treatment in Liquid Culture

Bacterial Strain	Media Supplemented with Crude Oil (g)*	Aliphatic Fraction Degraded (%)	Aromatic Fraction Degraded (%)	NSO Comps Degraded (%)
Control	1.86	12.6±0.8	10.3±0.6	04.7±0.2
P. aeruginosa OBP1	1.86	72.8±0.2	25.2±0.6	11.5±0.5
OBP2	1.86	67.5±0.7	12.3±0.9	07.3±0.1
OBP3	1.86	71.8±0.3	31.8±0.8	13.5±0.4
OBP4	1.86	73.0±0.3	30.3±0.3	14.7±0.7

**Determination* based on crude oil 2.0 ml (1.857 g); Results mean of 3 experiments ± SD.

As shown in Table 12.1, the strain OBP4 followed by OBP1 exhibited high-level degradation of aliphatic hydrocarbons with 73.0 and 72.8%, respectively. On the other hand, the strain OBP3 and OBP2 possessed 71.8 and 67.5% degradation of aliphatic hydrocarbons, respectively. In the case of aromatic fraction, the bacterial strain OBP3 followed by OBP4, OBP1, and OBP2 showed better degradation of 31.8, 30.3, 25.2, and 12.3%, respectively. The strain OBP4 exhibited the highest degradation of NSO compounds with 14.7% while OBP3, OBP1, and OBP2 degraded 13.5, 11.5, and 7.3%, respectively. The degradation of crude oil by the bacterial strains was further confirmed by gas chromatography (GC). The GC profiles of the saturated fractions of the crude oil after 30 days treatment with the bacterial strains along with the control were determined and are presented in Figure 12.3(A)–(E).

A total of 11 different combinations of bacterial strains were cultured in MSM supplemented with 2% (v/v) crude oil for a period of 96 h to determine their ability to grow on crude oil. The final dry biomass yield was determined, and the data are presented in Table 12.2.

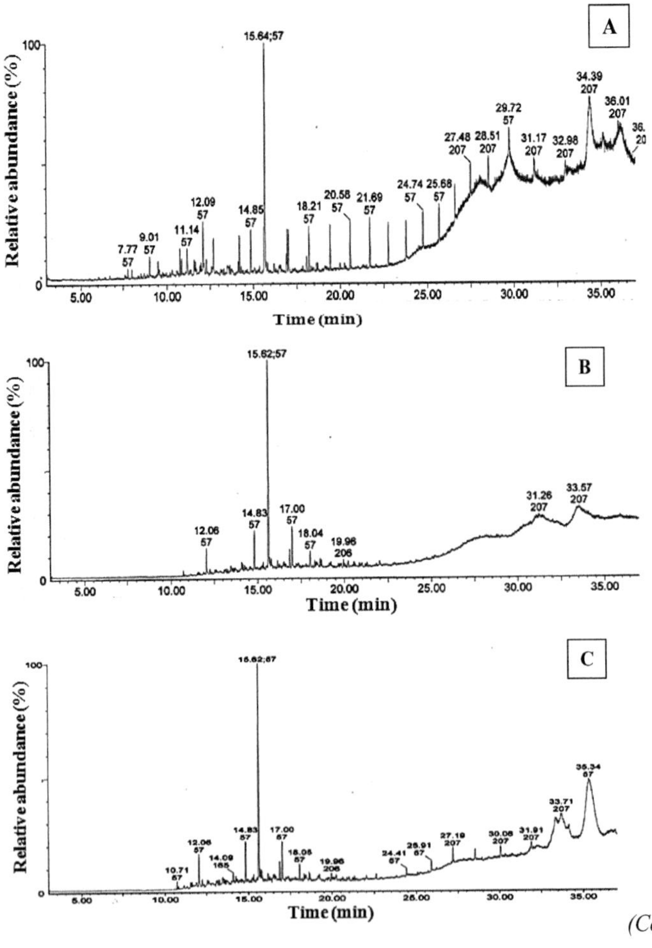

(Continued)

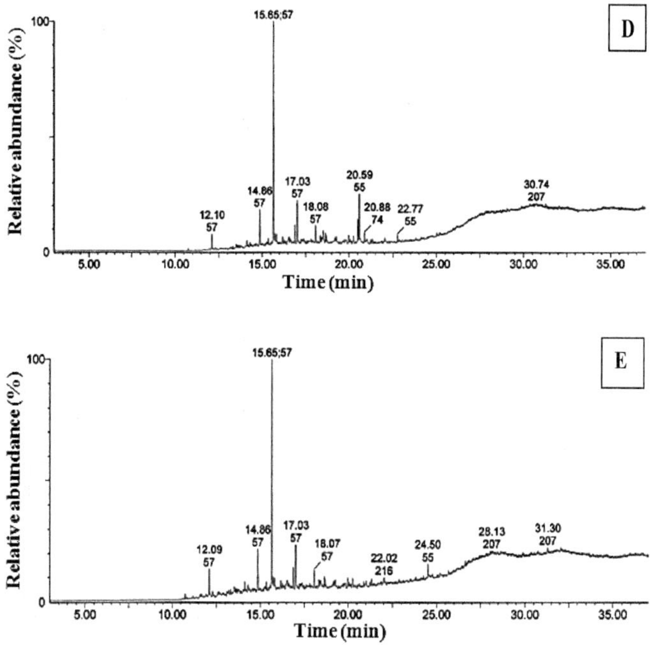

FIGURE 12.3 Gas chromatographic analysis of the saturate fraction of crude oil after treatment with bacterial strains. (A) Control without treatment; (B) *P. aeruginosa* OBP1; (C) *P. aeruginosa* OBP2; (D) *P. aeruginosa* OBP3; and (E) *P. aeruginosa* OBP4.

TABLE 12.2 Biomass of Bacterial Consortia in Mineral Salt Medium Supplemented with Crude After 96 h of Culture [Mean of 3 Experiments ± SD]

SL. No.	Bacterial Consortium	Dry Biomass (g.l^{-1})
1.	OBP1 + OBP2	1.45± 0.01
2.	OBP1 + OBP3	1.35± 0.04
3.	OBP1 + OBP4	0.85± 0.1
4.	OBP2 + OBP3	1.34± 0.2
5.	OBP2 + OBP4	0.71± 0.1
6.	OBP3 + OBP4	0.86± 0.1
7.	OBP1 + OBP3+ OBP4	2.72± 0.3
8.	OBP2 + OBP3+ OBP4	1.50± 0.4
9.	OBP1 + OBP2+ OBP4	1.90± 0.1
10.	OBP1 + OBP2+ OBP3	3.34± 0.1
11.	OBP1 + OBP2+ OBP3+ OBP4	3.33± 0.1

Based on growth performance as depicted by the dry biomass yield, bacterial consortia 7 and 11 designated as consortium I and consortium II were selected to assess their ability to degrade crude oil in MSM. The residual crude oil in the culture medium after 30 days of treatment by the two bacterial consortia separately was assayed and is presented in Table 12.3.

TABLE 12.3 Degradation of Aliphatic, Aromatic, and NSO Fractions of Crude Oil by Bacterial Consortia and in the Presence of Biosurfactant After 30 Days*

Bacterial Strain	Culture Media with Crude Oil (g)	Aliphatic Fraction Degraded (%)	Aromatic Fraction Degraded (%)	NSO Compound Degraded (%)
Control	1.85	12.2±0.5	08.9±0.5	05.2±0.7
Consortium I	1.85	78.6±0.5	42.7±0.7	21.60.3
Consortium II	1.85	80.4±0.8	42.4±0.4	19.2±0.4
Consortium I + biosurfactant	1.85	80.7±0.3	43.8±0.5	22.5±0.7
Consortium II+ biosurfactant	1.85	81.6±0.7	42.6±0.2	20.7±0.5

**Determination* based on crude oil 2.0 ml (1.849 g); Results mean of 3 experiments ± SD.

As shown in Table 12.2, the consortia were able to degrade 78.6–80.4% of aliphatic fractions, around 42.5% of aromatic fractions and 19.2–21.6% of NSO containing compounds of crude oil within 30 days. Similarly, the effect of BS on the degradation of crude oil by the bacterial consortia was estimated and the data are presented in Table 12.3. The GC profiles of the saturated fraction of crude oil in culture medium inoculated with consortium I, consortium II and in the presence of externally added BS after 30 days of incubation was determined and are presented in Figure 12.4(A)–(E).

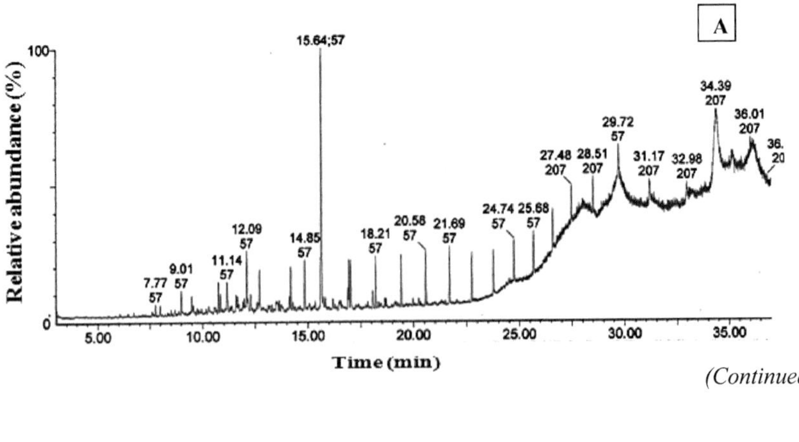

(Continued)

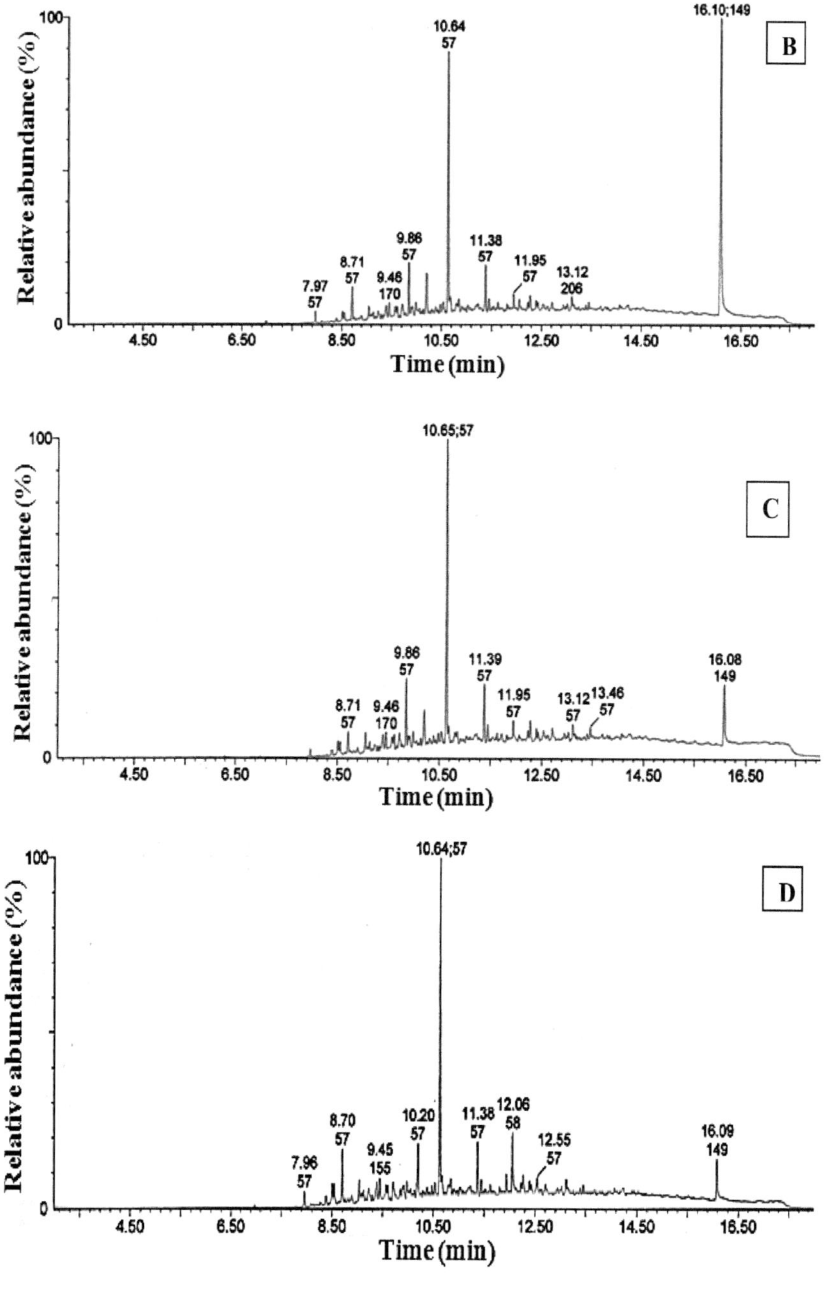

B
Relative abundance (%)
100
0
16.10;149
10.64
57
9.86
57
11.38
57
8.71
57
9.46
170
11.95
57
13.12
206
7.97
57
4.50
6.50
8.50
10.50
12.50
14.50
16.50
Time (min)
C
Relative abundance (%)
100
0
10.65;57
9.86
57
11.39
57
16.08
149
8.71
57
9.46
170
11.95
57
13.12
57
13.46
57
4.50
6.50
8.50
10.50
12.50
14.50
16.50
Time (min)
D
Relative abundance (%)
100
0
10.64;57
12.06
58
11.38
57
10.20
57
8.70
57
9.45
155
12.55
57
16.09
149
7.96
57
4.50
6.50
8.50
10.50
12.50
14.50
16.50
Time (min)

(Continued)

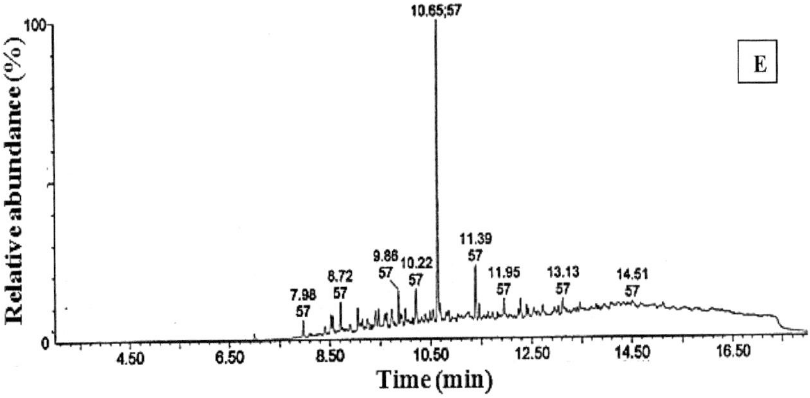

FIGURE 12.4 Gas chromatographic analysis of the saturate fraction of crude oil after treatment with (A) control (no treatment), (B) consortium I; (C) consortium II; (D) consortium I + bio-surfactant; and (E) consortium II+ biosurfactant.

KEYWORDS

- **chromatograph-mass**
- **dichloromethane**
- **gas chromatography**
- **hydrocarbon**
- **polyaromatic hydrocarbons**
- **sodium sulfate**

REFERENCES

Barkay, T., et al., (1999). Enhancement of solubilization and biodegradation of polyaromatic hydrocarbons by the bio-emulsifier Alasan. *Appl: Environ. Microbiol., 65*, 2697–2702.

Queiroga, C. L., et al., (2003). Evaluation of paraffins biodegradation and biosurfactant production by *Bacillus subtilis* in the presence of crude oil. *Brazil. J. Microbiol., 34*, 1321–1324.

Wentzel, A., et al., (2007). Bacterial metabolism of long-chain n-alkanes. *Appl. Microbiol. Biotechnol., 76*, 1209–1221.

CHAPTER 13

Separation of Crude Oil by Biosurfactant from Contaminated Sand and Petroleum Sludge

The capability of the aqueous biosurfactant (BS) solutions to remove crude oil from the crude oil-contaminated sand was investigated and the results are shown in Figure 13.1. The maximum crude oil removed by the BSs isolated from the bacterial strains was attained within their CMC, showing a total removal up to a range of 54–61.7%. However, increased concentration of BSs beyond CMC did not enhance further removal of crude oil from the contaminated sand.

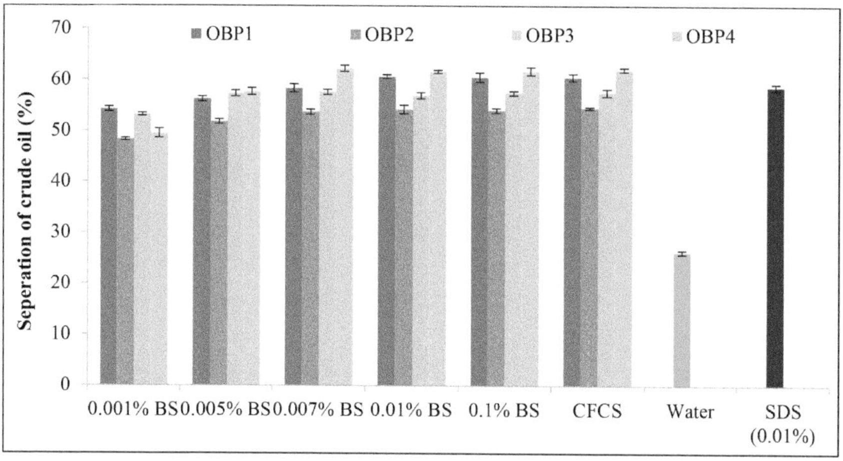

FIGURE 13.1 Removal of crude oil from the contaminated sand after washing with biosurfactant solution (BS) produced by *P. aeruginosa* strains. [CFCS: Cell-free culture supernatant, SDS: sodium dodecyl sulfate. Mean of three experiments ± SD].

The cell-free culture supernatants were also found to be efficient in separating the crude oil from the contaminated sand and exhibited almost similar efficiency as compared to the purified BS solutions. The control with only distilled water was able to remove only 26.2% while SDS could be able to separate 58.6% of crude oil from the contaminated sand.

13.1 RELEASE OF CRUDE OIL FROM SAND PACK COLUMN BY BIOSURFACTANTS (BSS)

The suitability of the isolated biosurfactant (BS) in microbial enhanced oil recovery (MEOR) was evaluated using the sand pack technique as described by Suthur et al. (294). Glass column of 20 mm × 25 mm × 85 mm dimensions with a sieve (100 μm pore size) was packed with 150 g of acid-washed sand of 140 μm particle size. The column was then saturated with 50 ml of crude oil supplied by ONGC, Assam, and India having a density of 0.86 $g.cc^{-1}$ at 15°C. A volume of 50 ml aqueous BS solution was pumped into the column and the amount of oil released was measured. Similarly, cell-free culture supernatant of the bacterial strain was also used in the recovery process. To determine the influence of temperature on the recovery process, the experiment was carried out at room temperature (RT), 50, 70, and 90°C. Sand pack column saturated with crude oil without the addition of aqueous BS was kept as control. The mean and standard deviation of triplicates for each treatment were calculated.

The crude oil-saturated sand pack columns were treated with individual cell-free culture supernatant of the bacterial strains and incubated at RT, 50, 70, and 90°C to determine the release of crude oil from the column. The released crude oil was quantified, and the data are shown in Figure 13.2. At RT the culture supernatant of the bacterial strains could recover 9.3–11.4% of crude oil from the saturated sand pack column, 6.4–8.7% at RT, 7.8–9.7% at 50°C, 8.4–10.5% at 70°C and 9.3–11.4% at 90°C.

The control with the sterilized culture medium could be able to recover only 1.1–3.1% crude oil. The BSs produced in the culture media by the bacterial strain while exposed to higher temperature between 70 and 90°C caused higher recovery of crude oil from the saturated sand pack column. This also proved the stability of BSs in the recovery process being subjected to higher temperatures. Further, addition of fresh medium separately with the bacterial strains could not enhance the further recovery of crude oil from the saturated sand pack column.

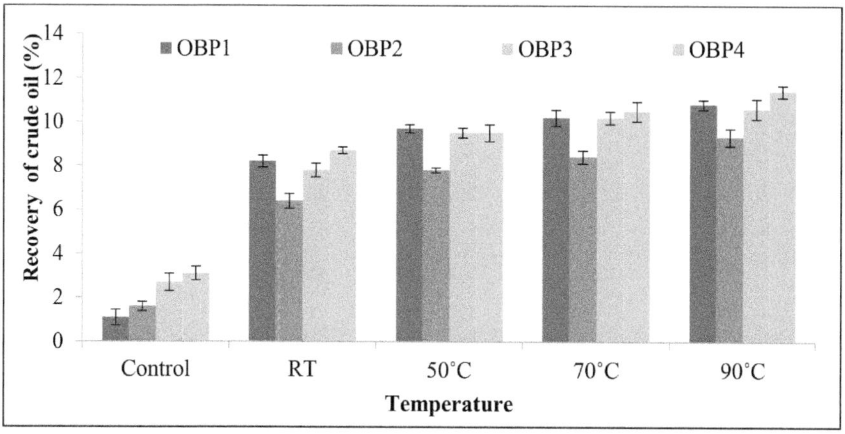

FIGURE 13.2 Recovery of crude oil (%) from the sand pack column at room temperature (RT), 50°C, 70°C, and 90°C after treatment with cell-free culture broth of *P. aeruginosa* strains. Mean of 3 experiments ± SD.

13.2 SEPARATION OF RESIDUAL CRUDE OIL FROM THE PETROLEUM SLUDGE BY BIOSURFACTANTS (BSS)

The petroleum sludge was mixed with acid-washed sterile sand to achieve a sludge concentration between 1 and 9% (w/w). The sludge samples weighing 20 g of the five concentrations were transferred to 250 ml Erlenmeyer flasks containing 100 ml of aqueous BS solution (0.001, 0.01 and 0.1% w/v) separately and kept at constant shaking (100–180 rpm) between 3 and 18 days at RT. After the treatment, the culture flasks could settle for few hours and the treated sludge samples were recovered. The total petroleum hydrocarbon (TPH) of the sludge sample after treatment was estimated and expressed as residual TPH. Flask receiving sludge sample with only water was kept as control (Joseph and Joseph, 2009).

The sticky solid sludge used in the present investigation was blackish-brown in color. The TPH present in the sludge was approximately 785 ± 130 g/kg. Results clearly explained that the suitability of the BSs for the removal of TPH from the petroleum sludge than those removed by water only. The BS solution of bacterial strain OBP1 could separate around 63.4–73.5% of residual crude oil from 7% petroleum sludge in 15 days of incubation and is presented in Figure 13.3(A)–(D).

The separation of the residual oil from the sludge gradually increased with the increase in BS concentration. The maximum recovery of crude oil by the BS solutions were achieved within their respective CMC

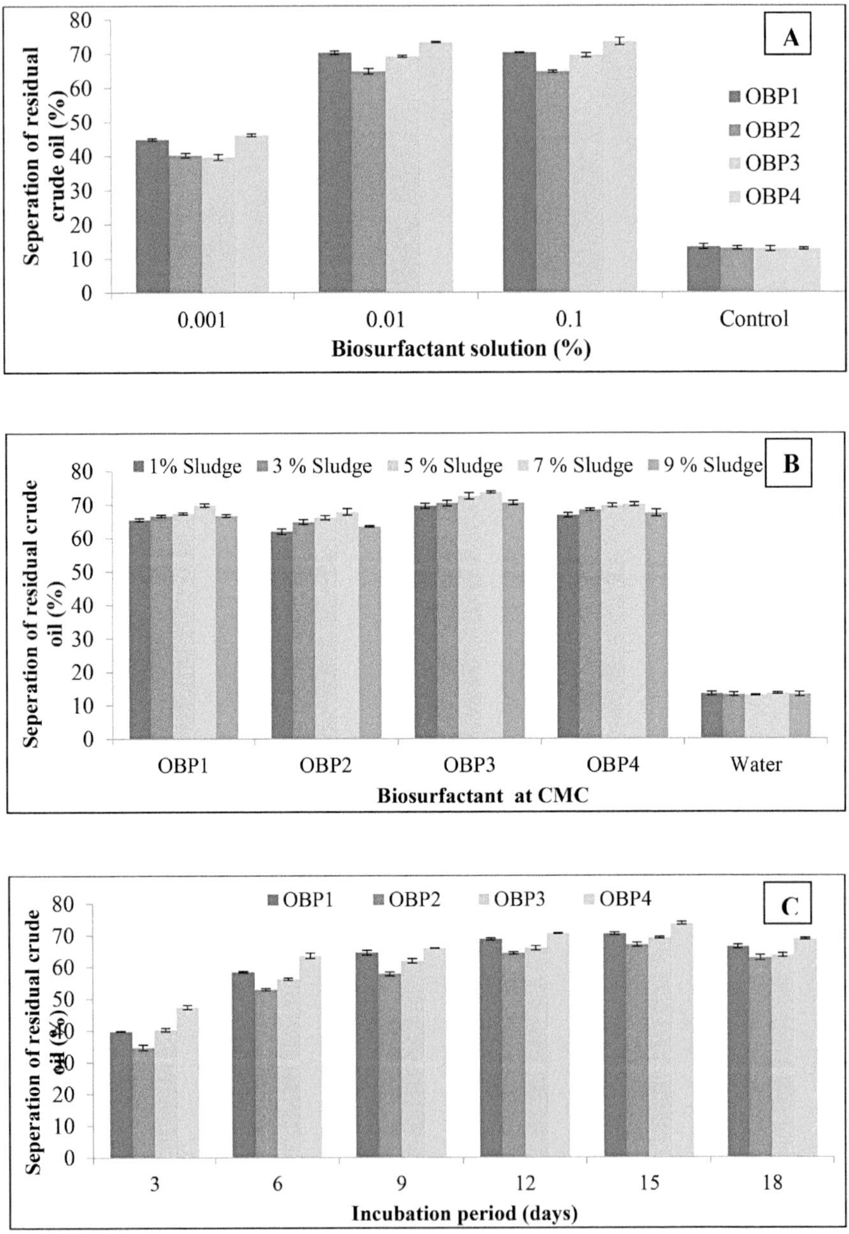
A
Seperation of residual crude oil (%)
80
70
60
50
40
30
20
10
0
OBP1
OBP2
OBP3
OBP4
0.001
0.01
0.1
Control
Biosurfactant solution (%)
B
1% Sludge
3 % Sludge
5 % Sludge
7 % Sludge
9 % Sludge
Seperation of residual crude oil (%)
OBP1
OBP2
OBP3
OBP4
Water
Biosurfactant at CMC
C
OBP1
OBP2
OBP3
OBP4
Seperation of residual crude oil (%)
3
6
9
12
15
18
Incubation period (days)

(Continued)

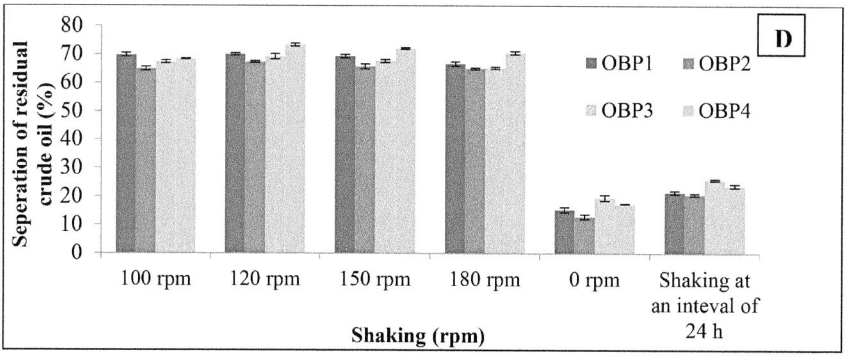

FIGURE 13.3 Effect of different parameters on the separation of residual crude oil from the petroleum sludge. (A) Biosurfactant concentration; (B) sludge concentration; (C) treatment period; and (D) shaking. Mean of 3 experiments ± SD.

values (p-value < 0.05). In the case of concentrations above CMC (Figure 13.3(A)), no enhancement in the release of oil from the sludge was observed (p-value > 0.05).

The amount of residual crude oil released gradually increased along with the treatment time (Figure 13.3(C)), but after 15 days of treatment there was no further increase (p-value > 0.05). Much more residual oil was released when the flasks were continuously shaked as compared to occasional (p-value < 0.05) and no shaking (p-value < 0.05). The release of oil increased with the increase in the rpm value (Figure 13.3(D)), however above 120 rpm there was a significant decrease in the separation of residual oil (p-value > 0.05).

The maximum recovery of crude oil by the BS solutions were achieved within their respective CMC values. In the case of concentrations above CMC, no enhancement in the release of oil from the sludge was observed. Similarly, cell-free culture supernatants of the bacterial strains were also efficient in removing the residual crude oil from the sludge and were almost comparable to their respective BS solutions. Separation of oil was up to 7% (w/w) sludge concentrations, but above this concentration, there was no further increase in the release of the residual crude oil from the sludge, as they could not form homogenous slurry.

Dry acid washed sand was mixed with crude oil (10%, w/w) and left at RT for 12 days. The crude oil-contaminated sand samples weighing 20 g was transferred to each of 250 ml Erlenmeyer flasks containing 100 ml of aqueous BS solution at various concentrations (0.001, 0.005, 0.007, 0.01 and 0.1% w/v) and kept at 150 rpm for 24 h at RT. The percentage of oil removed was calculated using the equation:

$$\text{Crude oil removed in \%} = \frac{(\text{Oi} - \text{Or})}{\text{Oi}} \times 100$$

where; Oi is the initial oil in the sand sample (g) before washing and Or is the oil remaining in the sand sample (g) after washing (Costa et al., 2010). The mean and standard deviation of triplicates for each treatment were calculated.

The TPH of the sludge sample after treatment was estimated and expressed as residual TPH. Flask receiving sludge sample with only water was kept as control.

13.3 DEGRADATION OF BACTERIAL BIOSURFACTANTS (BSS)

BSs isolated from the bacterial strains were significantly degraded by the bacterial strains *P. aeruginosa* (MTCC8165) and *B circulans* (MTCC8167) in liquid culture. The degradation patterns of BSs are shown in Figure 13.4(A)–(C). In both liquid cultures of *P. aeruginosa* (MTCC8165) and *B circulans* (MTCC8167) no inhibitory effect of the tested BSs was observed. The bacterial strains showed normal growth behavior. Both strains could utilize all four types of BSs efficiently, as revealed by the decrease in the BS concentration and increase in biomass content in the culture medium with the increase in the incubation period. However, the rate of degradation in all four types of BSs was more in the case of *B circulans* (MTCC8167) as compared to *P. aeruginosa* (MTCC8165).

The biodegradation assay of the isolated BSs was carried out using the procedure of by Zeng et al. (432). The experiment was carried out in 250 ml Erlenmeyer flasks containing 100 ml of culture media consisting of NH_4Cl (1.0 g), K_2HPO_4 (1.0 g), KH_2PO_4 (1.0 g), $MgSO_4.7H_2O$ (0.05 g), $CaCl_2$ (0.02 g), $FeSO_4.7H_2O$ (250 μg), glucose (1.0 g) and BS sample (1 g). The pH of the culture medium was adjusted to 7.2 and sterilized by autoclaving at 121°C for 15 min. Gram-negative bacteria *P. aeruginosa* (MTCC8165) and Gram-positive bacteria *B. circulans* (MTCC8167) were selected for the biodegradation assay. Pure cultures of the bacterial strains were prepared using Luria Bertani (LB) broth and incubated overnight in

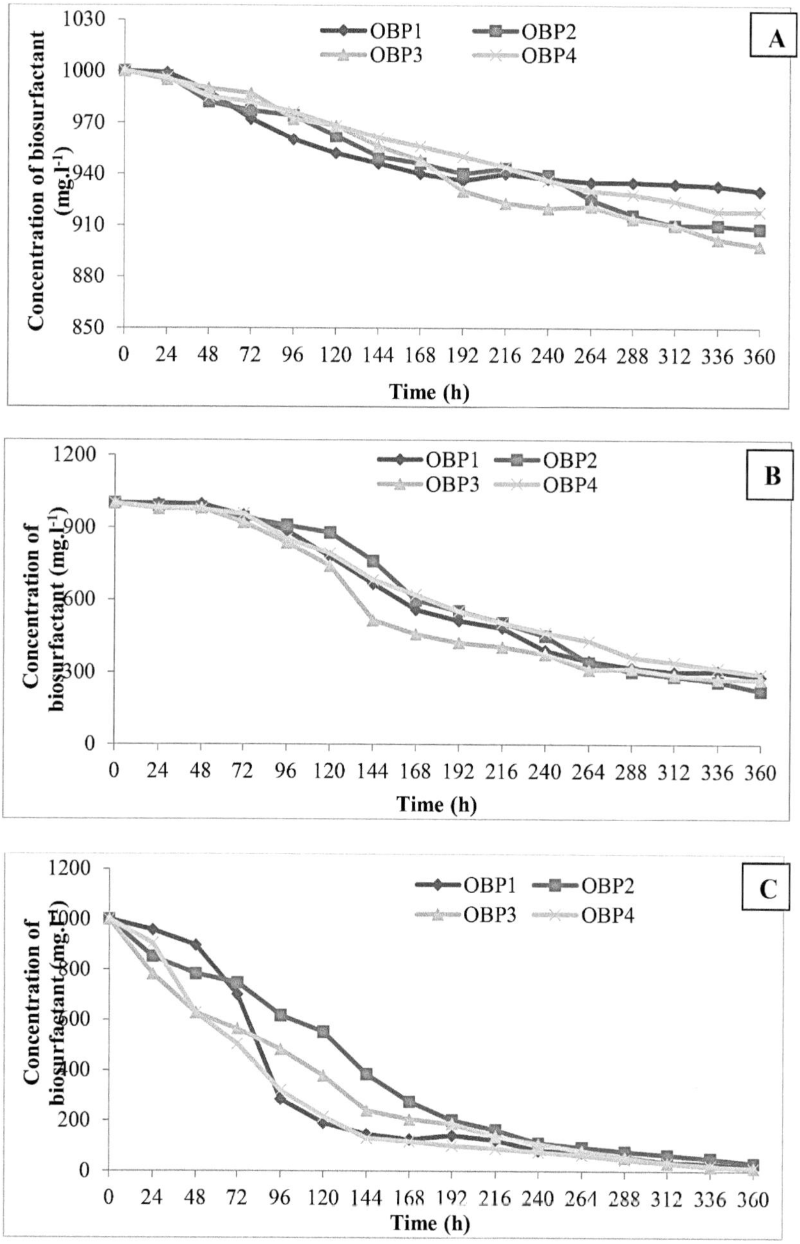

FIGURE 13.4 Degradation of biosurfactants produced by the selected bacterial strains by (A) *P. aeruginosa* (MTCC8165); (B) *B. circulans* (MTCC8167); and (C) in garden soil Mean of 3 experiments ± SD.

an orbital incubator shaker at 37°C with 180 rpm. An aliquot of 100 μl of overnight grown culture broth containing 1×108 ml^{-1} cells (McFarland turbidity method) was inoculated to the 250 ml volume Erlenmeyer flask containing 100 ml of MSM.

The degradation experiment was carried out at 37°C in an orbital incubator shaker at 180 rpm. For analytical use, 2 ml of the culture broth was removed from the culture flask at time intervals of 24 h. After centrifuged at 8,000 rpm for 15 min, the cell-free culture supernatant was separated and used for the determination of rhamnolipid (RL) concentration by using orcinol as described by Chandrasekaran and BeMiller (1980). The top surface soil was collected from the Departmental garden and distributed equally in several earthen pots. RL solution was added into the earthen pots and the ratio of soil and RL was adjusted to 800:1 (w: w). Water was added to earthen pots and the water content was adjusted to about 60%. The earthen pots were kept open air in a shed. The water content in the earthen pots was determined at an interval of 24 h and was maintained between 50 and 60% with fresh sprinkling. For analytical use, the soil sample was mixed up with water at the ratio of 1: 20 and incubated on an orbital incubator shaker at 100 rpm for 30 min until the suspension was formed. The residual suspension was centrifuged at 8,000 rpm for 15 min to recover the supernatant and used for the determination of RL content using the standard procedure as described above. Each experiment was repeated thrice to determine the mean and standard deviation.

Biodegradation of BSs produced by four *P. aeruginosa* strains confirmed their biodegradability nature when co-cultured with glucose in the culture medium. Several reports demonstrated co-degradation or sole utilization as a source of carbon and energy by various bacterial monocultures (Chaillan et al., 2004; Das and Mukherjee, 2007). Moreover, the degradation behavior of BS depends on the type of bacterial species involved. In the present investigation, the bacterial strains used for the degradation study are potential degraders of complex hydrocarbons isolated from the petroleum hydrocarbon contaminated soil (Bordoloi and Konwar, 2007). Chrzanowski et al. (2010) suggested that RLs might be preferentially degraded by a consortium of hydrocarbon degraders due to structural similarities between lipid moieties of RLs and fatty acid moieties present in the biodiesel. In the present study, the degradation of BS produced by *P. aeruginosa* strain (MTCC8165) was comparatively lower than that of *B. circulans* strain (MTCC8167), which signifies that BSs can't be easily degraded by its source species. Providenti et al. (1995) investigated the degradation behavior of RL and reported that it could be easily utilized by bacterial consortia in sandy loam, silt loam, and creosote contaminated soil, but immune to *P. aeruginosa* when it was present as the sole source of carbon. Zeng et al. (2007) reported that RL was degraded efficiently in compost without creating any disturbance to the microbial community present in the composting matrix, indicating its potential compatibility in environmental applications. Recent studies clearly demonstrated that the application of synthetic surfactants in the bioremediation process influences the dynamics of microbial community and the hydrocarbon degradation rates (Volkering et al., 1998; Das et al., 2008). According to Chrzanowski et al. (2010), biodegradation of RLs did not favor the growth of any specific consortium member, which confirmed that the employed BS did not interfere with the microbial equilibrium during diesel/biodiesel biodegradation.

KEYWORDS

- **biosurfactant solution**
- **Luria Bertani**
- **microbial enhanced oil recovery**
- **room temperature**
- **specific consortium member**
- **total petroleum hydrocarbon**

REFERENCES

Bordoloi, N. K., & Konwar, B. K., (2007). Microbial surfactant-enhanced mineral oil recovery under laboratory conditions. *Colloids Surf. B. Biointerfaces, 63*, 73–82.

Chaillan, F., et al., (2004). Identification and biodegradation potential of tropical aerobic hydrocarbon-degrading microorganisms. *Res. Microbiol., 155*(7), 587–595.

Chandrasekaran, E. V., & BeMiller, J. N., (1980). Constituents' analysis of glycosaminoglycans. In: Whistler, R. L., (eds.), *Methods in Carbohydrate Chemistry* (p. 89). Academic, New York.

Chrzanowski, L., et al., (2010). Biodegradation of rhamnolipids in liquid cultures: Effect of biosurfactant dissipation on diesel fuel/B20 blend biodegradation efficiency and bacterial community composition. *Bioresour. Technol., 111*, 328–335.

Das, K., & Mukherjee, A. K., (2007). Crude petroleum-oil biodegradation efficiency of *Bacillus subtilis* and *Pseudomonas aeruginosa* strains isolated from a petroleum oil-contaminated soil from north-east India. *Bioresour. Technol., 98*, 1339–1345.

Das, P., et al., (2008). Improved bioavailability and biodegradation of a model polyaromatic hydrocarbon by a biosurfactant producing bacterium of marine origin. *Chemosphere, 72*, 1229–1234.

Joseph, P. J., & Joseph, A., (2009). Microbial enhanced separation of oil from petroleum refinery sludge. *J. Hazard. Mater., 161*, 522–525.

Providenti, M. A., et al., (1995). Effect of addition of rhamnolipid biosurfactants of rhamnolipid-producing *Pseudomonas aeruginosa* on phenanthrene mineralization in soil slurries. *FEMS Microbiol. Ecol., 17*, 15–26.

Suthur, H., et al., (2008). Evaluation of bioemulsifier mediated microbial enhanced oil recovery using sand pack column. *Microbiol. Methods, 75*, 225–230.

Volkering, F., et al., (1998). Microbiological aspects of surfactant use for biological soil remediation. *Biodegr., 8*, 401–417.

Zeng, G., et al., (2007). Co-degradation with glucose of four surfactants, CTAB, triton X-100, SDS and rhamnolipid, in liquid culture media and compost matrix. *Biodegr., 18*, 303–310.

FURTHER READING

Bharali, P., & Konwar, B. K., (2011). Production and physicochemical characterization of a biosurfactant produced by *Pseudomonas aeruginosa* OBP1 isolated from petroleum sludge. *Appl. Biochem. Biotechnol., 164*, 1444–1460.

Bharali, P., Das, S., Konwar, B. K., & Thakur, A. J., (2011). Crude biosurfactant from thermophilic *Alcaligenes faecalis*: Feasibility in Petro-spill bioremediation. *Inter. Biodeter. Biodegr., 65*, 682–690.

Das, S., Kalita, S. J., Bharali, P., Konwar, B. K., Das, B., & Thakur, A. J., (2013). Organic reactions in "green surfactant": An avenue to bisuracil derivative. *ACS Sustainable Chem. Eng.* (p. 301). doi: 10.1021/sc4002774, Publication Date (Web).

Pranjal, B., (2015). *Bioremediation of Crude Oil Contaminated Soil.* (Thesis: Supervisor B K Konwar), Dept. of Mol. Biol. and Biotechnology, Tezpur University (Central), Napaam – 784028, Assam, India.

CHAPTER 14

Bioremediation with Bacterial Biosurfactant and Its Biological Activity

The pure cultures of the individual bacterial strains possess limited preferred substrates and thereby are assisting less significantly in utilizing the complex hydrocarbon mixtures present in the crude oil. Lal and Khanna (1996) reported that degradation of crude oil by microbes habitually occurs with the utilization of alkanes or light aromatic fractions, while the higher molecular weight aromatics, resins, and asphaltenes are considered as recalcitrant. Adebusoye et al. (2007) reported that the individual organisms such as *Corynebacterium spp, Acinetobacter lwoffi* and *P. aeruginosa* could metabolize only a limited range of hydrocarbon substrates. Biodegradation studies conducted by Sharma and Pant (2001) showed that 50% of the aliphatic fractions of the crude oil of Assam were degraded by the isolates of *Rhodococcus*. Several studies pointed out that the extent of oil and total petroleum hydrocarbon (TPH) biodegradation are closely linked to the type of oil and its molecular composition (Balachandran et al., 2012). The present study showed that the presence of crude oil in the culture medium had no inhibitory effect on BS production by the bacterial strains and might assist in the biodegradation of high molecular weight n-alkanes (C_{12}-C_{18}).

Individual microorganisms could metabolize only a limited range of hydrocarbon substrates; hence the assemblages of mixed populations with overall broad enzymatic capacities are required to bring the rate and extent of petroleum biodegradation further (Etoumi et al., 2007). Out of the eleven different combinations tried, the combination 7 and 11, named as consortium I and II exhibited the highest dry biomass production of 2.73 ± 0.3 and 3.33 ± 0.1 g.l^{-1}, respectively in the crude oil supplemented MSM medium requiring 96 h of culture. Both the consortia comprising of all four different strains of *P. aeruginosa* except OBP2 in consortium I were found to efficiently degrade the crude oil under the shaking condition. Individually the bacterial strains were capable to utilize the different fractions of crude oil, especially the aliphatic fractions in the range of 67.5–73.0%. However, the present investigation clearly revealed that consortium I and II were efficient in degrading 78.6–80.4% of aliphatic, 42.4–42.7% of aromatic and 19.2–21.6% of NSO compounds in 30 days of culture.

Degradation of the aromatic fraction up to 42.4–42.7% could not be attained in the case of any single bacterium (Table 12.3). Such potency of consortia in degrading aromatic fraction supports the co-metabolism behavior between the bacterial strains. The metabolic intermediates produced by one bacterial strain could be utilized by the other members of the consortium as the substrate for their growth and BS production (Mulligan, 2005). Ghazali et al. (2004) reported that biodegradation of complex hydrocarbons usually required the cooperation of more than a single species. This is particularly true in the case of pollutants that are made up of many different compounds such as crude oil or petroleum. Addition of BS (45 $mg.l^{-1}$) produced by the bacterial strain OBP1 to both of the consortia leads to an enhancement in the degradation of crude oil indicating the effectiveness of BS in the biodegradation process. Further, the GC analysis of the crude oil confirmed the enhancement in the degradation process. Research conducted by Itoh and Suzuki (1974); Rahman et al. (2003); and Zhang et al. (2009) reported that the addition of the RL produced by *P. aeruginosa* enhances hydrocarbon degradation by the same organism.

Gas chromatographic profile of the saturated fraction of the crude oil inoculated with Consortia I and II exhibited a much-reduced noise level as compared to the non-inoculated medium. Presence of certain distinct unreduced peaks indicates the accumulation of bacterial by-products which are not degraded further by the members of the consortia. The results clearly established that both consortia along with the BS were efficient in degrading the different components of crude oil. In designing a consortium, solubility, and accessibility of the hydrophobic compounds available in the crude oil are the two key aspects. Since only 0.02% of crude oil is soluble in water, hence there is a need for emulsification of the crude oil in the medium (She et al., 2011).

The aqueous BS solution of the bacterial strains could efficiently separate the crude oil from the contaminated sand which clearly indicated their capability in the soil washing. The maximum removal of crude oil by the BSs was achieved within their CMC, but by increasing the concentration beyond the CMC, the removal of crude oil from the contaminated sand could not be enhanced. With the reduction in the interfacial tension (IFT) between crude oil and the sand particles, the capillary force that holds them together in the sand-oil mixture gets reduced. Such reductions in the tension further increase the contact angle between the oil and soil particles and change the wettability of the system. This ultimately results in the mobilization of the crude oil from the sand-oil mixture to the aqueous solution. Such effect is directly related to the BS concentration in the solution until it reaches the CMC, the concentration at which the surfactant molecules start to form micelles and show the lowest tensional force (Rodrigues et al., 2006). Differences in the separation behavior of the BSs produced by the bacterial strains in the washing experiment suggests that the BS mediated removal of oil from the sand particles is also dependent on the physicochemical properties of the BS and combined behavior of surfactant/crude oil/sand systems (Urum and Pekdemir, 2004). Efficient washing off crude oil from the contaminated sand could be achieved even with synthetic surfactant SDS. It is now well established that synthetic surfactants are more recalcitrant than the petroleum hydrocarbons and potentially toxic to the environment (Das et al., 2008). Hence, use of BS seems to be more advantageous.

The available residual oil in the sand pack column was mobilized during the passage of the cell-free culture broth containing BS and began to exude with the effluent. Nearly 6.4–11.4% of residual crude oil was recovered from the saturated sand pack column using cell-free culture supernatant of the bacterial strains as compared to the control. Results clearly indicated mobilization of crude oil by BS in the sand packed column. Parameters like IFT are important in the recovery of crude oil as the capillary number is dependent on the ratio of viscous to capillary forces (Xia et al., 2011). Capillary forces arise due to the IFT between oil and water phases, which oppose the externally applied viscous forces and responsible for large quantities of oil that are left behind after water flooding. With the introduction of BSs into the column system, it starts to decrease the IFT at oil/brine interface which in turn increases the capillary number of the system. Further, increase in capillary number lowers the residual oil saturation in the column and increases the recovery process. Suthur et al. (2008) reported reduction in the IFT directs the mobilization of irregular oil lump to form available oil banks in between the reservoir rocks.

Further, exposure of all four types of BSs to high temperature between 70 and 90°C increased the release of the residual crude oil by 8.4–10.5% and 9.3–11.4%, respectively from the column. When temperature was raised between 50 and 90°C, there was a reduction in the inherent viscosity of the crude oil. This might be because of the reduction in the compaction between the molecules of crude oil, which makes the crude oil more mobile as compared to the room temperature (RT). With the increase in the temperature, the air which was previously trapped in between the vacant spaces of the sand particles of the column starts to move out and provides more space for the crude oil to flow. Further, reduction in the capillary forces with the involvement of BS enhances the release of the crude oil from the column. Such behavior in the release of crude oil from the column with the enhancement of temperature in the investigation agreed with the previous reports of Bordoloi and Konwar (2009). The results also indicated the retention of surface properties of the BSs even after exposure to higher temperatures. Such thermo-stable property of BSs is very crucial for their application in MEOR where the BSs must withstand the prevailing extreme temperatures inside the oil reservoirs.

The BS of four bacterial strains can separate residual crude oil (63.4–73.5%) from the petroleum-sludge having up to 7% (w/w) of the sludge concentration, but above this concentration there was no further increase in the release of oil. In-fact it was reported that the preparation of homogeneous slurry could be a critical factor in the treatment of sludge which could limit the process (Volkering et al., 1998). The release of residual oil from the sludge gradually increases with the increase in the duration of treatment (up to 15 days) with continuous shaking (120 rpm). The disturbance caused by the continuous shaking further separated the loosely bound oil droplets from the soil particles due to the reduction of IFT. With the increase in the BS concentration, there was an increase in the removal of residual crude oil from the sludge. Such result was due to the reduction in the surface tension and IFT, which lead to a gradual decrease in the capillary force that hold soil and oil. This further enhanced the contact angle between soil and oil, resulting in a change in the wettability of the system. The IFT of the system decreased gradually until it reached its CMC, after which it remained constant. The same mechanism described in the case of soil washing experiment is probably responsible for the mobilization of residual crude oil. The physicochemical properties of BS and combined behavior of surfactant/crude oil/soil systems probably cause the effect (Urum and Pekdemir, 2004). A successful attempt in using BS to recover oil from the sludge was also reported by Banat et al. (2010). Joseph and Joseph (2009) separated the residual oil from the petroleum sludge generated from the crude oil refinery by directly inoculating the strains of *Bacillus spp.* and with the addition of the cell-free culture supernatant of the bacteria. Helmy et al. (2010) reported the application of the BS produced by *Azotobacter vinelandii* AV01 for enhanced oil recovery from the oil sludge and recovered up to 15% of oil from the sludge. SDS is also effective but hazardous to the environmental components (Das et al., 2008). Hence, the use of BS seems to be more advantageous over the chemical surfactants because of their biodegradability, less toxicity and effectiveness at extreme temperatures, pH, and salinity (Ilori et al., 2005; Cunha et al., 2004; Ko-Sin et al., 2010).

Biodegradation of BSs produced by four *P. aeruginosa* strains confirmed their biodegradability nature when co-cultured with glucose in the culture medium. Several reports demonstrated co-degradation or sole utilization as a source of carbon and energy by various bacterial monocultures (Zeng et al., 2007). In the present study, the degradation of BS produced by *P. aeruginosa* strain (MTCC8165) was comparatively lower than that of *Bacillus circulans* strain (MTCC8167), which signifies that BSs can't be easily degraded by its source species. Similar observation was reported by Zeng et al. (2007). Recent studies clearly demonstrated that the application of synthetic surfactants in the bioremediation process influences the dynamics of microbial community and the hydrocarbon degradation rates (Das et al., 2008).

14.1 BIOSURFACTANT (BS) AND PRODUCING BACTERIA IN FIELD BIOREMEDIATION

All four bacterial strains efficiently reduced the viscosity of the crude oil after 30 days of treatment under laboratory conditions. Results of the present investigation clearly indicated the ability of bacterial strains to degrade the heavy fractions of crude oil, which in turn changed the physicochemical properties of the crude oil (Hao et al., 2007). The use of microbes in degrading long-chain alkanes might have several benefits such as minimizing paraffin precipitation or deposition problem along the production flow line, reduction of the viscosity of crude oil, increase in API gravity value and finally reduction of both pour point and paraffin content of crude oil (Etoumi et al., 2007). Moreover, the biosurfactants (BSs) produced during the bacterial growth on crude oil additionally reduces its viscosity by altering the surface and interfacial energy of the system, thereby increasing the mobility of crude oil in the pipelines. She et al. (2011) used the BS-producing indigenous *Bacillus* strains to degrade the higher fractions of crude oil. They also reported the enhancement in the flow characteristics of crude oil after the treatment in a petroleum reservoir of Daqing Oilfield. According to Gudiňa et al. (2012) *Bacillus* strains were able to degrade large alkyl chains and reduce the viscosity of hydrocarbon mixtures. These reports supported our view of using the indigenous BS producing bacterial strains in reducing the viscosity of crude oil.

Water solubility of some of the hydrophobic organic compounds HOCs could be increased with the addition of surfactant or BS (Eddouaouda et al., 2011). The solubilization assays of anthracene, phenanthrene, and naphthalene clearly showed that with the increase in the concentration of BSs up to a range of 45–120 $mg.l^{-1}$ could reduce the amount of available undissolved PAHs in the reaction mixture. Such behavior of the BSs is since the concentration of BSs above the CMC enhances the formation of micelle causing the undissolved organic components to dissolve within the micelle structure facilitating microbial uptake and bioremediation (Yin et al., 2009). Das et al. (2008) showed that the solubilization of anthracene increases with the increase in RL concentration beyond 100 $mg.l^{-1}$. Yin et al. (2009) reported the solubilization of phenanthrene at 50 $mg.l^{-1}$ of RL produced by *P. aeruginosa* strain S6. Further, differences in the degree of solubilization by the BSs might be due to differences in the physicochemical characters of the tested PAHs and the types of RL congers present in the BSs (Salihu et al., 2009; Abdel-Mawgoud et al., 2009).

All four bacterial strains were found to be efficient in hydrocarbon degradation. Bacterial strain OBP4 followed by OBP3 and OBP1 appeared to be the best degraders and worked best at neutral or near-neutral pH. The genus *Pseudomonas* stands out as the most versatile group among the several genera of microorganisms that have the capability of hydrocarbon degradation and BS production (Bodour et al., 2003; Tuleva et al., 2002; Adebusoye et al., 2007). Das and Mukherjee (2007) reported that strains of *P. aeruginosa* were much capable of degrading crude oil components as compared to the strains of *B. subtilis*. Mehdi and Giti (2008) showed that the strain of *Pseudomonas* was more efficient than the strains of *Bacillus* and *Rhodococcus* in degrading crude oil. The growth dynamics could either be due to the constitutive nature of hydrocarbon assimilation capability in the organism or reflected the adaptation of the strains because of previous exposure to exogenous hydrocarbons. This could be followed by simultaneous development of the capability to use the oil and/or its catabolic products as carbon and energy sources (Smits et al., 2002). Gas chromatographic analyzes of the treated crude oil with the bacterial strains showed a similar response of degradation. Bioconversion of crude oil components leads to the enrichment of the lighter fractions of hydrocarbons having shorter retention time.

The bacterial strains were competent in degrading crude oil in the mineral salt medium (MSM) and efficiently degraded n-alkanes in the range of C_9 to C_{18} (Sugiura et al., 1997). The high molecular weight n-alkanes could have been preferentially utilized as carbon and energy sources. Previous studies showed that alkane with C_{14}–C_{20} carbon atoms permits abundant growth for most of the bacteria (Balachandran et al., 2012; Lotfabada et al., 2010). However, the bacterial strains were not much efficient in utilizing aromatic and polyaromatic compounds. The predominance of mineralization of aliphatic over aromatic hydrocarbons with a greater rate of degradation by the bacterial community was reported by several authors (Chaillan et al., 2004).

14.2 BIOLOGICAL ACTIVITY OF BIOSURFACTANT (BS)

14.2.1 ON SEED GERMINATION AND SEEDLING GROWTH

The phytotoxicity of the isolated BS samples was evaluated in static test by seed germination and root elongation as described by Tiquia et al. (1996). The isolated BS samples were dissolved in water at three different concentrations, i.e., below CMC, at CMC and above CMC. The bioassay was determined in 50 ml glass Petri dish containing Whatman No.1 filter paper. Mung bean (*Vigna radiate*) as dicotyledonous and rice (*Oryza sativa*) as monocotyledonous plant were selected for the bioassay. The seeds were pre-treated with 0.1% (w/v) $HgCl_2$ solution and 10 seeds were transferred into each Petri dish which was inoculated with 5 ml of the test BS solution at 25 ± 2°C. After 3–7 days of incubation in the dark, the germination of seed, root elongation (≥5 mm) and germination index (GI) were determined as follows:

$$\text{Relative seed germination } \% = \frac{\text{Number of seeds germinated in the biosurfactant}}{\text{Number seeds germinate in the control}} \times 100$$

$$\text{Relative root length } \% = \frac{\text{Mean root length in the control}}{\text{Total root length in the biosurfactant}} \times 100$$

$$\text{Growth Index (GI) } \% = \frac{\text{Percentage of seed germination}}{\text{Percentage of root growth}} \times 100$$

Petri dishes containing the seeds and water were kept as controls. The mean and standard deviation of triplicate samples from each concentration were calculated. The seed germination test was performed with the addition of BS at various concentrations to evaluate their toxicity towards the tested plants, and the results are presented in Figure 14.1(A) and (B).

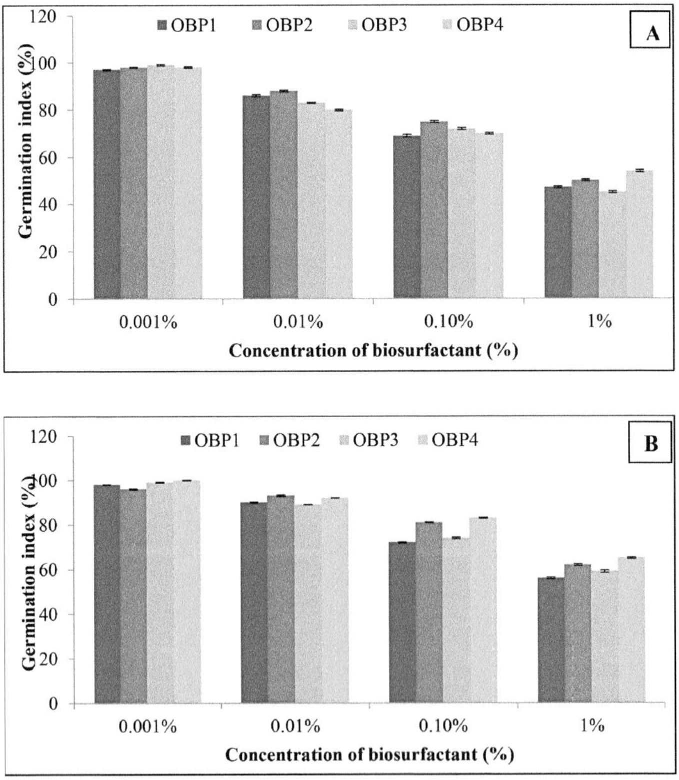

FIGURE 14.1 Effect of biosurfactants produced by *P. aeruginosa* strains on the germination index of (A) rice; and (B) mung bean. Mean of 3 experiments ± SD.

The BS solutions were tested at below CMC, at CMC and above CMC to determine the effect of BS concentrations on crop seed germination. The BS solutions at lower concentration did not exhibit any inhibitory effect on the seed germination and root elongation of mung bean and rice. Normal development of leaves and elongation of primary roots with root hairs were observed in both plants. The GI of mung bean (96–100%) was higher up to CMC concentration of the BS as compared to rice (97–99%). Increased concentrations of BSs above CMC values caused reduction in GI values to 55–65% and 45–54% in mung bean and rice, respectively.

14.2.2 INSECT LARVICIDAL ACTIVITY OF BIOSURFACTANTS (BSS)

The mosquito larvicidal activity of the isolated BS samples was evaluated against the third instar larvae of *Aedes albopictus* using the standard protocol approved by WHO (1996). The larvae used in the bioassay were obtained from the mosquito rearing facility section of the Defense Research Laboratory, Tezpur, India. A total of 20 numbers of larvae per replicate were transferred to 250 ml glass beakers containing 200 ml of distilled water. The BS samples were dissolved in ethanol ($\geq$ 99%, Merck) and added to the beakers at concentrations ranging from 100–1500 $mg.l^{-1}$. For each concentration, three replicates were made. The beakers with larvae and water were kept as negative control. The same procedure was also used for determining the larvicidal property of cell-free culture supernatant of the P. aeruginosa strains. The numbers of larvae killed were counted after 24 h, followed by counting the number of live larvae in each beaker. The mean and standard deviation of triplicates for each treatment were calculated.

The isolated BSs from all four strains showed no mortality of mosquito larvae at the recommended concentrations fixed by WHO. However, all four BSs exhibited almost negligible mortality (%) towards the 3rd instar *Aedes albopictus* larvae at a dose of 100–1,000 $mg.l^{-1}$ and the results are presented in Table 14.1.

TABLE 14.1 Mosquito Larvicidal Potency of Biosurfactants Produced by *P. aeruginosa* Strains

Biosurfactant (mg.l^{-1})	Mortality (%) of Mosquito Larvae After 24 h			
	OBP1	OBP2	OBP3	OBP4
Control	–	–	–	–
100	–	–	–	–
300	–	–	–	–
500	–	–	–	–
700	–	–	–	–
1000	–	–	10 ± 2.0	–
1500	10 ± 2.5	10 ± 1.5	15 ± 2.0	10 ± 1.5

Results mean of 3 experiments ± SD.

Mortality of mosquito larvae at much higher concentrations above the one recommended by the WHO indicated the ineffectiveness of the isolated BSs.

14.2.3 ANTIMICROBIAL ACTIVITY OF BIOSURFACTANTS (BSS)

The bacterial and fungal strains used in the research work were obtained from the Department of Molecular biology and Biotechnology, Tezpur, Assam, India, and Department of Plant Pathology, Assam Agricultural University, Jorhat, Assam, India, respectively. The studied bacterial strains were *E. coli* (MTCC40, MG1655), *B. subtilis* (MTCC121), *B. subtilis* (MTCC441), *Staphylococcus aureus* (MTCC737), *S. aureus* (MTCC3160), *Klebsiella pneumoniae* (MTCC618), *P. aeruginosa* (MTCC7815) and *P. diminuta* (AAU). The fungal strains used in the investigation include *Candida albicans* (MTCC227), *Fusarium oxysporium* (MTCC284), *Aspergillus niger* (AAU), *Colletotrichum capaci* (AAU) and *Alternaria solani* (AAU). The anti-bacterial activity of BS was evaluated by well diffusion method (Radhika et al., 2008). Briefly, 200 μl of the log phase culture of the test microbes (107–108 cells as per McFarland standard) were seeded on the surface of the Mueller Hinton agar (MHA) medium using a micropipette and spread over the medium uniformly using a sterile glass spreader. With the help of a sterile Cork Borer wells having 6 mm diameter each were made on MHA plates. The BS samples were dissolved in sterilized DMSO (10% v/v) and introduced into one of the wells. As the presence of 10% DMSO (v/v) had no detectable effect on bacterial growth, compounds at concentrations of 10 mg.ml-1 were prepared in 10% DMSO (v/v).

Streptomycin sulfate (1 mg.ml-1) was taken as a positive control and 10% DMSO (v/v) as negative one. After the incubation of the plates at 37°C for an overnight period, microbial growth was determined by measuring the diameter of inhibition zone using a transparent metric ruler. For antifungal investigation, the fungal strains (0.5–2.5 × 106. ml-1) were grown on Sabouraud dextrose agar (SDA). BS solution was introduced into the wells in the similar manner as described for bacteria. After incubation for 36 h at 25°C, the growth was determined by measuring the diameter of inhibition zone. Nystatin (1 mg.ml-1) was used as positive while 10% DMSO (v/v) was kept as negative control. The micro broth dilution method was performed to determine the minimum inhibitory concentration (MIC). The dissolved RL solution was diluted to a series of tenfold in Luria Bertani (LB) broth, seeded in a 96-well culture plate, and then inoculated with a fresh bacterial inoculum. Inoculated microplates were incubated at 37°C for 24 h. In the case of fungi, the RL solution was diluted in Sabouraud dextrose (SD) broth, and the plates were incubated at 25°C for 24 h. Each BS concentration was tested in duplicates for each organism. Two wells containing suspension test organism with no drug (growth control) and 2 wells containing only media (background control) were included in the microtiter plate. The viability of the treated cells was determined by MTT (3-[4,5-dimethylthiazol-2-yl]-2,5-Diphenyltetrazolium) assay and the absorbance was measured at 570 nm and 405 nm using a microtiter plate reader (Bio-Rad Model 680; Hercules, California) for bacterial and fungal strains respectively. The MIC was determined as the lowest concentration of BS required to inhibit the growth of each organism. The mean and standard deviation of triplicates for each treatment were calculated.

BS produced by the bacterial strains OBP1 exhibited the highest antibacterial activity against the tested microorganisms followed by OBP3, OBP2, and OBP4 as shown in Table 14.2. The BS showed high activity against *E. coli* (MG1655) and *Klebsiella pneumoniae* (MTCC618), but less against *P. aeruginosa* (MTCC7815), *B. subtilis* (MTCC441), *P. diminuta* (AAU) and *Staphylococcus aureus* (MTCC3160).

TABLE 14.2 Antimicrobial Properties of Biosurfactants Produced by *P. aeruginosa* Strains

Microorganisms	**Zone of Inhibition (mm)**			
	Purified Biosurfactants from *P. aeruginosa* Strains			
	OBP1	**OBP2**	**OBP3**	**OBP4**
Escherichia coli (MTCC40)	16 ± 0.2	15 ± 0.4	16 ± 0.3	14 ± 0.3
E. coli (MG1655)	18 ± 0.3	1 ± 0.8	18 ± 0.9	16 ± 0.4
Bacillus subtilis (MTCC441)	14 ± 0.4	10 ± 0.3	12 ± 0.2	10 ± 0.1
B. subtilis (MTCC121)	12 ± 0.1	12 ± 0.7	14 ± 0.4	12 ± 0.4
Staphylococcus aureus (MTCC737)	14 ± 0.3	14 ± 0.2	14 ± 0.5	12 ± 0.5
S. aureus (MTCC3160)	12 ± 0.3	12 ± 0.8	10 ± 0.2	10 ± 0.3
Klebsilla pneumoniae (MTCC618)	14 ± 0.4	16 ± 0.9	14 ± 0.8	14 ± 0.8
P. aeruginosa (MTCC7815)	10 ± 0.7	08 ± 0.2	08 ± 0.3	08 ± 0.4
P. diminuta (AAU)	12 ± 0.3	12 ± 0.7	10 ± 0.7	10 ± 0.4
Candida ablicans (MTCC227)	09 ± 0.7	07 ± 0.2	08 ± 0.6	07 ± 0.7
Fusarium oxysporium (MTCC284)	10 ± 0.2	12 ± 0.5	08 ± 0.7	07 ± 1.2
Aspergillus niger (AAU)	10 ± 0.8	08 ± 1.0	07 ± 0.2	–
Colleototricum capaci (AAU)	10 ± 0.1	08 ± 0.3	–	–
Alternaria solani (AAU)	10 ± 0.3	07 ± 0.8	–	–

Results mean of 3 experiments ± SD.

The BS displayed antifungal activity with higher MIC values against tested fungal strains as compared to bacterial strains and results are presented in Table 14.3.

TABLE 14.3 Antimicrobial Potency of Biosurfactants Produced by the *P. aeruginosa* Strains

Microorganisms	**Minimum Inhibitory Concentration ($mg.ml^{-1}$) of Purified Biosurfactants from *P. aeruginosa* Strains**			
	OBP1	**OBP2**	**OBP3**	**OBP4**
E. coli (MTCC40)	0.5 ± 0.3	0.5 ± 0.5	0.5 ± 0.1	0.5 ± 0.1
(MG1655)	0.25 ± 0.8	1.25 ± 0.2	0.25 ± 0.6	0.5 ± 0.5
B. subtilis (MTCC441)	0.5 ± 0.2	2.0 ± 0.6	1.0 ± 0.2	2.0 ± 0.2
(MTCC121)	1.0 ± 0.2	1.0 ± 0.2	0.5 ± 0.6	1.0 ± 0.7
S. aureus (MTCC737)	0.5 ± 0.5	0.5 ± 0.6	0.5 ± 0.4	1.0 ± 0.5
(MTCC3160)	0.5 ± 0.1	1.0 ± 0.3	2.0 ± 0.1	2.0 ± 0.2

TABLE 14.3 (Continued)

Microorganisms	**Minimum Inhibitory Concentration (mg.ml^{-1}) of Purified Biosurfactants from *P. aeruginosa* Strains**			
	OBP1	**OBP2**	**OBP3**	**OBP4**
K. pneumoniae (MTCC618)	0.25 ± 0.3	0.5 ± 0.6	0.5 ± 0.3	0.5 ± 0.8
P. aeruginosa (MTCC7815	2.0 ± 0.7	4.0 ± 0.5	4.0 ± 0.9	4.0 ± 0.3
P. diminuta (AAU)	1.0 ± 0.3	1.0 ± 0.1	2.0 ± 0.3	2.0 ± 0.8
Candida ablicans (MTCC227)	2.0 ± 0.1	4.0 ± 0.6	4.0 ± 0.6	4.0 ± 0.3
Fusarium oxysporium (MTCC284)	2.0 ± 0.5	1.0 ± 0.7	4.0 ± 0.2	4.0 ± 0.2
Aspergillus niger (AAU)	2.0 ± 0.2	4.0 ± 0.7	–	–
Colleototricum capaci (AAU)	2.0 ± 0.4	4.0 ± 0.2	–	–
Alternaria solani (AAU)	2.0 ± 0.2	2.0 ± 0.4	–	–

Results mean of 3 experiments ± SD.

14.2.4 CHEMO-ATTRACTANT PROPERTY OF BIOSURFACTANT (BS)

The chemotaxis property of the BS samples was examined by using chemical gradient motility agar (CGMA) method as described by Garg and Kanitkar (2006). Motility agar medium (MAM) containing 0.7% agar was used for the assay. After the solidification of the medium, three long rectangular wells were made. The well present on the right and left side were loaded with RL (0.5%, w/v) and streptomycin (1 mg.ml^{-1}), respectively, and kept undisturbed for 60 min at RT. The well present at the center was loaded with 100 μl of the fresh overnight grown test bacterial culture and incubated at 37°C for 24 h. Glucose 5% (w/v) was used as a positive control because it has an excellent chemo-attractant property. The BS (RL) used in the present investigation was isolated from *P. aeruginosa* OBP1 because this strain showed the highest antimicrobial property among the four selected bacterial strains of *P. aeruginosa*.

The BS produced by the bacterial strain OBP1 was found to be RL in nature and this BS at a concentration of 0.5% (w/v) was used to determine the chemo-attractant property on *Staphylococcus aureus* (MTCC3160) and *Klebsiella pneumoniae* (MTCC618). Glucose 5% (w/v) was the positive control. The cultures were incubated for 24 h. Both bacterial strains exhibited growth almost like that of the control (Glucose, 5% w/v); but they failed to show growth towards streptomycin (1 mg.ml^{-1}) containing wells and the same are shown in Figure 14.2. The tested bacterial strains showed positive chemotaxis response towards the RL BSs like that of glucose but negative chemotaxis response towards the streptomycin.

14.2.5 CELL CYTOTOXICITY OF BIOSURFACTANT (BS)

Primary mouse connective tissue cell line (L929) was obtained from the National Centre for Cell Science (Pune, India). The cells were maintained in Dulbecco's minimum essential medium (DMEM) containing 2 mM.l^{-1} glutamine, 1.5 g.l^{-1} sodium bicarbonate ($NaHCO_3$), 0.1 mM non-essential amino acids, and 1.0 mM sodium pyruvate, supplemented with 10% (v/v) fetal bovine serum and 1% antibiotic antimycotic solution (1000 U.ml^{-1} penicillin G, 10 mg.ml^{-1} streptomycin sulfate, 5 mg.ml^{-1} gentamicin and 25 μg.ml^{-1} amphotericin B). Cells were maintained at 37°C in a saturated-humidity atmosphere containing air 95%, CO_2 5%. A simple, non-radioactive, and colorimetric MTT (3-[4, 5-dimethylthia-zole-2-yl]-2, 5-diphenyl tetrazolium bromide) dye conversion assay was used (Mossman, 1983) to quantitatively measure the cell toxicity. For MTT assay viability studies, mouse fibroblast cell line L929 was cultured at a density of 1×10^4 cells per well in a 100 μl volume of cell culture medium (DMEM supplemented with 10% fetal bovine serum) in a 96-well cell culture plate. After 24 h, cultured cells were treated with a series of different BS concentrations (10,

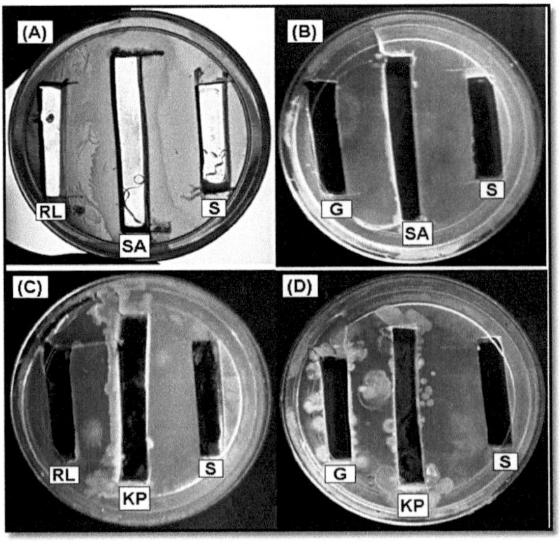

FIGURE 14.2 Chemotaxis activity of (A) *Staphylococcus aureus* (SA) with RL and streptomycin (S); (B) *S. aureus* (SA) with glucose (G) and streptomycin (S); (C) *Klebsiella pneumoniae* (KP) with RL and streptomycin (S); and (D) *K. pneumoniae* (KP) with glucose (G) and streptomycin (S).

20, 30, 40, 50, 60, 70, 80, 90, and 100 μg.ml^{-1}) of OBP1, OBP2, OBP3, and OBP4 dispersed in 100 μl per well DMEM without serum and phenol red and incubated further for 4 h with MTT dye. After the incubation, an aliquot of 100 μl of dimethyl sulfoxide (DMSO) was added to each well to dissolve blue formazan precipitate, and absorbance was measured at 570 nm using a microtiter plate reader (Bio-Rad Model 680; Hercules, California). The cell viability was expressed as a percentage of the control by the following equation:

$$\text{Viability \%} = \frac{Nt}{Nc} \times 100$$

where; Nt is the absorbance of the compound treated cells and Nc is the absorbance of the untreated cells. All experiments were performed in quadruplets.

The BSs of the bacterial strains did not exhibit any inhibitory effect on the primary mouse fibroblast cell line (L929). Data are presented in Figure 14.3(A)–(D). BS of OBP1, OBP2, OBP3, and OBP4 up to a concentration of 100 μg.ml^{-1} failed to prevent the growth of mouse fibroblast cells, referring to lack of cell cytotoxicity at higher concentrations.

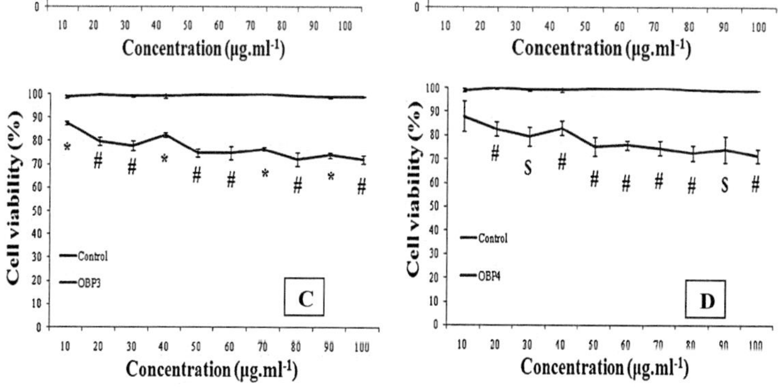

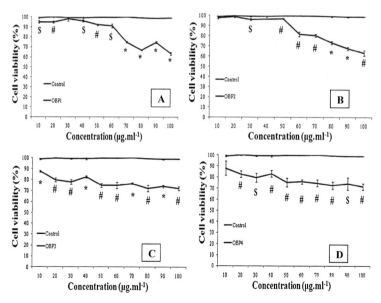

FIGURE 14.3 Effect of biosurfactants on the viability of mouse fibroblast cell line (L929) grown in Dulbecco's minimum essential medium (DMEM) supplemented with 10% fetal bovine serum. (A) *P. aeruginosa* OBP1; (B) *P. aeruginosa* OBP2; (C) *P. aeruginosa* OBP3; and (D) *P. aeruginosa* OBP4. Mean of 3 experiments ± SD.

14.2.6 ACUTE DERMAL IRRITATION STUDY WITH THE ISOLATED BIOSURFACTANTS (BSS)

The acute dermal irritation study showed no effect till 72 h, no erythema or edemas as compared to their negative control after the application of BSs on the shaved region of rabbits and the same are shown in Figure 14.4(A)–(I).

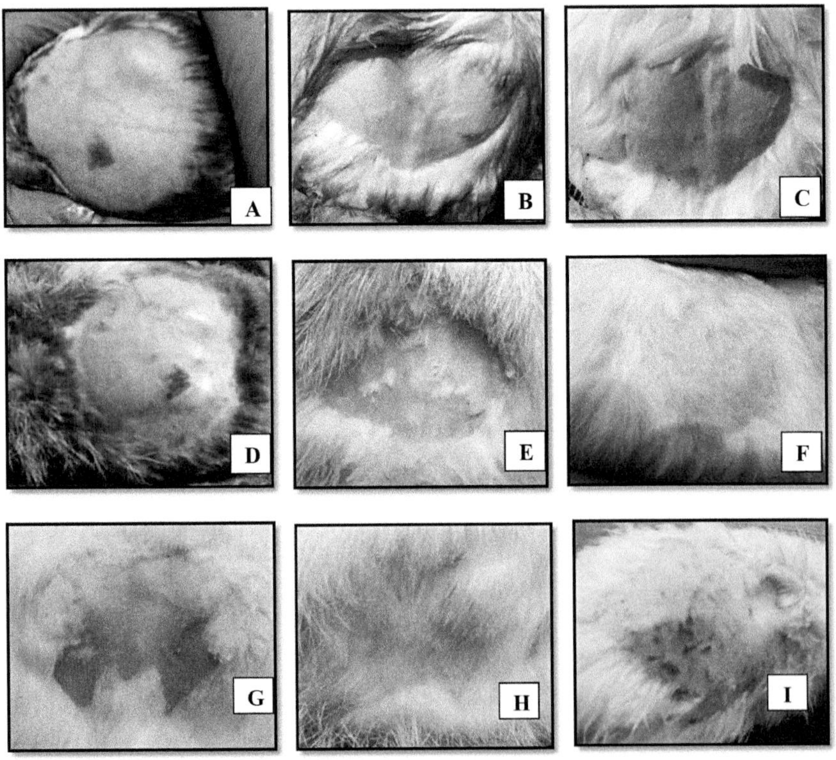

FIGURE 14.4 The acute dermal irritation study of transdermal patch in rabbits. (A) at 0 h and (D) after 72 h of treatment with biosurfactant from OBP1 strain; (B) at 0 h and (E) after 72 h of treatment with biosurfactant from OBP2 strain; (C) at 0 h and (F) after 72 h of treatment with biosurfactant from OBP3 strain; (G) at 0 h and (H) after 72 h of treatment with biosurfactant from OBP4 strain; (I) Control: after 72 h of treatment with 0.8% HCHO (v/v).

The primary skin irritation potential of isolated BSs was studied on rabbits. The isolated BS samples were dissolved in ethanol (≥99%, Merck) at a concentration of 1 mg.ml- 1 and were evenly applied to the shaved skin of the rabbits under a patch. After the treatment of 24 h, the patch was removed, and the compound applied site was wiped off with distilled water to remove the residual test substance. Treated skin sites of the animals were examined for erythema, edema, and eschar at 72 h after the application of the test compound. The blood samples were also collected from the test animals at 72 h to determine the biochemical changes in the blood after the application of the test compound (Geetha et al., 2010; Banerjee et al., 2013).

No significant reductions in the total body weight in the treated groups of rabbits were observed as compared to the control. The concentration of BSs above their CMC was non-toxic to the skin of rabbit. The acute dermal irritation study with the isolated BSs also showed no adverse effect on the hematological parameters of the treated rabbits and is presented in Table 14.4.

TABLE 14.4 Results of Hematological Test After Treatment with Biosurfactants Produced by *P. aeruginosa* Strains

Components	Treatment 1	Treatment 2	Treatment 3	Treatment 4
	OBP1	**OBP2**	**OBP3**	**OBP4**
WBC(K)	6.11 ± 0.98	6.79 ± 0.35	6.24 ± 0.26	7.21 ± 0.05
Neutrophil (%)	11.16 ± 1.7	19.06 ± 1.0	22.32 ± 2.8	12.51 ± 1.2
Lymphocyte (%)	68.72 ± 6.2	81.25 ± 2.2	61.06 ± 0.4	71.25 ± 0.2
Monocyte (%)	6.23 ± 0.15	1.08 ± 2.66	5.77 ± 1.29	3.11 ± 2.16
Eosinophil (%)	8.33 ± 0.49	3.19 ± 1.38	2.03 ± 0.91	1.28 ± 1.31
Basophil (%)	0.87 ± 0.01	0.75 ± 0.01	0.65 ± 0.01	0.83 ± 0.01
RBC (M)	10.02 ± 0.1	7.98 ± 0.15	7.62 ± 0.76	5.36 ± 0.81
Hb ($g.dl^{-1}$)	36.19 ± 8.0	18.76 ± 1.0	22.29 ± 3.7	21.06 ± 5.0
Hct (%)	34.07 ± 1.0	42.21 ± 3.2	49.39 ± 0.2	32.31 ± 1.2
MCV (fl)	76.33 ± 3.9	51.79 ± 2.5	70.86 ± 2.0	41.62 ± 2.1
MCH (pg)	16.02 ± 0.5	13.07 ± 1.0	14.96 ± 1.3	13.37 ± 2.2
MCHC ($g.dl^{-1}$)	37.18 ± 3.9	41.07 ± 0.5	29.10 ± 2.4	42.17 ± 0.2
Reticulocyte (%)	2.16 ± 0.98	5.33 ± 2.13	1.91 ± 0.88	5.73 ± 3.11
PLT(K)	1578 ± 156.5	1708 ± 118.3	1611 ± 179.9	1428 ± 104.1
PT (sec)	13.75 ± 0.3	13.19 ± 3.8	9.98 ± 0.95	12.29 ± 1.2
APTT (sec)	17.61 ± 0.9	46.79 ± 5.5	20.50 ± 3.3	26.09 ± 1.5
Components	**Positive Control**		**Negative Control**	
WBC (K)	5.02 ± 0.66		7.11 ± 0.78	

TABLE 14.4 *(Continued)*

Components	Treatment 1 OBP1	Treatment 2 OBP2	Treatment 3 OBP3	Treatment 4 OBP4
Neutrophil (%)	19.36 ± 2.66		16.26 ± 2.4	
Lymphocyte (%)	51.69 ± 1.90		52.69 ± 0.9	
Monocyte (%)	4.04 ± 0.97		4.00 ± 0.87	
Eosinophil (%)	2.28 ± 0.16		3.88 ± 0.63	
Basophil (%)	0.85 ± 0.007		0.65 ± 0.01	
RBC (M)	8.36 ± 0.66		5.96 ± 0.78	
Hb ($g.dl^{-1}$)	19.86 ± 2.24		17.06 ± 4.0	
Hct (%)	49.66 ± 4.0		42.16 ± 2.1	
MCV (fl)	49.36 ± 3.2		48.96 ± 3.6	
MCH (pg)	11.55 ± 5.2		10.35 ± 5.9	
MCHC ($g.dl^{-1}$)	44.98 ± 4.5		41.98 ± 0.5	
Reticulocyte (%)	3.99 ± 0.71		3.63 ± 0.26	
PLT(K)	1620 ± 111.5		1360 ± 107.3	
PT (sec)	16.98 ± 1.2		13.58 ± 0.2	
APTT (sec)	23.71 ± 5.5		27.51 ± 5.9	

Mean of 3 experiments ± S.D.

Abbreviations: WBC: white blood cell; RBC: red blood cell; Hb: hemoglobin; Hct: hematocrit; MCV: mean corpuscular volume; MCH: mean corpuscular hemoglobin; MCHC: mean corpuscular hemoglobin concentration; PLT: platelet; PT: prothrombin time; APTT: activated partial thromboplastin time.

KEYWORDS

- **chemical gradient motility agar**
- **dimethyl sulfoxide**
- **germination index**
- **Luria Bertani**
- **minimum inhibitory concentration**
- **motility agar medium**

REFERENCES

Abdel-Mawgoud, A. M., et al., (2009). Characterization of rhamnolipid produced by *Pseudomonas aeruginosa* isolate BS20. *Appl. Biochem. Biotechnol., 2*, 329–345.

Adebusoye, S. A., et al., (2007). Microbial degradation of petroleum in a polluted tropical stream. *World J. Microbiol. Biotechnol., 23*, 1149–1159.

Balachandran, C., et al., (2012). Petroleum and polycyclic aromatic hydrocarbons (PAHs) degradation and naphthalene metabolism in *Streptomyces* sp. (ERI-CPDA-1) isolated from oil-contaminated soil. *Bioresour. Technol., 112*, 83–90.

Banat, I. M., (1993). The isolation of a thermophilic biosurfactant producing *Bacillus* sp. *Biotech Letters, 15*(6), 591–594.

Banat, I. M., (1995). Biosurfactants production and possible uses in microbial enhanced oil recovery and oil pollution remediation: A review. *Bioresour. Technol., 51*, 1–12.

Banat, I. M., et al., (1991). Biosurfactant production and use in oil tank clean-up. *World J. Microbiol. Biotechnol., 7*, 80–88.

Banat, I. M., et al., (2000). Potential commercial applications of microbial surfactants. *Appl. Microbiol. Biotechnol., 53*, 495–508.

Banat, I. M., et al., (2010). Microbial biosurfactants production applications and future potential. *Appl. Microbiol. Biotechnol., 87*, 427–444.

Banerjee, S., et al., (2013). Acute dermal irritation, sensitization, and acute toxicity studies of a transdermal patch for prophylaxis against (+) Anatoxin-A poisoning. *Inter. Toxicol., 32*, 4308–4313.

Bodour, A. A., et al., (2003). Distribution of biosurfactant-producing bacteria in undisturbed and contaminated arid southwestern soils. *Appl. Environ. Microbiol., 69*, 3280–3287.

Bordoloi, N. K., & Konwar, B. K., (2007). Microbial surfactant-enhanced mineral oil recovery under laboratory conditions. *Colloids Surf. B. Biointerfaces, 63*, 73–82.

Bordoloi, N. K., & Konwar, B. K., (2009). Bacterial biosurfactant in enhancing solubility and metabolism of petroleum hydrocarbons. *J. Hazardous Materials, 170*(1), 495–505.

Chaillan, F., et al., (2004). Identification and biodegradation potential of tropical aerobic hydrocarbon-degrading microorganisms. *Res. Microbiol., 155*(7), 587–595.

Chrzanowski, L., et al., (2010). Biodegradation of rhamnolipids in liquid cultures: Effect of biosurfactant dissipation on diesel fuel/B20 blend biodegradation efficiency and bacterial community composition. *Bioresour. Technol., 111*, 328–335.

Cunha, C. D., et al., (2004). Serratia sp. SVGG 16: A promising biosurfactant producer isolated from tropical soil during growth with ethanol-blended gasoline. *Process Biochem., 39*, 2277–2282.

Das, K., & Mukherjee, A. K., (2007). Crude petroleum-oil biodegradation efficiency of *Bacillus subtilis* and *Pseudomonas aeruginosa* strains isolated from a petroleum oil-contaminated soil from north-east India. *Bioresour. Technol., 98*, 1339–1345.

Das, P., et al., (2008). Improved bioavailability and biodegradation of a model polyaromatic hydrocarbon by a biosurfactant producing bacterium of marine origin. *Chemosphere, 72*, 1229–1234.

Eddouaouda, K., et al., (2011). Characterization of a novel biosurfactant produced by *Staphylococcus* sp. strain 1E with potential application on hydrocarbon bioremediation. *J. Basic Microbiol., 51*, 1–11.

Etoumi, A., et al., (2007). The reduction of wax precipitation in waxy crude oils by *Pseudomonas* species. *J. Ind. Microbiol. Biotechnol., 35*, 1241–1245.

Garg, A. D., & Kanitkar, D. V., (2006). *12th National Symposium Biotechcellence*. Center for Biotechnology, Anna University, Chennai.

Geetha, I., et al., (2010). Identification and characterization of a mosquito pupicidal metabolite of a *Bacillus subtilis* subsp. subtilis strain. *Appl. Microbiol. Biotechnol., 86*, 1737–1744.

Ghazali, F. M., Rahman, R. N. Z. A., Salleh, A. B., & Basri, M., (2004). Biodegradation of hydrocarbons in soil by microbial consortium. *Int. Biodet. Biodegr., 54*, 61–67.

Gudina, E. J., et al., (2012). Isolation and study of microorganisms from soil samples for application in microbial enhanced oil recovery. *Inter. Biodeter. Biodegrad., 68*, 56–64.

Hao, R., et al., (2004). Effect on crude oil by thermophilic bacterium. *J. Petrol. Sci. Eng., 43*, 247–258.

Helmy, Q., et al., (2010). Application of biosurfactant produced by *Azotobacter vinelandii* AVOI for enhanced oil recovery and biodegradation of oil sludge. *Inter. J. Civil Environ. Eng., 10*, 7–14.

Ilori, M. O., et al., (2005). Factors affecting biosurfactant production by oil-degrading *Aeromonas spp.* isolated from a tropical environment. *Chemosphere, 61*, 985–992.

Itoh, S., & Suzuki, T., (1972). Effect of rhamnolipids on growth of *Pseudomonas aeruginosa* mutant deficient in n-paraffin-utilizing ability. *Agric. Biol. Chern., 36*, 2233–2235.

Joseph, P. J., & Joseph, A., (2009). Microbial enhanced separation of oil from petroleum refinery sludge. *J. Hazard. Mater., 161*, 522–525.

Ko-Sin, N., et al., (2010). Evaluation of jatropha oil to produce poly(3-hydroxybutyrate) by *Cupriavidus necator* H16. *Polym. Oegrad. Stab., 95*, 1365–1369.

Lal, B., & Khanna, S., (1996). Degradation of crude oil by *Acinetobacter calcoaceticus* and *Alcaligenes odorans*. *J. Appl. Bacteriol., 81*, 355–362.

Lotfabada, T. B., et al., (2010). Structural characterization of a rhamnolipid-type biosurfactant produced by *Pseudomonas aeruginosa* MR01: Enhancement of dirhamnolipid proportion using gamma irradiation. *Colloids Surf. B. Biointerfaces, 81*, 397–405.

Mehdi, H., & Giti, E., (2008). Investigation of alkane biodegradation using the microtiter plate method and correlation between biofilm formation, biosurfactant production and crude oil degradation. *Inter. Biodeter. Biodegr., 62*, 170–178.

Mossman, T., (1983). Rapid colorimetric assay for cellular growth and survival: Application to proliferation and cytotoxicity assays. *J. Immunol. Methods, 16*, 55–63.

Mulligan, C. N., (2005). Environmental applications for biosurfactants. *Environ. Pollut., 133*, 183–198.

Providenti, M. A., et al., (1995). Effect of addition of rhamnolipid biosurfactants of rhamnolipid-producing *Pseudomonas aeruginosa* on phenanthrene mineralization in soil slurries. *FEMS Microbiol. Ecol., 17*, 15–26.

Radhika, P., et al., (2008). Antimicrobial screening of *Andrographis paniculata* (Acanthaceae) root extracts. *Res. J. Biotech., 3*, 62–63.

Rahman, K. S. M., et al., (2003). Enhanced bioremediation of n-alkane petroleum sludge using bacterial consortium amended with rhamnolipid and micronutrients. *Bioresour. Technol., 90*, 159–168.

Rahman, P. K. S. M., & Gakpe, E., (2008). Production, characterization and application of biosurfactants-review. *Biotechnol., 7*, 360–370.

Rodrigues, L. R., et al., (2006). Biosurfactants: Potential applications in medicine. *J. Antimicrob. Chemother., 57*, 609–618.

Salihu, A., et al., (2009). An investigation for potential development on biosurfactants. *Biotechnol. Mol. Biol. Rev., 3*, 111–117.

Sharma, S. L., & Pant, A., (2001). Biodegradation and conversion of alkanes and crude oil by a marine *Rhodococcus* sp. *Biodegra., 11*, 289–294.

She, Y. H., et al., (2011). Investigation of biosurfactant-producing indigenous microorganisms that enhance residue oil recovery in an oil reservoir after polymer flooding. *Appl. Biochem. Biotechnol., 163*, 223–234.

Smits, T. H. M., et al., (2002). Functional analysis of alkane hydroxylases from gram-negative and gram-positive bacteria. *J. Bacteriol., 184*, 1733–1742.

Sugiura, K., et al., (1997). Physicochemical properties and biodegradability of crude oil. *Environ. Sci. Technol., 31*, 45–51.

Suthur, H., et al., (2008). Evaluation of bio-emulsifier mediated microbial enhanced oil recovery using sand pack column. *Microbiol. Methods, 75*, 225–230.

Tiquia, S. M., et al., (1996). Effects of composting on phytotoxicity of spent pig-manure sawdust litter. *Environ. Pollut., 93*, 249–256.

Tuleva, B. K., et al., (2002). Biosurfactant production by a new *Pseudomonas putida* strain. *Z. Naturforsch., 57C*, 356–360.

Urum, K., & Pekdemir, T., (2004). Evaluation of biosurfactant for crude oil contaminated soil washing. *Chemosphere, 57*, 1139–1150.

Volkering, F., et al., (1993). Effect of microorganisms on the bioavailability and biodegradation of crystalline naphthalene. *Appl. Microbiol. Biotechnol., 40*, 535–540.

Volkering, F., et al., (1998). Microbiological aspects of surfactant use for biological soil remediation. *Biodegr., 8*, 401–417.

WHO Health Organization, (1996). Protocols for laboratory and field evaluation of insecticides and repellents. *Report of the WHO Informal Consultation on the Evaluation and Testing of Insecticides.* Ref.: CTDIWHOPES/IC/96.1, Geneva.

Xia, W., et al., (2011). Comparative study of biosurfactant produced by microorganisms isolated from formation water of petroleum reservoir. *Colloids Surf. A Physicochem. Eng. Aspects, 392*, 124–130.

Yin, H., et al., (2009). Characteristics of biosurfactant produced by *Pseudomonas aeruginosa* S6 isolated from oil-contaminating wastewater. *Process Biochem., 44*, 302–308.

Zeng, G., et al., (2007). Co-degradation with glucose of four surfactants, CTAB, triton X-100, SDS and rhamnolipid, in liquid culture media and compost matrix. *Biodegr., 18*, 303–310.

Zhang, H., et al., (2009). Enhanced treatment of waste frying oil in an activated sludge system by addition of crude rhamnolipid solution. *J. Hazard. Mater., 167*, 217–223.

Zhang, Y., & Miller, R. M., (1992). Enhanced octadecane dispersion and biodegradation by a *Pseudomonas rhamnolipid* surfactant (biosurfactant). *Appl. Environ. Microbiol., 58*, 3276–3282.

Zhang, Y., & Miller, R. M., (1994). Effect of a *Pseudomonas* rhamnolipid biosurfactant on cell hydrophobicity and biodegradation of octadecane. *Appl. Environ. Microbiol., 60*, 2101–2106.

FURTHER READING

Bharali, P., & Konwar, B. K., (2011). Production and physicochemical characterization of a biosurfactant produced by *Pseudomonas aeruginosa* OBP1 isolated from petroleum sludge. *Appl. Biochem. Biotechnol., 164*, 1444–1460.

Bharali, P., Das, S., Konwar, B. K., & Thakur, A. J., (2011). Crude biosurfactant from thermophilic *Alcaligenes faecalis*: Feasibility in Petro-spill bioremediation. *Inter. Biodeter. Biodegr., 65*, 682–690.

Das, S., Kalita, S. J., Bharali, P., Konwar, B. K., Das, B., & Thakur, A. J., (2013). Organic reactions in "green surfactant": An avenue to bisuracil derivative. *ACS Sustainable Chem. Eng.* (p. 301). doi: 10.1021/sc4002774, Publication Date (Web).

Pranjal, B., (2015). *Bioremediation of Crude Oil Contaminated Soil.* (Thesis: Supervisor B K Konwar), Dept. of Mol. Biol. and Biotechnology, Tezpur University (Central), Napaam – 784028, Assam, India.

CHAPTER 15

Biosurfactants in Nanostructures

The biosurfactants (BSs) are emerging as a green surfactant alternate to their chemically synthesized counterparts for the synthesis and stabilization of nanoparticles. Rhamnolipid (RL) BSs produced by the bacterial strain OBP1 exhibited stabilization of silver and iron oxide nanoparticles. Such behavior of BSs is due to the solubilization or incorporation of the reactant species and/or synthesized particles into the micellar phase and adsorption of nanoparticles on the surface of the BS micelles (Sanchez-Peinado et al., 2010). It is possible that the positive charge on Ag^{+}/Fe^{2+} leads to the formation of the ion pair with the negative head of RL micelles which concentrate the Ag^{+}/Fe^{2+} within the small volume through the electrostatic interactions into the reaction sites (Sanchez-Peinado et al., 2010). Kiran et al. (2010) suggested that the presence of BS in the colloidal solution of nanoparticles would act as the stabilizing agent that prevents the formation of aggregates, and favors the production and stability of nanoparticles under the experimental conditions. Silver nanoparticles (SNP) synthesized in the RL colloid were found to be stable for more than one month. The SNP in RL colloid (SNPRL) was not affected by the addition of external NaCl up to 60 $mg.ml^{-1}$ and this also prevented the destruction of SNP. The inference has suggested that the RL shall undergo vesicle formation to prevent SNP' exposure to NaCl (Rosenberg and Ron, 1996). Xie et al. (2006) also reported the use of RL BS as stabilizing agent of SNP. Biswas and Raichur (2008) used the RL BS to evaluate its effect on the synthesis and stabilization of nano zirconia particles.

Iron oxide nanocrystal-rhamnolipid (ION-RL) and silver nanoparticle rhamnolipid (SNP-RL) nanocomposites were found to be effective against gram-positive and Gram-negative bacteria suggesting their broad-spectrum antibacterial properties. Different strains belonging to the same species of bacteria exhibited different susceptibility towards the BS, ION-RL, and SNP-RL. The excellent antibacterial efficacy of the surfactant-adsorbed nanoparticles could be envisaged as a complex interplay of the following factors:

1. Better accessibility and as such greater activity of the surface adsorbed BS moieties. The immobilization of bioactive molecules onto nanomaterials displays enhanced activity, stability, and reusability (Saikia et al., 2013; Konwarh, 2010).
2. Triangular nanoplates are much more competent than spherical counterparts (Konwarh, 2010). The evolution of hierarchical structure with more of strained facets or planes in the iron oxide nanostructure could be ascribed to the antimicrobial action. The differential action against gram-positive and Gram-negative bacteria may be credited to the varied cell wall architecture and the surface moieties that interact differentially with the RL and the BS assisted nanocomposites of iron oxide and SNP.

15.1 IRON OXIDE NANOCRYSTALS (IONRLS)

IONRLs were prepared using the protocol as described by Konwarh et al. (2010) with slight modification. Briefly, 0.1 M $FeCl_2 \cdot 4H_2O$ (≥99.0%, BDH) solution in 50:50 (v/v) Millipore water: methanol containing 2% (w/v) biosurfactant (BS) of bacterial strain OBP1 was subjected to sonication for 3 min with 60% amplitude and 0.5 cycles instead of constant magnetic stirring, as reported previously. The NaOH (Merck) solution 1 M was added drop-wise to the above mixture with constant stirring at room temperature (RT) till pH 8.4 was attained. Addition of 3% (v/v) H_2O_2 (Ranbaxy) to the resultant dark green suspension yielded a black dispersion that was attracted by a permanent magnet. The mixture was then subjected to sonication at the same parameters mentioned previously. After separating the BIONPs by centrifugal decantation, they were washed with distilled water followed by washing with acetone and re-dispersion in water.

15.2 CHARACTERIZATION OF IONRLS

Water-dispersed iron oxide nanoparticles were casted on carbon-coated copper grids, and their morphology was observed using a JEOL, JEM 2100 transmission electron microscope (TEM) at an operating voltage of 200 KV. The morphology of the magnetite crystals depends more on the ultrasonic irradiation than on the growth temperature. It was interesting to note that the randomly distributed BSs could stabilize iron oxide nanorods and nanoplates in contrast to the spherical nanoparticles when PEG was used as the supporting matrix. The diameter of nanorods was distributed within a spectrum of (9.5–34.0) nm with an average diameter of 23 nm. Nanorods were if 91 nm along their long axis. On the other hand, the nano-discs appeared with strained geometry of 4–6 faces.

15.3 WITH SILVER NANOPARTICLES (SNP)

The RL produced by *P. aeruginosa* OBP1 was dispersed in distilled water at its CMC concentration (45.0 $mg.l^{-1}$). The SNP was synthesized in the RL dispersed distilled water following the method as described by Phukon et al. (2011). Briefly, 30 ml of RL (45.0 $mg.l^{-1}$) solution was taken in a 250 ml Erlenmeyer flask and sodium borohydride ($NaBH_4$) was dissolved into the solution to get a concentration of 0.002 M. The Erlenmeyer flask was placed into an ice bath and could cool for 20 min. The assembly was stirred gently using a magnetic stirrer. Now, 10 ml 0.001 M $AgNO_3$ solution was added drop-wise, about 1 drop per second, until the whole amount was used up. After the addition of the $AgNO_3$, the solution turned light yellow in color and the silver nanoparticles (SNP) were synthesized in the RL solution. The synthesized colloidal composite (SNPRL) was analyzed using SEM. The particle size of SNPRL is distributed from 40–200 nm having the majority of them within the range of 40–100 nm. The uniformity in the size may be attributed to the application of RL. Detailed morphological study of the nanoparticles was performed using TEM.

The TEM micrographs represented naked and RL-coated SNPs. The RL coated were almost spherical and well separated from each other with a particle size distribution of 20–40 nm. The micrograph also revealed uniform thickness coating (marked with arrow) of the RL molecules on the SNP. The small clusters of SNP (particle size 20 nm) in RL molecules might be due to drying of the colloidal sample during sample preparation on the carbon-coated TEM-grid used for analysis.

15.4 CHARACTERIZATION OF SNP/SNP-RL

The morphology and size of the particles were investigated by scanning electron microscopy (SEM) using model no. JSM-6390LV of JEOL,

Japan. The samples were directly observed under SEM without platinum or gold coating. Transmission electron microscopy (TEM) analysis was done using a 200 KV system of JEOL JEM 2100, Japan.

15.5 PROTECTION OF SILVER NANOPARTICLES (SNP) BY BIOSURFACTANT (BS) AGAINST SALT TREATMENT

The SNPRL colloid was subjected to UV-visible scan (Thermo Scientific, UV-10 model) from 300 to 700 nm at 0, 10, 16, 22-, 27-, 32-, and 33-days interval from the day of synthesis. After finding the SNPRL colloid samples to be stable for more than 1 month, they were exposed to 0, 2, 20, 60 mg NaCl.ml^{-1} of the colloid. A freshly prepared SNP negative control was also put in contact with 2 mg NaCl.ml^{-1} of the colloid. All spectra were recorded with a gap of 5 min between NaCl addition and spectra recording (Saikia et al., 2013).

In Figure 15.1, the intensity comparison of the SNPRL and RL colloids can be observed. The magnification of the overlapping peaks from day 10 to 33 clearly showed changes in the peak position and intensity with respect to time.

As seen in Figure 15.2, the salt stability of the SNPRL was very high as compared to freshly prepared SNP colloid without RL. On application of NaCl to attain the final concentration of 2.0 mg.ml^{-1} of the colloids, the SNPRL did not show any color change.

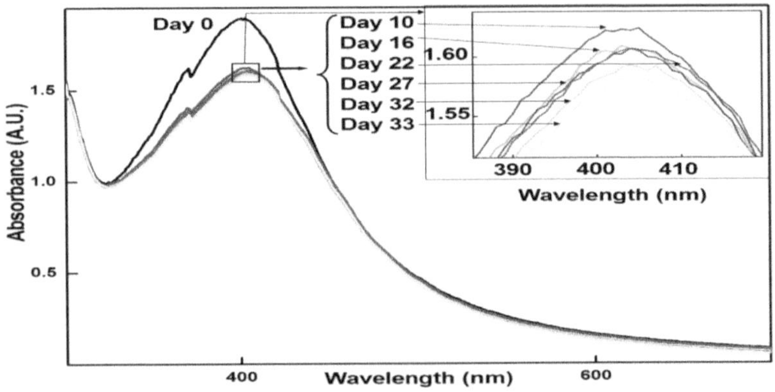

FIGURE 15.1 UV-Vis spectroscopic analysis of the silver nanoparticle colloid in rhamnolipid suspension with respect to time in days. The details on the absorption peaks during days 10 to 33 were shown in the right-side corner.

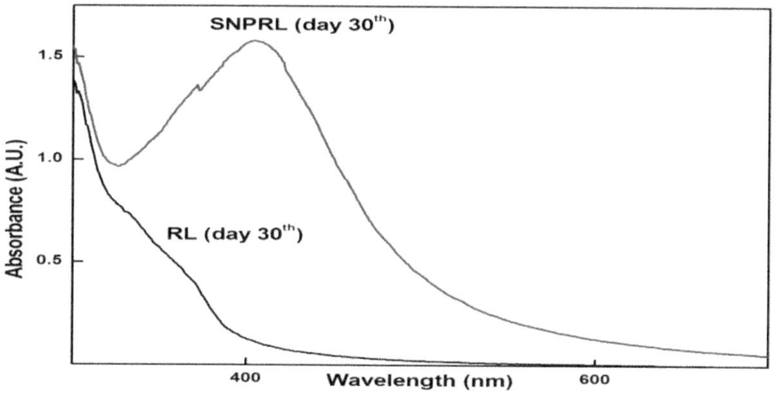

FIGURE 15.2 Graph showing the relative absorbance of the RL solution and SNPRL solution after synthesis on the 30th day.

As evident in Figure 15.3, there is consistency in the UV-Vis analysis of silver nanoparticles rhamnolipid (SNPRL) colloid for 10–33 days period as compared to the rapid fall for 0–10 days. The loss of intensity without broadening on day 10 as compared to day 0 suggested a decrease in the number of SNP in the SNPRL colloid. During the period of study, the peak of the SNPRL shifted from 401 nm on day 0 to 404 nm on day 33rd.

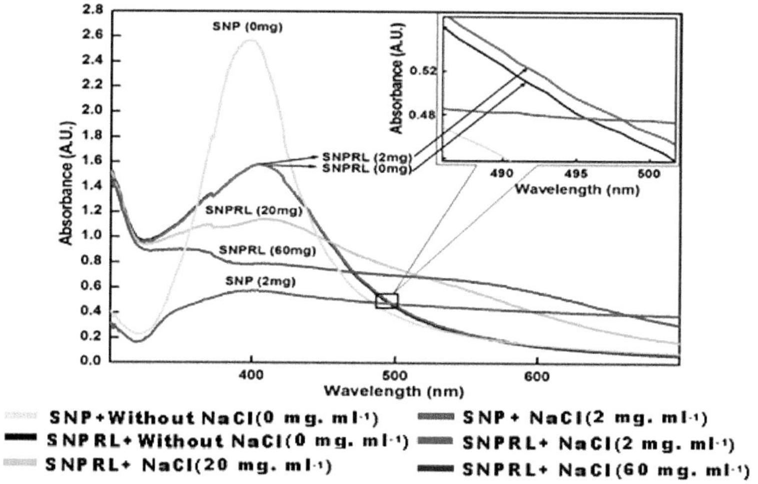

FIGURE 15.3 UV-Vis spectroscopic evidence of silver nanoparticle rhamnolipid (SNPRL) stability during the treatment with NaCl (2 mg.ml^{-1}) and further degradation of SNPRL at an exposure of 60 mg.ml^{-1} of NaCl. Freshly prepared silver nanoparticles (SNP) was used as negative control.

When examined the same using UV-Vis spectrophotometer, there was no change in the SNPRL (2 mg) as compared to blank SNPRL (0 mg). Broadening of the slight insignificant peak was observed in SNPRL near 490 nm, suggesting a minor amount of SNP clumping. On the contrary, when the same concentration of NaCl was applied to the freshly prepared SNP the intensity collapsed signifying that all the nanoparticles diminished in SNP (0 mg) and SNP (2 mg). The SNPRL was found to be dominant in the protection of SNP from NaCl as it took NaCl concentration of 60 $mg.ml^{-1}$ to diminish all SNP in the colloid. The surface plasmon resonance property of the SNPRL was retained after addition of NaCl, suggesting the presence of silver in the form of nanoparticles.

15.6 ANTIMICROBIAL ACTIVITY OF RHAMNOLIPID (RL) NANOCOMPOSITES OF IONRL AND SNPRL

The agar well diffusion method was used for the determination of antibacterial activities of the IONRL, SNPRL, ION, SNP, and RL. The RL used in the present investigation was isolated from *P. aeruginosa* OBP1 strain. The tested bacterial strains, *S. aureus* (MTCC3160), *E. coli* (MTCC40), *P. aeruginosa* (MTCC8163) and *B. subtilis* (MTCC441) were grown on Mueller-Hinton agar plate. Wells were made with the help of sterile Cork Borer and 100 μl of SNPRL, SNP, and RL along with positive control streptomycin (1 $mg.ml^{-1}$) were added. The plates were incubated at 37°C for 24 h and the observed zones of inhibition were measured using a transparent metric ruler. The mean and standard deviation of triplicates data from each sample were calculated. The antibacterial activity of SNP, ION, RL, SNPRL, and IONRL was depicted in Figure 15.4.

However, SNP, ION, and RL when applied individually did not possess lethal effect except in *E. coli*. Both SNPRL and IONRL possessed positive antibacterial activity against all four tested bacterial strains. The nanocomposites exhibited enhanced antibacterial activity as compared to that of pristine SNP, ION, and RL in the same concentrations.

15.7 DEGRADATION OF HYPERBRANCHED EPOXY CLAY NANOCOMPOSITES BY BIOSURFACTANT OF BACTERIA

MSM was used for assessing the biodegradation of modified hyperbranched epoxy (MHBE) clay nanocomposites as described by Dutta et al. (2009). Vegetable oil-based branched polyester MHBE nanocomposites were

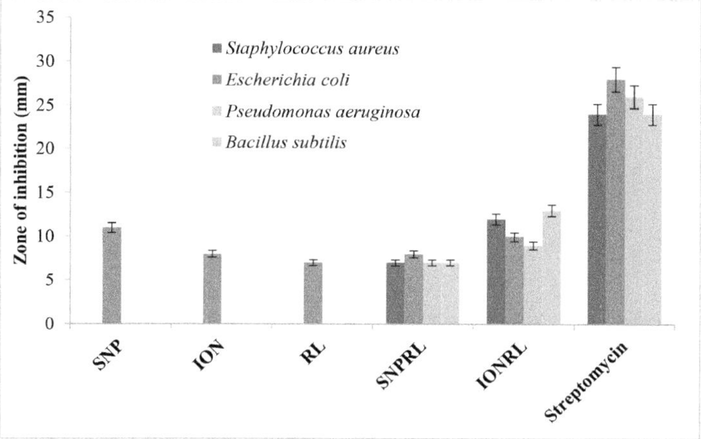

FIGURE 15.4 Antibacterial activity of (a) pristine SNP, ION, and RL; (b) SNPRL and IONRL nanocomposite; and (c) streptomycin against different bacterial strains. Mean of 3 experiments ± SD.

prepared with organically modified montmorillonite clay (OMMT) at different clay loadings (1, 2.5 and 5 wt.% with respect to the total amount of resin). The prepared nanocomposite films were introduced into the sterilized MSM under the laminar air hood. The BS-producing bacterial strain *P. aeruginosa* (OBP1) was selected for the biodegradation assay. The nanocomposite films were the only sole carbon source in the culture medium required for the growth of the bacterial strain. An aliquot of 100 μl of overnight grown culture both containing 1×10^8 ml^{-1} cells (calculated from McFarland turbidity method) was inoculated to the 250 ml volume Erlenmeyer flask containing 100 ml of MSM. The culture flasks were kept at 37°C with 180 rpm on an orbital incubator shaker for 5 weeks. The growth of the bacterial strain was monitored by measuring the optical density at 630 nm at a time interval of 7 days. Culture media without any polymer film was kept as control.

KEYWORDS

- **iron oxide nanocrystal-rhamnolipid**
- **organically modified montmorillonite clay**
- **rhamnolipid**
- **scanning electron microscopy**
- **silver nanoparticle rhamnolipid**
- **synthesized colloidal composite**

REFERENCES

Biswas, M., & Raichur, A. M., (2008). Electrokinetic and rheological properties of nano zirconia in the presence of rhamnolipid biosurfactant. *J. Am. Ceram. Soc., 91*, 3197–3201.

Dutta, S., et al., (2009). Biocompatible epoxy modified bio based polyurethane nanocomposites: Mechanical property, cytotoxicity and biodegradation. *Bioresour. Technol., 100*, 6391–6397.

Kiran, G. S., et al., (2011). Biosurfactants as green stabilizers for the biological synthesis of nanoparticles. *Critical-Rev. Biotechnol., 31*, 354–364.

Kiran, S. G., et al., (2010). Optimization and characterization of a new lipopeptide biosurfactant produced by marine *Brevibacterium aureum* MSA 13 in solid-state culture. *Bioresour. Technol., 101*, 2389–2396.

Konwarh, R., (2010). Magnetically recyclable, antimicrobial, and catalytically enhanced polymer-assisted "green" nanosystem-immobilized *Aspergillus niger* amyloglucosidase. *Appl. Microbiol. Biotechnol., 87*, 1983–1992.

Rosenberg, E., & Ron, E. Z., (1996). Bioremediation of petroleum contamination. In: Ronald, L. C., & Don, L. C., (eds.), *Bioremediation: Principles and Applications* (pp. 100–124). Cambridge University Press, UK.

Saikia, J. P., et al., (2013). Possible protection of silver nanoparticles against salt by using rhamnolipid. *Colloids Surf. B. Biointerfaces, 104*, 330–332.

Sanchez-Peinado, M. D. M., et al., (2010). Influence of linear alkylbenzene sulfonate (LAS) on the structure of *Alphaproteobacteria, Actinobacteria*, and *Acidobacteria* communities in a soil microcosm. *Environ. Sci. Pollut. Res. Int., 17*, 779–790.

Xie, Y., et al., (2006). Synthesis of silver nanoparticles in reverse micelles stabilized by natural biosurfactant. *Colloids Surf. A Physicochem. Eng. Aspects, 279*, 175–178.

FURTHER READING

Bharali, P., & Konwar, B. K., (2011). Production and physicochemical characterization of a biosurfactant produced by *Pseudomonas aeruginosa* OBP1 isolated from petroleum sludge. *Appl. Biochem. Biotechnol., 164*, 1444–1460.

Bharali, P., Das, S., Konwar, B. K., & Thakur, A. J., (2011). Crude biosurfactant from thermophilic *Alcaligenes faecalis*: Feasibility in Petro-spill bioremediation. *Inter. Biodeter. Biodegr., 65*, 682–690.

Das, S., Kalita, S. J., Bharali, P., Konwar, B. K., Das, B., & Thakur, A. J., (2013). Organic reactions in "green surfactant": An avenue to bisuracil derivative. *ACS Sustainable Chem. Eng.* (p. 301). doi: 10.1021/sc4002774, Publication Date (Web).

Pranjal, B., (2015). *Bioremediation of Crude Oil Contaminated Soil.* (Thesis: Supervisor B K Konwar), Dept. of Mol. Biol. and Biotechnology, Tezpur University (Central), Napaam – 784028, Assam, India.

CHAPTER 16

Industrial Applications of Biosurfactant-Producing Bacteria

16.1 SYNTHESIS OF BIS-URACIL DERIVATIVES

The rhamnolipid (RL) produced by *P. aeruginosa* OBP1 was dispersed in distilled water at its CMC (45 $mg.l^{-1}$). 6-[(dimethylamino)-methylene]1,3-aminouracil (210 mg, 1 mM) was added to the 8 ml of the above biosurfactant (BS) solution in a 100 ml round-bottomed flask and the mixture was stirred at room temperature (RT) until all the 6-[(dimethylamino) methylene]1,3-aminouracil got dissolved. To the above mixture, 15% (w/v) of p-toluene sulfonic acid (PTSA) was added to enhance the organic chemical reaction. Further, benzaldehyde (108 mg, 1 mmol) was added drop-wise to the above mixture solution with constant stirring at RT and the product was formed within 5–10 min. The product was separated from the reaction mixture through TLC, dissolved in distilled ethanol (≥98%) and warmed. The ethanol solution of the product was then filtered, allowed to cool, and evaporated at RT to obtain the square-shaped yellow shining transparent crystal product. The crystals were collected, dried, and stored in glass vials for further characterization. The same procedure was followed for the synthesis of other derivatives of bis-uracil. For comparison, the same experiment was conducted in water with the addition of BS.

The compound 6-[(dimethylamino) methyleneamino]-1,3-dimethyl-pyrimidine-2, 4(1*H*, 3*H*)-dione[4] reacts with aldehydes in solution with water. But, the reaction did not proceed at all under this reaction condition. The use of *p*-toluene sulfonic acid (*p*-TSA) as the catalyst or the application of heat as well did not support the reaction to proceed. Interestingly, when the reaction was carried out using the BS produced by *P. aeruginosa* OBP1 in water, the reaction took place, but produced a very low yield. However, when *p*-TSA (15 mol%) was added as catalyst, the yield of the product was found to increase. To optimize the condition, the reaction between 6-[(dimethylamino) methyleneamino]-1, 3-dimethyl pyrimidine-2, 4(1*H*, 3*H*)-dione $_4$ with benzaldehyde (2a; R= $-C_6H_5$) was chosen as the model reaction. Results are presented in Table 16.1.

TABLE 16.1 Optimization of the Reaction for the Model Reaction between 6-[(Dimethylamino) Methylene-Amino]-1, 3-dimethyl-Pyrimidine-2, 4(1*H*, 3*H*)-Dione [4] and Benzalaldehyde [2a]

Entry	Reaction Condition	Time (h)	Yield [%][a] of (5a)
1	Water, rt	48	No product found
2	Water, reflux	48	No product found
3	Water, reflux, 15 mol% *p*-TSA	12	Trace amount
4	CH_3CN, reflux	12	No product found
5	EtOH, reflux	12	No product found
6	MeOH, reflux	12	No product found
7	DMF, reflux	12	No product found
8	Toluene, reflux	12	No product found
9	$PhNO_2$, reflux	12	No product found
10	Water, SDS, [b] rt	12	56
11	Water, biosurfactant, rt	12	25
12	Water, biosurfactant, rt, 5 mol% *p*-TSA	12	40
13	Water, biosurfactant, rt, 10 mol% *p*-TSA	8	60
14	Water, biosurfactant, rt, 15 mol% *p*-TSA	4	75

[a] Yield [%] is referred to isolated yields and calculated from (mol of product)/(mol of initial substrate) ×100.

[b] SDS=Sodium dodecyl sulfate.

To study the scope and limitation of the reaction, the above scheme was extended to other differently substituted aromatic, aliphatic, and heterocyclic aldehydes and ketones (2a-r) under the same optimized reaction conditions. The results are summarized in Table 16.2(a–r) along with comparison in the presence or absence of the BS.

TABLE 16.2 Synthesis of Bisuracil Derivatives (3) in the Presence of Biosurfactant (BS) vide Scheme 1

Entry	Carbonyl Compounds (2)	Time (h)		Yield [%][a]	
		In Presence of BS	In Absence of BS	In Presence of BS	In Absence of BS
a	C_6H_5CHO	0.25	1	99	95
b	$C_6H_5CH{=}CHCHO$	0.91	4	95	93
c	*p*-$OMeC_6H_4CHO$	1.83	4	97	91
d	*p*-ClC_6H_4CHO	0.25	0.41	99	99
e	*p*-OHC_6H_4CHO	2	7	91	75
f	*o*-OHC_6H_4CHO	2	7	90	73
g	*p*-MeC_6H_4CHO	0.33	0.41	99	99
h	*p*-$NO_2C_6H_4CHO$	2.33	10	93	73
i	*m*-$NO_2C_6H_4CHO$	2.25	5	91	82
j	2-furaldehyde	0.33	1	99	99
k	Thiophene-2-carbaldehyde	0.75	6	99	87
l	Paraformaldehyde	0.41	1	93	91
m	CH_3CHO	0.75	3	94	92
n	$CH_3(CH_2)_2CHO$	0.83	3	95	93
o	$CH_3(CH_2)_3CHO$	0.83	3	95	90
p	$(CH_3)_2CO$	48	48	–[b]	–[b]
q	CH_3COPh	48	48	–[b]	–[b]
r	PhCOPh	48	48	–[b]	–[b]

[a]*Yield* (%) is referred to isolated yields and calculated from (mol of product)/(mol of initial substrate) ×100.

[b]*No* product formation.

All the aldehydes [entries 2 (a–o), Table 16.2] reacted with equal ease within the short times, furnishing the products, bis-uracils possessed excellent yields (90–99%) without the formation of by-products. However, under the present condition, also, the reaction with ketones (entries 2p-r) failed. The yield of products increased, and the reaction time reduced drastically along with a reduction in the amount of water used from 30–35 ml but without BS to 8 ml.

In the next set of experiment, as shown in Table 16.3 (Scheme 2) instead of 1, 3-dimethyl-6-aminopyrimidine-2, 4(1H, 3H)-dione$_1$, the reaction between 6-[(dimethylamino)-methyleneamino]-1,3-dimethylpyri-midine-2, 4(1H, 3H)-dione $_4$ and benzaldehyde (2a; R=—C_6H_5) was studied using the BS in water. Accordingly, stirring of 1, 3-dimethyl-6-aminopyrimidine-2,

4(1H, 3H)-dione (1) and benzaldehyde (2a; R=Ph) together in the presence of BS solution at its CMC resulted in the formation of 5, 5'-phenylmethylenebis (1, 3-dimethyl-6-amino-pyrimidine-2, 4-dione) (3a) within 15 min in 99% yield (Scheme 2). The reaction was monitored by TLC. Accordingly, stirring of 1, 3-dimethyl-6-aminopyrimidine-2, 4(1H, 3H)-dione $_1$ and benzaldehyde (2a; R=Ph) together in the presence of BS solution at its CMC resulted in the formation of 5, 5'-phenylmethylenebis (1, 3-dimethyl-6-amino-pyrimidine-2, 4-dione) (3a) within 15 min in 99% yield.

The reaction was monitored by TLC. The reaction was clean, providing only one product that is (3a). To study the scope and limitation of the reaction, the above scheme was extended to other differently substituted aromatic, aliphatic, and heterocyclic aldehydes, and ketones [2 (a-r)] under the same optimized reaction conditions.

TABLE 16.3 Synthesis of Bisuracil Derivatives (5) and (6) in the Presence of Biosurfactant Vide Scheme 2

Entry	Carbonyl Compounds (2)	Time (h)	Yield of Product (5) [%][a]	Yield of Product (6) [%][a]
a	PhCHO	3.5	75	Trace
b	PhCH=CHCHO	10	25	Trace
c	*p*-OMePhCHO	3	trace	93
d	*p*-ClPhCHO	2	95	Nil
e	*o*-OHC_6H_4CHO	5	trace	75
f	*p*-OHPhCHO	5	trace	85
g	*p*-MePhCHO	4	15	87
h	*p*-NO_2PhCHO	4.5	96	Trace
i	*m*-NO_2PhCHO	4.5	60	37
j	2-furaldehyde	3	73	25
k	Thiophene-2-carbaldehyde	3.5	trace	93
l	Paraformaldehyde	10	trace	Nil
m	CH_3CHO	10	trace	Nil
N	$CH_3(CH_2)_2CHO$	10	trace	Nil
O	$CH_3(CH_2)_3CHO$	10	trace	Nil
P	$(CH_3)_2CO$	48	-[b]	-[b]
Q	CH_3COPh	48	-[b]	-[b]
r	PhCOPh	48	-[b]	-[b]

[a] *Yield* [%] is referred to isolated yields and calculated from (mol of product)/(mol of initial substrate)×100.

[b] No product formation.

Towards the completion of the reaction, a mixture of products, bisuracils (5) and (6) were obtained. The yield of the product(s) was also not satisfactory. In view of poor yield, 15 mol% of *p*-TSA was added to increase the yield of the products. Remarkably, the reaction was very clean providing 5a (R=—C_6H_5) as the major product and trace amount of 6a.

The reaction was extended to other differently substituted aromatic, aliphatic, and heterocyclic aldehydes, and ketones (2a-r) under the same optimized reaction condition. Out of two, one product was obtained always predominantly or exclusively. The results are summarized in Table 16.3 [entries (a-r)]. All aldehydes [Table 16.2, entries 2 (a-o)] reacted with equal ease within short duration, furnishing the products, 6-amino-6'-(dimethylamino) methyleneamino-1, 1,' 3, 3'-tetramethyl-5, 5'-(phenylidene)-bis-[pyrimidine-2, 4(1H, 3H)-dione] derivatives (5a-k) and (6a-k) in good yields except (2b) and with no by-product formation. It was observed that the same reaction did not proceed in water in the absence of BS, thereby proving the necessity of BS. Like Scheme 1 (entries p-r), here also the reaction failed with ketones. However, with aliphatic aldehydes entries l-o), the reaction was not at all satisfactory.

The results are summarized in Table 16.3(a–r) along with comparison in the presence or absence of the BS. All the aldehydes [Table 16.2 (entries 2a-o)] reacted with equal ease within the short times, furnishing the products, bis-uracils (3a-o) possessed excellent yields (90–99%) without the formation of by-products. However, under the present condition, also, the reaction with ketones [entries 2 (p-r)] failed. The yield of products increased, and the reaction time reduced drastically along with a reduction in the amount of water used from 30–35 ml but without BS to 8 ml. In the next set of experiment, instead of 1, 3-dimethyl-6-aminopyrimidine-2, 4(1H, 3H)-dione $_1$, the reaction between 6-[(dimethylamino)-methyleneamino]-1,3-dimethylpyrimidine-2,4 (1H, 3H)-dione $_4$ and benzaldehyde (2a; R=—C_6H_5) was studied using the BS in water (Scheme 2). Towards the completion of the reaction, a mixture of products, bisuracils (5) and (6) were obtained. The yield of the product(s) was also not satisfactory. In view of poor yield, 15 mol% of *p*-TSA was added to increase the yield of the products and the data obtained are presented in Table 16.4. Remarkably, the reaction was very clean providing 5a (R=—C_6H_5) as the major product and trace amount of 6a.

To extend the scope and limitation of the reaction further, the reaction between 6-[(dimethylamino)methyleneamino]-1,3-dimethylpyrimidine-2,

4(1H, 3H)-dione $_4$ and dicarbaldehydes (Kosswig, 2005) (Scheme 3) was studied. The results are summarized in Table 16.3 (entries a and b). In the case of an aliphatic dicarbaldehyde, glutaraldehyde the bis-uracil adduct (8b) as major one (25%) was obtained, but in the case of p-benzene dicarbaldehyde (entry a) bis-uracil adduct (8a) was obtained predominantly along with little amount of N-formylated bis-uracil derivatives (9). Un-reacted amount of 4 was recovered.

TABLE 16.4 Synthesis of Bisuracil Derivatives (8) and (9) in the Presence of Biosurfactant Vide Scheme 3

Entry	Dicarbonyl Compounds (7)	Time (h)	Yield of Product (8) [%][a]	Yield of Product (9) [%][a]
a	*p*-(CHO)C_6H_4(CHO)	5	65	28
b	CHO$(CH_2)_5$CHO	10	25	Trace

[a] *Yield* [%] is referred to isolated yields and calculated from (mol of product)/(mol of initial substrate) ×100.

16.2 BACTERIAL BIOSURFACTANT (BS) AND THEIR VARIOUS ACTIVITIES

At low concentrations, the BSs did not show phytotoxicity, whereas at the increased concentration of BSs beyond their CMC caused decrease in the GI of both seeds. Tiquia et al. (1996) reported that GI 80% could be an indicator for the absence of phytotoxicity. The increased concentrations of the BS solutions above the CMC caused reduction in the GI, clearly indicating the inhibitory effect on seed germination and root elongation. The reduction in GI of mung bean was moderate to 55–65% but it was prominent in rice to 45–54%. The progressive decrease in the GI with the increasing concentration of BSs might be due to the alternations in the permeability of the cellular membrane induced by BSs (Sanjeet et al., 2004). According to Millioli et al. (2009), the sole presence of RLs may influence the GI of lettuce (*Lactuca sativa*) and the GI dropped linearly with the increasing concentration of RLs. Silva et al. (2010) reported that cabbage (*Brassica oleracea*) tolerated the presence of RLs without showing any significant reduction in the GI up to the CMC of the tested BS which was 700 mg.l^{-1}. Marecik et al. (2012) reported the appearance of notable changes in the GI value of alfalfa (*Medicago sativa*), mustard (*Brassica nigra*) and sorghum (*Sorghum saccharatum*) at two different concentrations of RL, the first at CMC and the second far above it. They also reported the presence of inhibitory effect of RLs on the germination of alfalfa and mustard and most prominent for monocotyledonous plant-like sorghum while cuckoo flower species (*Cardamine pratensis*) remained unaffected suggesting that the phytotoxicity of RLs maybe plant-specific.

In the present study, BSs produced by *P. aeruginosa* strains did not show any larvicidal potency against *Aedes albopictus* at normal concentrations as recommended by WHO (1996). The same was observed at much higher concentrations. This clearly indicated that the tested BSs were not lethal to the mosquito larvae. Several studies reported that only certain cellular proteins secreted by *P. aeruginosa* have virulence effect on various insects. Mostakim et al. (2012) reported the larvicidal activity of *P. aeruginosa* against the 3rd instar larvae of *Bactrocera oleae*, the most serious pest in olive cultivation. Among the known BSs, the cyclic lipopeptide surfactin, produced by *B. subtilis* subsp. *Subtilis* was reported to exhibit potential larvicidal and mosquitocidal activities (Geetha et al., 2010). Since the BSs from *P. aeruginosa* strains possessed negligible mosquito larvicidal potency as compared to many other currently available preparations, the application of such BSs in contaminated environments during the remediation process might not cause any undesirable effects to the beneficial insects or their larvae.

The MIC values were much less for Gram-positive bacteria, indicating their effectiveness at low concentrations except for few bacteria such as *Staphylococcus aureus*. However, the MIC values were much higher for Gram-negative bacteria because of the surface proteins and lipopolysaccharides (LPSs), which are the two main constituents of the cell wall. The LPS either acts as a barrier or provides protection to the inner sensitive membrane and cell wall from the toxic compounds (Sharma and Pant 2001; Adams and Jackson, 1996). The present study suggested that the RL molecules having both hydrophobic and hydrophilic groups could insert their fatty acid components into the cell membrane that cause considerable alteration in the ultra-structure of the cell, such as the ability of the cell to interiorize plasma membrane. Alternately, it might also be possible that insertion of the shorter acyl tails of the RL into the cell membrane causes disruption between cytoskeleton elements and the plasma membrane, allowing the membrane to lift away from the cytoplasmic constituents (Yalcin and Ergene, 2009). Gram-negative bacteria are known to have intrinsic resistance against a variety of antibiotics due to the transenvelope multidrug resistance (MDR) pump (Girish and Satish, 2008). Therefore, treatment of such MDR strains with RLs seems to be more advantageous as it interacts only with the bacterial cell surface. Several workers reported enhancement of the cytoplasmic membrane permeability by RLs with the consequential alteration leading to cell damage (Bharali et al., 2013).

The concentration gradient motility agar (CGMA) assay showed positive chemoattractant property of RLs at their CMC towards *Staphylococcus aureus* (MTCC3160) and *Klebsiella pneumoniae* (MTCC618) confirming chemo-attractant property. Such chemo-attractant property of the test compounds enhances its antibacterial property towards the targeted pathogen (Devi et al., 2010). It has been suggested that when the microbes get RL molecules, they might cause disruption of the cell surface of bacteria (Jarvis and Johnson, 1949) by being integrated in between the phospholipids of the plasma membrane. RLs can also enhance permeability of the cytoplasmic membrane with the consequential alteration leading to cell damage (Sotirova et al., 2008). Such unique property of RLs suggests their application in increasing the efficacy of several drug molecules such as nisin. According to Magalhaes et al. (2012), the combinations of nisin, an antimicrobial peptide produced by *Lactococcus lactis* and RLs have a strong synergistic effect on the cytoplasmic membrane of *Listeria monocytogenes*, a serious food born pathogen.

The BSs of *P. aeruginosa* strains were found to be cytotoxic to growing mouse fibroblast cells (L929) at concentrations above 100 $\mu g.ml^{-1}$, probably by directly perturbing cell membranes (Hauler et al., 2003). The hemolytic and cytotoxic activities of the RL are due to the detergent-like properties, and cell membranes are possibly perturbed by the introduction of fatty acid chains into the organized lipid layers of cells. Häuler et al. (2003) reported the time and dose-dependent cytotoxicity against non-phagocytic HeLa and phagocytic HL60 cells by the RL of *Burkholderia pseudomallei*. No inhibitory effect of the purified RL fractions RL-a and RL-b isolated from *P. aeruginosa* MR01 was observed on the normal Vero cell line at concentrations up to 50 $\mu g.ml^{-1}$. RL, especially the diRL (50 $\mu g.ml^{-1}$) in the presence of serum, favors the keratinocyte differentiation and inhibits the proliferation of fibroblasts thus helping in the tissue repair (Stipcevic et al., 2005). These features broaden the application of RLs in the new advanced field of medicine as wound healers. Several others reported the non-toxic and antiproliferative properties of RLs (Lotfabada et al., 2010), suggesting that the exposure of BSs to the human skin during the field operation process of bioremediated sites doesn't cause any hazard.

Acute dermal irritation study of the transdermal patch in rabbits showed no dermal responses such as erythema or edema as compared to the negative control against the application of BSs. The total body weight in the transdermal patch-treated groups of rabbits did not differ significantly from the control group. Loss of body weight is an important indicator of gross toxicity. Severe toxicity or interference with absorption of nutrients is reflected by the reduced body weight (Banerjee et al., 2013). Hence, it could be concluded that the application of RL BSs have neither the potency to produce severe tissue destruction nor does it seem to interfere with the absorption of nutrients. Further, the study showed that the patch-treated group did not show any hematological and biochemical changes confirming their non-toxicity to the mammalian skin. Maier and Soberon-Chavez (2000) reported that RL produced by *P. aeruginosa* have extremely low irritancy and even anti-irritating effects, as well as compatibility with human skin. Clinical trials for the treatment of psoriasis, lichen planus, neurodermatitis, and human burn wound healing confirmed excellent ameliorative effect of RLs as compared to the conventional therapy using corticosteroids (Stipcevic et al., 2005, 2006). RLs also exhibited differential effects on human keratinocyte and fibroblast cultures (Stipcevic et al., 2005). The innovative application of BSs has appeared with the suggestion that BSs may aid wound healing, hence opening up new avenues for incorporating BSs into a wide range of skincare products in place of chemical surfactants and this could lead to healing of minor skin lesions (Suneel et al., 1996).

16.3 DEGRADATION OF COMPLEX NANO-BIOCOMPOSITE POLYMERS BY BIOPOLYMER

The growth of *P. aeruginosa* strain OBP1 on various nanocomposite films such as hyperbranched epoxy (HBE), modified hyperbranched epoxy (MHBE) and their clay nanocomposites along with the corresponding pristine polymeric films can be realized, as shown in Figure 16.1.

The growth of the bacterial strain on the tested polymers increased with the increase in exposure time to the bacterial strain. The growth rate in all the tested nanocomposites as well as on the pristine polymers was not significant up to two weeks of bacterial exposure. But after 2–3 weeks of treatment, the biodegradation rate increased sharply as could be realized from the bacterial population density determination by McFarland turbidity method. The biodegradation process caused severe damage to almost all the nanocomposites, which could be observed from SEM micrographs of the recovered nanocomposite films following five weeks of bacterial exposure.

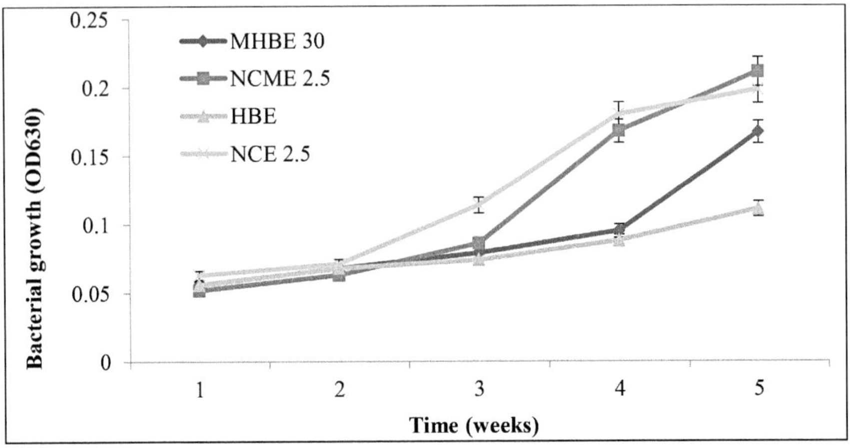

FIGURE 16.1 Growth curves of bacterial strain OBP1 on HBE, MHBE30, NCE2.5, and NCME2.5 their nanocomposites. Mean of 3 experiments ± SD.

The same was further supported by the decrease in the weight of the nanocomposite films after biodegradation. However, the extent of biodegradation was found to be low in the case of pristine hyperbranched epoxy/clay nanocomposite in comparison to the MHBE nanocomposites, which could be seen clearly from the bacterial population as observed in SEM micrographs. Moreover, the rate of biodegradation was higher in the case of MHBE and its nanocomposites in comparison to the pristine HBE and its nanocomposite. The same are presented in Figure 16.2.

When examined the same using UV-Vis spectrophotometer, there was no change in the SNPRL (2 mg) as compared to blank SNPRL (0 mg), as shown in Figures 15.2 and 15.3. Broadening of the slight insignificant peak was observed in SNPRL near 490 nm, suggesting a minor amount of SNP clumping. On the contrary, when the same concentration of NaCl was applied to the freshly prepared silver nanoparticles (SNP) the intensity collapsed signifying that all the nanoparticles diminished in SNP (0 mg) and SNP (2 mg).

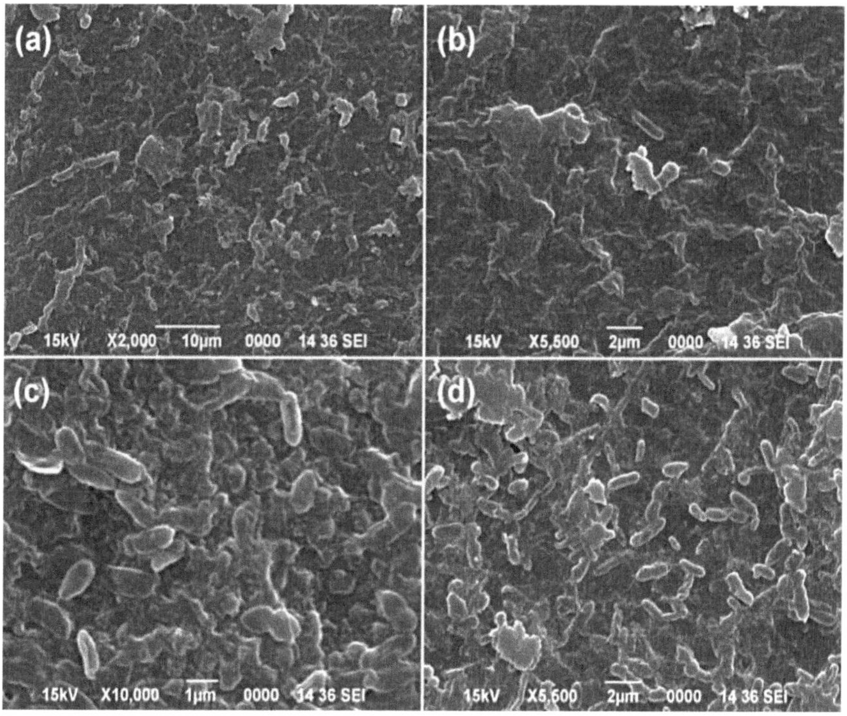

FIGURE 16.2 SEM micrographs of biodegraded nanocomposite films (a) HBE; (b) NCE2.5; (c) NCME 2.5; and (d) MHBE30.

16.4 USE OF BIOSURFACTANT (BS) IN INDUSTRIAL PROCESSES

A highly efficient and environmentally benign nucleophilic addition of 6-amino-1, 3-dimethyl pyrimidine-2, 4(1H, 3H)-dione and 6 [(dimethylamino) methyleneamino]-1, 3-dimethylpyrimidine-2, 4(1H, 3H)-dione with aldehydes (aromatic, aliphatic, and heterocyclic) using BS isolated from OBP1 achieved in water at the RT with higher yield of products as compared to their counterparts without the application of BSs. There could be two factors responsible for the increased yield of products; first, the equilibrium of the nucleophilic reaction is shifted far to the right in the presence of the BS. Gradual addition of surfactant into the reaction mixture to an extent of CMC led to the formation of hydrophobic pockets within the bulk water solvent due to hydrophobic interior of micelle, which paves the way for bringing the reactants to proximity facilitating product formation. The BS usually interferes with the existing solubility of the solvation layer, which led to the other probable interactions and chemical events that might be responsible for the release of binding energy. The released binding energy, in turn, helps in expediting the reaction towards completion in the presence of catalysts. Secondly, from the plausible reaction mechanism, the nucleophilic 5th position of 4 attacks of the carbon center belonging to the aldehyde, followed by the elimination of water molecule, a second molecule of uracil derivative then attacks via its nucleophilic 5th position affording the product. Dehydration has successfully been achieved in water due to the hydrophobic nature of the BS interior (Wyrwas et al., 2012). The BS has been recovered and recycled for several times and used repeatedly.

P. aeruginosa was reported to be a versatile bacterial strain capable of utilizing various types of carbon sources ranging from simple glucose to complex petroleum-based hydrocarbons (Vasileva-Tonkova and Galabova, 2003; Vasileva-Tonkova et al., 2006). From biodegradation studies of synthetic hyperbranched epoxy/OMMT clay nanocomposites-based polymers, it is clear that *P. aeruginosa* strain OBP1 is capable of utilizing the synthesized polymers as is evident by an increase in bacterial population density. The rate of bacterial cell density increases with the duration of the treatment time. This was particularly observed in the case of MHBE nanocomposites which might be due to the catalytic role of clay in the hydrolysis of the ester groups present in the modified systems (Dutta et al., 2009). The presence of terminal hydroxyl groups in the clay layers can cause heterogeneous hydrolysis after absorbing water in the presence of microbes. This process is known to require some time for the initiation, which might be the reason for the remarkable improvement of biodegradation after 2–3 weeks of bacterial treatment (Wyrwas et al., 2012). The extent of biodegradation was comparatively low in the case of pristine hyperbranched epoxy/clay nanocomposite in comparison to the MHBE nanocomposites. This was mainly due to the presence of polar ester groups and fatty acid chain moieties in the vegetable oil-based polyester MHBE/clay nanocomposites, which attract microbes.

The BS produced by the bacterial strains was able to reduce the surface tension of the culture medium significantly from 68.5 to 31.1 mNm^{-1}. However, the behavior of BS production by the bacterial strains was quite different (Tables 9.12–9.15). The CMC values of the BSs produced by the bacterial strains were in the range of 45–105 $mg.l^{-1}$ (Figure 10.1). Previously, a range of CMC values between 10 and 230 $mg.l^{-1}$ were reported for the RLs isolated from the different microbial sources (Nitschke et al., 2005). The RL homologs could also differ with the bacterium, medium, and cultivation conditions (Nitschke and Costa, 2007). The CMC values obtained in the present study were much lower than the chemical surfactants such as sodium dodecyl sulfate (SDS) having a CMC value of 2,100 $mg.l^{-1}$ (Monterio et al., 2007).

It was reported that concentration of the surfactant below their CMC level reduces the surface and interfacial tension (IFT) between air/water, oil/water and soil/water systems. Capillary number is determined by the ratio of viscous force to the capillary force (Xia et al., 2011).

Critical micelle dilution (CMD) is an indirect means of measuring the surfactant production related to the range of the critical micelle concentration (CMC) (Reddy et al., 1983). Therefore, it is very much important to determine the surface activity of the surfactant at different dilutions (Tables 10.2–10.3) levels such as CMD^{-1} (1:10) and CMD^{-2} (1:100). Reduction in the surface tension at CMD^{-1} was almost like that of normal, whereas the CMD^{-2} caused a slight increase in the surface tension of the system due to the higher dilution.

KEYWORDS

- **biosurfactant**
- **concentration gradient motility agar**
- **critical micelle dilution**
- **hyperbranched epoxy**
- **interfacial tension**
- **lipopolysaccharides**
- **modified hyperbranched epoxy**

REFERENCES

Adams, P., & Jackson, P. P., (1996). Bioremediation of oil spills, theory and practice. In: *The Proceedings of International Seminar on Petroleum Industry and the Nigerian Environment* (pp. 30–42). Nigerian National Petroleum Corporation (NNPC), Lagos.

Bailey, J. E., & Ollis, D. F., (1986). *Biochemical Engineering Fundamentals* (2nd edn.). New McGraw-Hill, York.

Banerjee, S., et al., (2013). Acute dermal irritation, sensitization, and acute toxicity studies of a transdermal patch for prophylaxis against (+) anatoxin-a poisoning. *Inter. Toxicol., 32*, 4308–4313.

Bharali, P., et al., (2011). Crude biosurfactant from thermophilic *Alcaligenes faecalis*: Feasibility in Petro-spill bioremediation. *Int. Biodeter. Biodegr., 65*, 682–690.

Bharali, P., et al., (2013). Colloidal silver nanoparticles/rhamnolipid (SNPRL) composite as novel chemotactic antibacterial agent. *Inter. J. Biol. Macromol., 61*, 238–242.

Choi, Y. J., et al., (2009). Enhancement of aerobic biodegradation in an oxygen-limiting environment using a saponin-based microbubble suspension. *Environ. Pollut., 157*, 2197–2202.

Devi, K. P., et al., (2010). Eugenol (an essential oil of clove) acts as an antibacterial agent against Salmonella typhi by disrupting the cellular membrane. *J. Ethnopharmacol., 130*, 107–115.

Dutta, S., et al., (2009). Biocompatible epoxy modified bio based polyurethane nanocomposites: Mechanical property, cytotoxicity and biodegradation. *Bioresour. Technol., 100*, 6391–6397.

Geetha, I., et al., (2010). Identification and characterization of a mosquito pupicidal metabolite of a *Bacillus subtilis* subsp. subtilis strain. *Appl. Microbiol. Biotechnol., 86*, 1737–1744.

Girish, H. V., & Satish, S., (2008). Antibacterial activity of important medicinal plants on human pathogenic bacteria. *Comparative Analysis, 5*, 267–271.

Hauler, S., et al., (2003). Structural and functional cellular changes induced by *Burkholderia pseudomallei* rhamnolipid. *Infect. Immun., 71*, 2970–2975.

Jarvis, F. G., & Johnson, M. J., (1949). A glycolipid produced by *Pseudomonas aeruginosa. J. Am. Oil Chem. Soc., 71*, 4124–4126.

Kosswig, K., (2005). Surfactants. In: *Ullmann's Encyclopedia of Industrial Chemistry*. Wiley-VCH, Weinheim.

Lee, K. M., et al., (2004). Rhamnolipid production in batch and fed-batch fermentation using *Pseudomonas aeruginosa* BYK-2 KCTC 18012P. *Biotech. Bioprocess Engg., 9*, 267–273.

Lotfabada, T. B., et al., (2010). Structural characterization of a rhamnolipid-type biosurfactant produced by *Pseudomonas aeruginosa* MR01: Enhancement of dirhamnolipid proportion using gamma irradiation. *Colloids Surf. B. Biointerfaces, 81*, 397–405.

Magalhaes, L., & Nitschke, M., (2012). Antimicrobial activity of rhamnolipids against *Listeria* monocytogenes and their synergistic interaction with nisin. *Food Cont., 29*, 138–142.

Maier, R. M., & Soberon-Chavez, G., (2000). *Pseudomonas aeruginosa* rhamnolipids: biosynthesis and potential applications. *Appl. Microbiol. Biotechnol., 54*, 625–633.

Marecik, R., et al., (2012). Rhamnolipids increase the phytotoxicity of diesel oil towards four common plant species in a terrestrial environment. *Water Air Soil Pollut., 223*, 4275–4282.

Millioli, V. S., et al., (2009). Bioremediation of crude oil-bearing soil: Evaluating the effect of rhamnolipid addition to soil toxicity and to crude oil biodegradation efficiency. *Global NEST J., 11*, 181–188.

Monterio, S. A., et al., (2007). Molecular and structural characterization of the biosurfactant produced by *Pseudomonas aeruginosa* DAUPE 614. *Chem Phys Lipids, 147*, 1–13.

Mostakim, M., et al., (2012). Biocontrol potential of a *Pseudomonas aeruginosa* strains against *Bactrocera oieae. Afr. J. Microbiol. Res., 6*, 5472–5478.

Nitschke, M., & Costa, S. G. Y. A. O., (2007). Biosurfactants in food industry. *Trends Food Sci. Technol., 18*, 252–259.

Nitschke, M., et al., (2005). Rhamnolipid Surfactants: An update on the general aspects of these remarkable biomolecules. *Biotechnol. Prog., 21*, 1593–1600.

Reddy, P. G., et al., (1983). Isolation and functional characterization of hydrocarbon emulsifying and solubilizing factors produced by a *Pseudomonas species. Biotechnol. Bioeng., 25*, 387–401.

Sanjeet, M., et al., (2004). Crude oils degradation efficiency of a recombinant *Acinetobacter baumannii* strain and its survival in crude oil-contaminated soil microcosm. *FEMS Microbiol. Lett., 235*, 323–332.

Sharma, S. L., & Pant, A., (2001). Biodegradation and conversion of alkanes and crude oil by a marine *Rhodococcus* sp. *Biodegra., 11*, 289–294.

Silva, S. N. R. L., et al., (2010). Glycerol as substrate for the production of biosurfactant by *Pseudomonas aeruginosa* UCP 0992. *Colloids Surf. B. Biointerfaces, 79*, 174–183.

Sotirova, A. V., et al., (2008). Rhamnolipid-biosurfactant permeabilizing effects on gram-positive and gram-negative bacterial strains. *Curr. Microbiol., 56*, 639–6441.

Stipcevic, T., et al., (2005). Di-rhamnolipid from *Pseudomonas aeruginosa* displays differential effects on human keratinocyte and fibroblast cultures. *Dermat. Sci., 40*, 141–143.

Stipcevic, T., et al., (2006). Enhanced healing of full-thickness burn wounds using dirhamnolipid. *Bums, 32*, 24–34.

Suneel, C., et al., (1996). Bacterial consortia for crude oil spill remediation. *J. Water Sci. Technol., 34*, 187–193.

Tiquia, S. M., et al., (1996). Effects of composting on phytotoxicity of spent pig-manure sawdust litter. *Environ. Pollut., 93*, 249–256.

Vasileva-Tonkova, E., & Galabova, D., (2003). Hydrolytic enzymes and surfactants of bacterial isolates from lubricant-contaminated wastewater. *Z. Naturforsch. C., 58*, 87–92.

Vasileva-Tonkova, E., et al., (2006). Production and properties of biosurfactants from a newly isolated *Pseudomonas fluorescens* HW-6 growing on hexadecane. *Z. Naturforsch. C, 61*, 553–559.

Wyrwas, B., et al., (2012). Utilization of triton X-I00 and polyethylene glycols during surfactant mediated biodegradation of diesel fuel. *Hazard. Mater., 197*, 97–103.

Xia, W., et al., (2011). Comparative study of biosurfactant produced by microorganisms isolated from formation water of petroleum reservoir. *Colloids Surf. A Physicochem. Eng. Aspects, 392*, 124–130.

Yalcin, E., & Ergene, A., (2009). Screening antimicrobial activity of biosurfactants produced by microorganisms isolated from refinery wastewaters. *J. Appl. Biol. Sci., 3*, 148–153.

CHAPTER 17

Conclusions

1. The potential biosurfactant (BS) producing bacterial isolates belong to the genus *Pseudomonas*. The nucleotide frequency count shows ATGC, C+G, and A+T compositions of the bacterial strains to have almost similar frequency (~99%) distribution as compared to the other reported strains of *P. aeruginosa*.
2. The bacterial isolates could utilize a wide spectrum of hydrocarbons as the sole source of carbon and energy. All four bacterial strains exhibited a preference for high molecular weight aliphatic hydrocarbons as compared to aromatic and polyaromatic hydrocarbons (PAHs).
3. Hemolytic and CTAB agar assay confirm the BS producing ability of the bacteria and provide a criterion to produce rhamnolipid (RL).
4. The potential *P. aeruginosa* strains possessed presumptive production of extracellular anionic BSs.
5. In the case of BS producing *P. aeruginosa*, n-hexadecane was the most suitable carbon source. Diesel is also an efficient carbon source as well in terms of reduction in surface tension. The level of BS production in the diesel-supplemented medium is almost like that of n-hexadecane.
6. Optimization of media components including carbon, nitrogen, macro-micro nutrients, and culture conditions, including temperature, pH, and agitation, leads to the enhancement of BS production in all four bacterial strains.
7. Among the low-cost and renewable carbon substrates screened to produce BS by these four bacterial strains, waste glycerol followed by kitchen-waste oil (*Sesamum indicum*) and Nahor seed oil (*Mesua ferrea*) were found to be suitable.

8. Physical parameters like reduction in surface tension (ST), interfacial tension (IFT), CMC, emulsification activity ($E_{24}\%$), and foaming index ($F_{24}\%$) of the BSs isolated from the bacterial strains clearly confirmed their effective surface-active properties.

9. BSs produced by the bacterial strains exhibited excellent surface properties and remained stable while exposed to extreme conditions like high temperature, pH, salinity, and metal ion concentration. The stability of BSs in higher dilutions (CMD^{-1} and CMD^{-2}) further confirmed their intact surface properties.

10. Chemical characterization of isolated BSs with TLC, FTIR, and MS confirmed their glycolipidic nature. Further, the mass spectroscopic studies confirmed the production of RLs by the bacterial strains; however, di-rhamnolipids were found to be predominant over mono-rhamnolipids.

11. The isolated BSs were efficient within their CMCs in solubilizing PAHs and thereby could remove the crude oil from the contaminated sand. This effected recovery of residual crude oil from the petroleum sludge. The same was true in the case of crude oil recovery from the crude oil-saturated sand pack column. Further, the increase in the release of crude oil at a higher temperature of 70–90°C confirmed the thermo-stability of BS in crude oil separation and recovery processes.

12. All four bacterial strains could grow in crude oil supplemented media utilizing them as the sole source of carbon. The efficiency of each bacterial strain in degrading the test hydrocarbons within 30 days was established by liquid chromatography followed by gravimetric analysis. Further, GC analysis of the saturated fractions of the hydrocarbons confirmed the degradation ability of each individual bacterial strain.

13. Bacterial Consortia I, comprising of OBP1, OBP3, OBP4, and Consortia II, comprising of OBP1, OBP2, OBP3, and OBP4 were found to be more effective in biodegradation of crude oil. Further, the addition of BS to the Consortia enhanced the biodegradation process.

14. BS solutions below the CMC were found to be non-toxic on germinating seeds of rice and mung bean, but at concentrations above the

CMC exhibited inhibitory effect on seed germination and root elongation which was more pronounced in the case of rice.

15. BS of these four bacterial strains showed no larval mortality of the insect *Aedes albopictus* at almost all the recommended concentrations of WHO. However, at higher concentrations of 1000–1500 $mg.l^{-1}$ could kill only about 3% of the larvae.

16. BS of the bacterial strains OBP1 exhibited the highest antibiotic activity against the test microorganisms. Further, this BS exhibited an excellent chemotactic response towards *Staphylococcus aureus* (MTCC3160) and *Klebsiella pneumoniae* (MTCC618) strains.

17. BSs did not exhibit inhibitory effect on the mouse fibroblast cell line-1929 up to 100 $\mu g.ml^{-1}$. BSs concentrations above CMCs revealed non-toxicity to the skin of rabbit and showed no adverse effect on the hematological parameters of the treated rabbits.

18. Two different types of nanoparticles such as iron oxide nanocrystal (IONRL) and silver nanoparticles (SNPRL) were synthesized in the presence of BS of the bacterial strain OBP1 and characterized. The SNPRL was found to stable up to 31 days and protected the silver nanoparticles (SNP) from NaCl up to a concentration of 60 $mg.ml^{-1}$.

19. Both IONRL and SNPRL nanocomposites exhibited considerable antibacterial properties against a wide range of bacterial strains.

20. BS of OBP1 was found to be highly efficient in nucleophilic addition reactions of 6-amino-1, 3-dimethyl pyrimidine-2, 4 (1H, 3H)-dione and 6[(dimethylamino)ethylene amino]-1, 3-dimethyl pyrimidine-2, 4(1H, 3H)-dione with aldehydes in water at room temperature (RT) to give higher yield of products.

21. The bacterial strain OBP1 exhibited growth on various nanocomposite films such as hyperbranched epoxy (HBE), modified hyperbranched epoxy (MHBE), and their clay nanocomposites along with the corresponding pristine polymeric films and could utilize vegetable oil-based polymer films as carbon substrate. Further, gravimetric analysis and SEM micrographs confirmed the degradation of the tested polymer.

Index

C

D

E

I

J

K

L

M

Q

R

S

T